International Federation of Automatic Control

RELIABILITY
OF
INSTRUMENTATION SYSTEMS
FOR SAFEGUARDING AND CONTROL

RELIABILITY
OF
INSTRUMENTATION SYSTEMS
FOR SAFEGUARDING AND CONTROL

Proceedings of the IFAC Workshop
The Hague, The Netherlands
12–14 May 1986

Edited by

J. P. JANSEN
NIRIA, The Netherlands

and

L. BOULLART
University of Gent, Belgium

Published for the

INTERNATIONAL FEDERATION OF AUTOMATIC CONTROL

by

PERGAMON PRESS

OXFORD · NEW YORK · BEIJING · FRANKFURT
SÃO PAULO · SYDNEY · TOKYO · TORONTO

U.K.	Pergamon Press, Headington Hill Hall, Oxford OX3 0BW, England
U.S.A.	Pergamon Press, Maxwell House, Fairview Park, Elmsford, New York 10523, U.S.A.
PEOPLE'S REPUBLIC OF CHINA	Pergamon Press, Room 4037, Qianmen Hotel, Beijing, People's Republic of China
FEDERAL REPUBLIC OF GERMANY	Pergamon Press, Hammerweg 6, D-6242 Kronberg, Federal Republic of Germany
BRAZIL	Pergamon Editora, Rua Eça de Queiros, 346, CEP 04011, Paraiso, São Paulo, Brazil
AUSTRALIA	Pergamon Press Australia, P.O. Box 544, Potts Point, N.S.W. 2011, Australia
JAPAN	Pergamon Press, 8th Floor, Matsuoka Central Building, 1-7-1 Nishishinjuku, Shinjuku-ku, Tokyo 160, Japan
CANADA	Pergamon Press Canada, Suite No. 271, 253 College Street, Toronto, Ontario, Canada M5T 1R5

First edition 1987

British Library Cataloguing in Publication Data

Reliability of instrumentation systems for safeguarding & control:
proceedings of the IFAC workshop, The Hague, The Netherlands, 12-14 May 1986.
1. Engineering instruments
I. Jansen, J. P.   II. Boullart, L.   III. International Federation of Automatic Control
620'.0044   TA165
ISBN 0-08-034063-6

These proceedings were reproduced by means of the photo-offset process using the manuscripts supplied by the authors of the different papers. The manuscripts have been typed using different typewriters and typefaces. The lay-out, figures and tables of some papers did not agree completely with the standard requirements: consequently the reproduction does not display complete uniformity. To ensure rapid publication this discrepancy could not be changed: nor could the English be checked completely. Therefore, the readers are asked to excuse any deficiencies of this publication which may be due to the above mentioned reasons.

The Editors

Printed in Great Britain by A. Wheaton & Co. Ltd., Exeter

IFAC WORKSHOP ON RELIABILITY OF INSTRUMENTATION SYSTEMS FOR SAFEGUARDING AND CONTROL

Organized by
Netherlands Association of Engineers NIRIA
Royal Institution of Engineers in the Netherlands (KIvI)

Sponsored by
IFAC Technical Committee on Applications

International Programme Committee
L. Boullart, Belgium (Chairman)
P. Andow, UK
O. A. Asbjornsen, Norway
G. Bello, Italy
D. R. Bristol, USA
R. Genser, Austria
R. Goarin, France
P. Inzelt, Hungary
E. O'Shima, Japan
A. Poucet, Belgium
H. U. Steusloff, FRG
A. Work, USSR

National Organizing Committee
J. P. Jansen (Chairman)
L. Winkel (Secretary)
D. Kortlandt
G. van Reijen
C. P. Willig
Chr. Wilmering

FOREWORD

This volume contains the papers and the major discussions presented at the IFAC Workshop on Reliability of Instrumentation Systems for Safeguarding and Control.

This Workshop, which was sponsored by the IFAC Applications Committee (APCOM), was the first of its kind.

It was organised as a cooperation between the Netherlands Association of Engineers and the Royal Institute of Engineers in the Netherlands.

The aim of the Workshop was to present and discuss the various reliability aspects of modern instrumentation systems for industrial processes. The programme was divided into a number of sessions covering the following topics: System design, Reliability modelling, Field data and maintenance and Human factors.

In addition, invited tutorial papers were given to introduce the subject in more general terms. Due to some dramatic events in industrial processes in the weeks before the Workshop, the awareness of the reliability problems in general and hence instrumentation systems is steadily increasing. This was felt during the informal talks throughout the Workshop.

Although a significant amount of literature is available, this has tended to emphasize more the mathematical and analytical aspects. Many presentations emphasized that the practical aspects of reliability and availability and its assessment are in many circumstances still a large problem. Software and Human reliability aspects are only slightly covered areas. Reliability engineering is a science where a large number of disciplines strongly interact. As most participants came from industry, it reflects the high impact and need in today's industrial life. This was also high-lighted by the fact that the special "Industrial Problem Session" organized on one of the evenings was attended by almost all participants.

The aim was to provide a platform to discuss "real life" problems without the necessity to write a full paper. Nevertheless papers and notes were presented and are recorded in these proceedings under the session headed: "Industrial Problems".

We would like to thank the Netherlands Association of Engineers (NIRIA) for their constant support and assistance in making this first Workshop a success-ful one, the International Program Committee for their effort in the selection of papers and the members of the National Organizing Committee for their support in the organisation.

We hope that by the publication of these papers, which came from specialists of 12 different countries, reliability engineering will find its way into the design, engineering and management of industrial instrumentation systems.

J P Jansen
L Boullart

CONTENTS

INTRODUCTION TO RELIABILITY MODELING

E. Schrüfer

Lehrstuhl für Elektrische Messtechnik, Technische Universität München, FRG

Abstract. Reliability predictions refer to the future behaviour of items and therefore can only be expressed as probabilities. With wearout failures the lifetimes of devices are normally distributed and with random ones they are exponentially distributed. The latter distribution is determined by the failure rate λ, which is independent on time but dependent on operational and environmental conditions (MIL HDBK 217).

Also the reliability of equipments which are composed by parts can be described by a failure rate. The failure effect analysis indicates whether the failures are detectable or not detectable, whether they are safe or unsafe. By means of failure detection, combined with repair and restauration, fault tolerant equipments can be realized. Their availability depends upon failure rate, failure detecting rate and repair rate.

In order to establish high reliable systems the principles of redundancy, diversity and physical and electrical separation are used as preventive measures against random and common mode failures. By this way the availability of the systems is higher than that of the components. It can be numerically predicted by means of a fault tree analysis.

Keywords. Failure rate; failure effect analysis; fault tree analysis; Markov processes; quality control; reliability theory; system failure and recovery.

INTRODUCTION

Recent intrumentation and control systems have not only to fulfill the operational conditions; they have to fulfill them reliably, a goal which is not achieved by chance but only by a careful design /1-6/. The paper deals with some aspects useful for the design of parts, equipments (composed by parts) and systems (composed by equipments). It explains some models suitable for a numerical prediction of failure probability and unavailability. As those quantities describe the future behaviour of items they can only be expressed as probabilities. Therefore this introductory lecture will start with a short chapter about probability distributions. It is followed then by three others in which reliability models of parts, units and systems are discussed.

1 LIFETIME DISTRIBUTIONS

1.1 Definitions

In order to define some terms needed in the reliability theory the following experiment is considered: A lifetime test is started with a number of n_0 items operating at time t = 0; after the time t the number n(t) is still working; the number $n_0-n(t)$ has failed. If the fractional number $n(t)/n_0$ is graphically represented with respect to time, then a curve is obtained like that in Fig. 1.1a. At the beginning all items were operating, at the end all failed. At a given time t the ratio $n(t)/n_0$ will have survived. Its lifetime is equal to or longer than the corresponding time t.

This ratio can be used as an estimate of future behaviour based on the outcomes of a previous series of events. The ratio $n(t)/n_0$ is interpreted as the probability p(t) that an item will reach or

exceed the lifetime t. It is called reliability: The reliability R(t) of an item is the probability that the item will perform a required function under a specified condition for a stated period of time t,

$$R(t) = \frac{n(t)}{n_0} = p(\text{lifetime } X \geq t) . \qquad (1.1)$$

The complement of the reliability is the failure probability F(t). This is the probability that the lifetime of an item is shorter than or equal to a given time t:

$$F(t) = 1 - R(t) = \frac{n_0-n(t)}{n_0} = p(\text{lifetime } X \leq t). \quad (1.2)$$

The failure probability (Fig. 1.1b), but not the reliability, is a cumulative probability distribution function, which can range in value from zero to unity.

If the failure probability function is differentiated with respect to time then the probability density function f(t) is obtained (Fig. 1.1c)

$$f(t) = \frac{dF(t)}{dt} = - \frac{dR(t)}{dt} . \qquad (1.3)$$

This function divided by the reliability R(t) gives the hazard function or failure rate function z(t) (Fig. 1.1d)

$$z(t) = \frac{f(t)}{R(t)} = - \frac{1}{R(t)} \frac{dR(t)}{dt} . \qquad (1.4)$$

As characteristic marks of probability distributions the median and the mean are used. The median t_{50} is the lifetime of the middle item; the mean lifetime $\bar{t}$ is given by

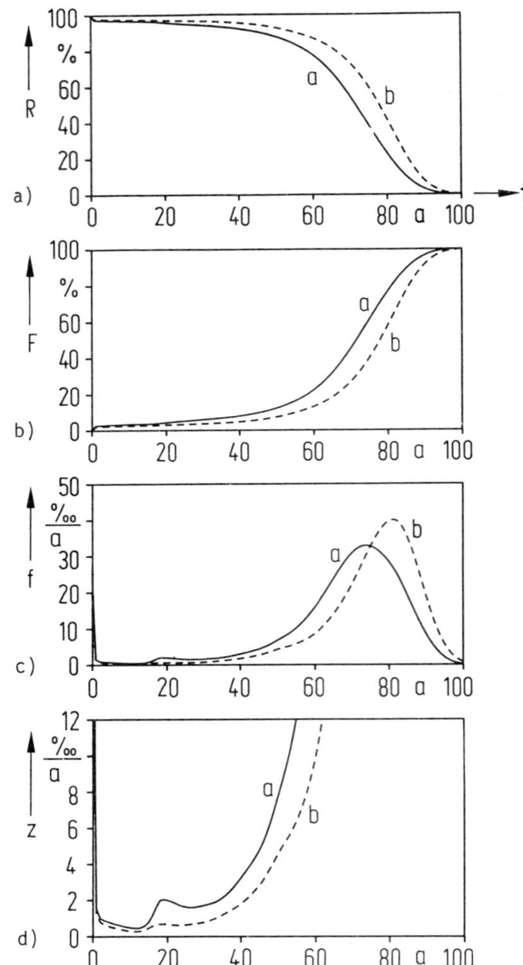

Fig. 1.1 Lifetime Distribution of the People in
         Western Germany /5/
         R reliability, F failure probability,
         f failure probability density function,
         z failure rate function, a male, b female

$$\overline{t} = \int\limits_{0}^{\infty} t \cdot f(t)\, dt,$$

which after some steps yields

$$\overline{t} = \int\limits_{0}^{\infty} R(t)\, dt. \tag{1.5}$$

Fig. 1.1 shows the lifetime distribution of the
people in Western Germany. Curve a refers to males,
curve b to females. Owing to the infant mortality
the "reliability" decreases within the first year
of life, remains fairly constant over a longer period of time and begins to fall down at an age of
50. At any age the mortality of men is higher than
that of women. The probability density function
f(t) has its maximum at 70, respectively 80 years.
The failure rate function z(t) clearly indicates
the "early failures". Furthermore it has an un-
expected peak for boys at the age of about 20
years. This is due to "annatural" accidents mostly
caused by cars and sports. Yet the death rate is
low, that of the infants is not reached until an
age of about 60 years.

Half of the population gets an age of 71 (men),
resp. 77 (women) years, while the mean lifetimes of
67 and 73 years are a bit shorter.

In the following three distributions are briefly
discussed which are important in engineering. These
are

   - the normal distribution, describing the lifetime of items with wear out failures,

   - the exponential distribution effected by random failures and

   - the Weibull distribution applicable for both,
the wear out and random failure mode.

<u>1.2 Normal Distribution</u>

Wear out failures occur as a result of deteriora-
tion processes or mechanical wear. The lifetimes of
items which have failed by wear outs correspond to
the normal or Gaußian distribution. The failure
probability F(t) is defined as

$$F(t) = \frac{1}{\sigma\sqrt{2\pi}} \int\limits_{v=-\infty}^{t} e^{-\frac{1}{2}\left(\frac{v-\overline{t}}{\sigma}\right)^2} dv. \tag{1.6}$$

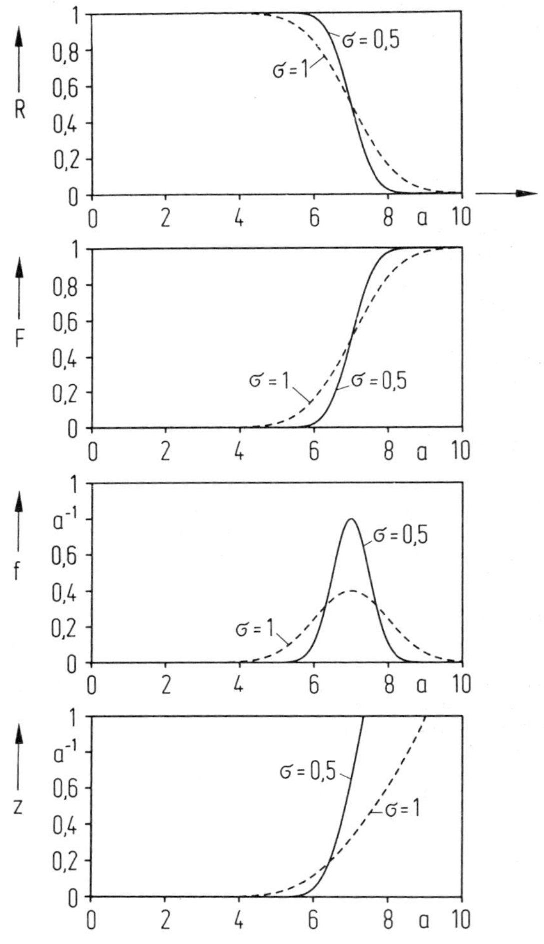

Fig. 1.2 Normal Distribution

It is determined by 2 parameters. The one is the
mean lifetime $\overline{t}$ and the other is the standard deviation σ (Fig. 1.2). The graph of the probability
density function is a bell-shaped curve with a maximum at $t = \overline{t}$. The most likely lifetime t_{max} has
the same value as the median t_{50} and the mean lifetime $\overline{t}$. The standard deviation σ is a shape parame-

ter. It expresses how far the different lifetimes are grouped around the mean.

With the normal distribution the reliability, the probability of success, does not decrease at the beginning of an item's application. There is a period of time in which items do not fail and in which the reliability remains unity. In this region the failure probability and the failure rate are zero. If the latter starts increasing, it increases rather fast and in a short period of time all items will fail.

An example of items with normally distributed lifetimes are candles, light-bulbs or tires. The useful lifetime of these items is limited by wear out. If the items are replaced before entering the region with decreasing reliability, then an operation without breakdowns is possible. Therefore a <u>preventative maintenance</u> is assumed to be very useful.

1.3 Exponential Distribution

Another group of failures are the random ones, which are predictable only in a probabilistic or statistical sense. Such failures predominate in electronic devices. Their lifetimes are exponentially distributed with the reliability $R(t)$ and the failure probability $F(t)$ (Fig. 1.3)

$$R(t) = e^{-\lambda t} \tag{1.7}$$

$$F(t) = 1 - e^{-\lambda t} \tag{1.8}$$

$$\approx 1 - (1 - \lambda t) = \lambda t \quad \text{if } \lambda t \ll 1. \tag{1.9}$$

Differentiating of Eq. (1.8) yields the probability density function $f(t)$

$$f(t) = \lambda e^{-\lambda t} \tag{1.10}$$

and the failure rate function $z(t)$ is obtained by dividing Eq. (1.10) by the reliability $R(t)$

$$z(t) = \frac{\lambda e^{-\lambda t}}{e^{-\lambda t}} = \lambda. \tag{1.11}$$

The failure rate function has a time independent constant value λ. It is called "failure rate" and it is the only parameter determining the exponential distribution.

The time independent failure rate is connected with another important property of the exponential distribution. It has no memory. This results from the fact that the distribution is related to the Poisson resp. Markov process. Asking for the probability that an item will reach the lifetime t, given it has reached the lifetime t_1 with $t_1 < t$ it can be shown that this conditional failure probability $F(t|t_1)$ is obtained by

$$F(t|t_1) = \frac{\frac{F(t)-F(t_1)}{1-F(t_1)}}{}$$

$$= \frac{1-e^{-\lambda t}-(1-e^{-\lambda t_1})}{1-(1-e^{-\lambda t_1})} = F(t-t_1). \tag{1.12}$$

The corresponding conditional failure density function $f(t|t_1)$ then becomes

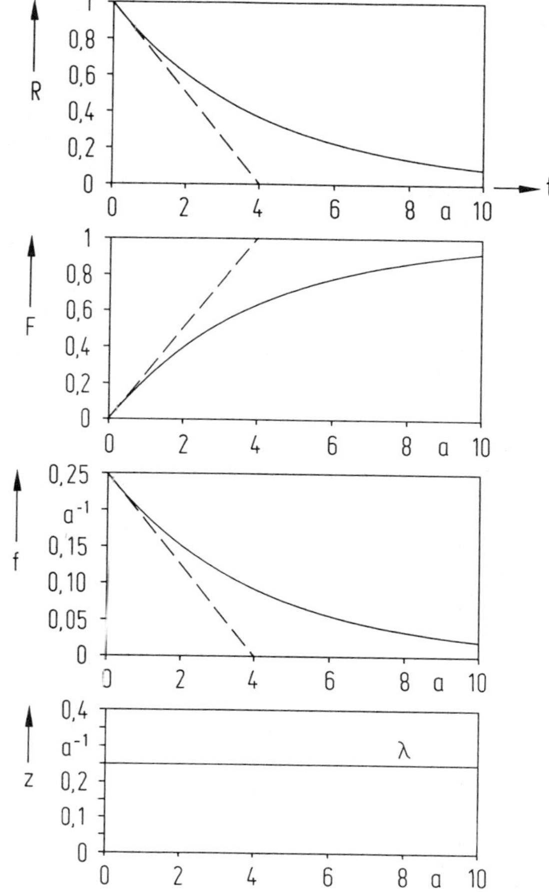

Fig. 1.3 Exponential Distribution

$$f(t|t_1) = \frac{dF(t|t_1)}{dt} = \frac{f(t)}{1-F(t_1)} = \frac{e^{-\lambda t}}{1-(1-e^{-\lambda t_1})}$$

$$= \lambda e^{-\lambda(t-t_1)} = f(t-t_1). \tag{1.13}$$

That means that the functions $F(t|t_1)$ and $f(t|t_1)$ are the original functions $F(t)$ and $f(t)$ but shifted for the time t_1. Has an item survived until the time t_1 then its failure probability starts at this time with $F(t=t_1|t_1) = 0$. The history before t_1 is no longer relevant.

For these reasons a <u>preventative maintenance does not improve</u> the reliability. It is not only not useful, but can even deteriorate the performance. This shall be explained by means of the bath-tub curve in the following section.

1.4 Bath-tub curve

It may be replied that the time independent failure rate is only a mathematical idealization which does not represent the real situation. It should be better to describe the failure rate with respect to time by a curve like that of Fig. 1.4, the well-known bath-tub curve. That is true, but not a serious argument.

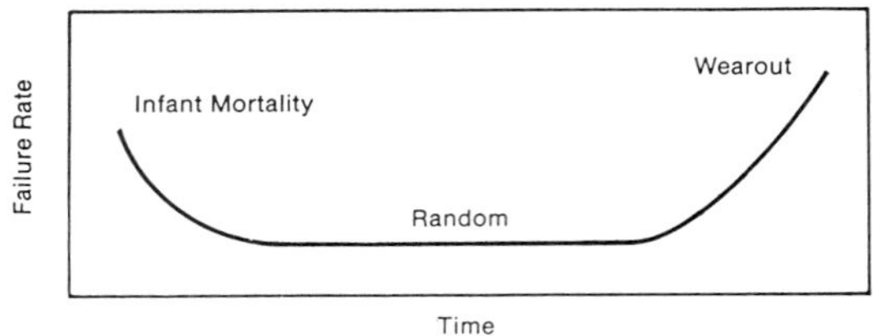

Fig. 1.4 Bath-tub Failure Rate Curve

The resulting characteristic can be seen falling into three distinct phases representing the infant mortality, the random failures and the wear out failures. The infant mortality is concerned with the early-life failure of a part. These failures are usually associated with material or manufacturing defects. The failure rate decreases rapidly and stabilizes when the weak units have died out. That may occur after some hundred hours for discrete semiconductors and after some months for integrated circuits. Are the parts used in larger control systems then the preoperational tests generally last so many hours that the early failures happened before the actual start of the plant. Therefore they do not affect the reliability. But in these cases in which operated parts get replaced by new ones, the possiblity exists that the new devices may early fail. For this reason a preventative renewal is not recommended.

The useful portion of life is the region with the time independent failure rate caused by random events. The low failure rate depends upon operational and environmental stresses. It remains constant for a period up to 30 - 50 years. Then the wear out failures occur and the failure rate increases rapidly. The end of the useful life is reached.

In most industrial applications the parts are not applied for such a long time. After about 10 or 20 years the devices do no longer represent the state of the art and have to be replaced before the wear out failures become essential. The end of the bath-tub is not be reached. Only the period with the time independent failure rate is important for industrial applications.

1.5 Weibull Distribution

The Weibull distribution (Fig. 1.5) can be used to describe all three regions of the bath-tub curve. The distribution has two parameters,

- the scale parameter or characteristic life time T
- the shape paramter a.

The reliability functions are given by

$$R(t) = e^{-(t/T)^a} \qquad (1.14)$$

$$F(t) = 1 - e^{-(t/T)^a} \qquad (1.15)$$

$$f(t) = \frac{at^{a-1}}{T^a} \, e^{-(t/T)^a} \qquad (1.16)$$

$$z(t) = \frac{at^{a-1}}{T^a} . \qquad (1.17)$$

In the region $0 < a < 1$ the failure rate decreases and the infant mortality can be described. With a = 1 the Weibull distribution corresponds to the exponential distribution for $\lambda = 1/T$. With $a > 1$ the failure rate increases. The wear out failures are characterized as with the normal distribution.

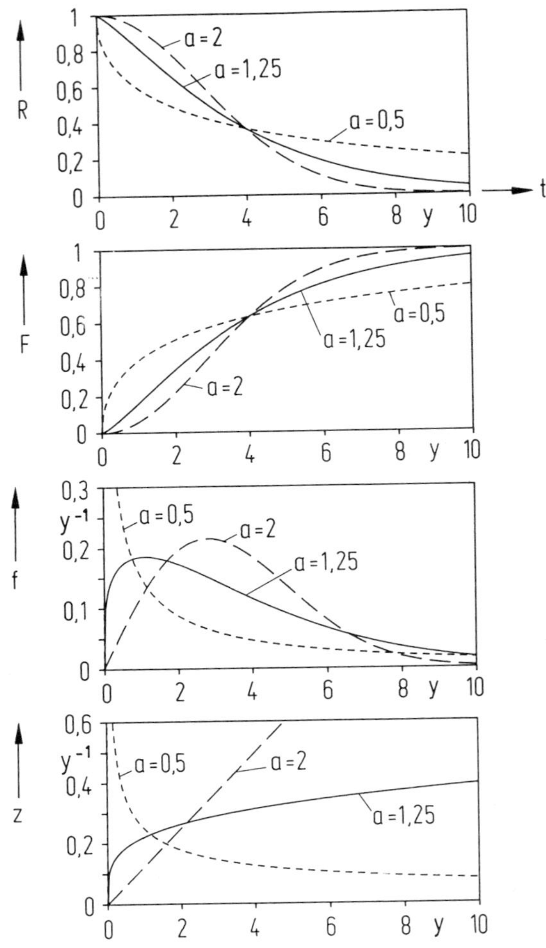

Fig. 1.5 Weibull Distribution

In the region $3 \le a \le 5$ the Weibull distribution approaches the normal distribution especially well.

2 PART FAILURE RATES

2.1 Random and Systematic Failures

We refer to the useful life phase of the parts which is connected with a constant time independent failure rate function. The corresponding life times are exponentially distributed. This model describes not only the behaviour of electronic and electric components such as diodes, transistors, IC's, resistors, inductive devices, capacitors but also that of mechanical parts such as wrapped, soldered, crimped or welded joints, relays and switches.

The failure rate includes only random catastrophic failures (sudden and complete), but not deterministic or systematic ones which occur if an item is not handled under the specified conditions. Such a misuse includes

- the application of a device in environmental stresses beyond those intended
- the human errors resulting in an improper installation, operation, maintenance and transportation and
- the use in a service never intended for the device.

If a transistor fails e.g., because it is operated with too large power or at too large an ambient temperature then the failure is not a random event but caused by the unskillness of the operator. Such failures are not included within the failure rate. They are found nearly as numerous as the random ones. Fig. 2.1 gives the result of an investigation of failed integrated circuits. 47 % of the failures are due to material and manufacturing weakness, but 49 % are caused by the customers' misuse.

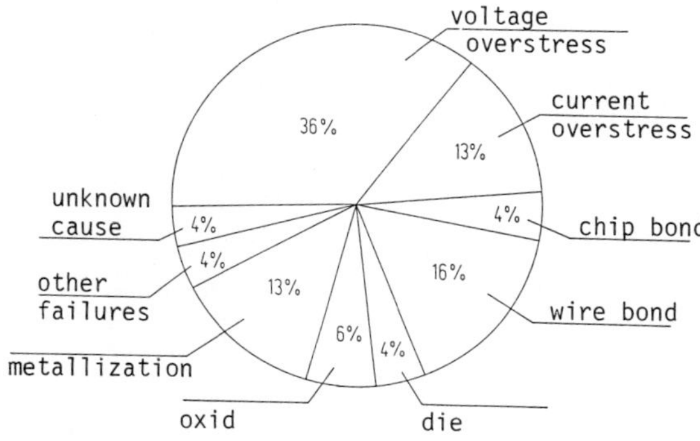

Fig. 2.1 Failures of integrated circuits /4/

2.2 Failure Rate Data

Mostly the failure rates are obtained by analysing data from the field use of operated systems. Manufacturers and particularly the customers of automatic control systems have detailed operation-, failure- and repair-listings. These inform about the appliances installed and the used parts. The failed and repaired devices are reported, too. These are large samples in a mathematical sense and an estimate $\hat{\lambda}$ for the failure rate may be obtained by the number k of failed devices divided by the product NT, N representing the number of parts operated and T the operating time,

$$\hat{\lambda} = \frac{k}{NT} . \qquad (2.1)$$

The probably most detailed failure rate data bank is the Military Handbook MIL HDBK 217 published by the United States Department of Defense /6/. In this report the lifetimes of the equipment used by the US forces - army, navy, air force - and the National Aeronautics and Space Administration are collected and interpreted. The edition A of this handbook was published more than 20 years ago in 1962, the edition D in 1982. In the handbook the failure rates are modeled as the product of a base failure rate λ_b and a number of adjustment factors that relate to the manufacturing process and to the anticipated stresses. Such factors are e.g.

  - the quality factor π_Q, accounting for effects of different quality levels
  - the learning factor π_L
  - the temperature factor π_T, accounting for effects of temperature
  - the electrical stress factors π_V, π_R, π_{S2}, accounting for voltage and power ratings
  - the environmental factor π_E, accounting for environmental effects others than temperature; it is related to application categories
  - the application factor π_A, accounting for effect of application in terms of circuit function
  - the complexity factor π_C, accounting for effect of multiple devices in a single package.

The different failure rates are tabulated in the MIL HDBK. They allow to predict the failure rate of a discrete semiconductor e.g. as

$$\lambda = \lambda_b \cdot \pi_Q \cdot \pi_E \cdot \pi_A \cdot \pi_R \cdot \pi_{S2} \cdot \pi_C \cdot 10^{-6} \ h^{-1}, \qquad (2.2)$$

where the base failure rate λ_b depends upon the temperature. The failure rate of monolithic bipolar and MOS digital SSI/MSI-devices (less than 100 gates) follows from

$$\lambda = \pi_Q [C_1 \pi_T \pi_V + (C_2 + C_3) \pi_E] \pi_L \cdot 10^{-6} \ h^{-1}, \qquad (2.3)$$

In this equation C_1 and C_2 are circuit complexity failure rates based upon gate count and C_3 is the package complexity failure rate.

Some of these factors affecting the part failure rate and particularly relating to the natural enemies of electronic part such as heat, excess voltage and environmental stress are discussed in the following sections.

2.3 Quality System

The manufacturing of a typical integrated circuit requires 35 - 40 processing steps that must be performed with sufficient accuracy to ensure both the reliability and the reasonable cost of the product. Thus the first key to quality is a good process control. In order to ensure the anticipated low failure rate a lot of tests are carried out during and after the production process. The objectives are

  - the elimination of early failures
  - the elimination of weak or potentially weak devices randomly present in a lot of components
  - the elimination of lots having too large a proportion of unstable products.

Beginning 20 years ago, test procedures and very strong and comprehensive screening specifications were developed. Among the US-specification the MIL-STD-883 is applied to IC's and the MIL-STD-750 to discrete semiconductors. Within the European CECC-System the IC's are processed according to CECC-90000 and the discretes according to CECC-50000 /7, 8/. Both systems include the following tests

  - _Internal Visual Inspection_ of the device prior to sealing to screen out defects such as insufficient metallization or oxide and bond defects and to detect the presence of foreign material.

  - _High Temperature Storage_ without electrical power applied to stabilize electrical characteristics.

  - _Temperature Cycling_ to check the thermal compatability of dissimilar materials (die attachment on mounting base).

  - _Constant Acceleration_ to detect mechanically weak devices, particularly loose connections.

  - _Leak Test_ (gross leak and fine leak) to detect faulty seals and to assure package hermeticity.

  - _Burn-in_, operating a device at an elevated temperature with electrical biases applied, to detect excessive parametric drift and to eliminate early failures.

  - _External Visual Inspection_ to assure that mechanical characteristics and visual aspects are within specifications.

  - _Final Electrical Quality Conformance_ to assure that devices are within electrical paramter limits.

Among these the burn-in is the most effective procedure as Table 2.1 shows:

TABLE 2.1 Effectiveness of Screening Tests of hermetically sealed IC's /9/

High Temperature Storage	0 - 15 %
Temperature Cycling	5 - 15 %
Constant Acceleration	0 - 5 %
Burn-in	60 - 80 %
Hermetic Seal	5 - 15 %

In many cases it is technically feasible to produce components which are virtually infinitely reliable. The costs of such parts and of equipment constructed from them would, however, be prohibitive unless - as it is the case with aerospace equipment - any failure cannot be tolerated under any circumstances. On the other hand in applications easily repairable an extremely high price due to high screening costs is less acceptable than a lower commercial grade reliability. Therefore many parts are covered by specifications that have several quality levels. Hereby the highest reliability - the lowest failure rate - is related to the highest price and vice versa. Table 2.2 shows some quality levels and the corresponding relative failure rates. A transistor e.g. can be delivered in 5 different grades. The transistors encapsulated or sealed with organic materials have the lowest quality level P1. That failure rate is a 100 times higher than the failure rate of JAN-TXV-quality.

TABLE 2.2 Quality levels and relative quality factors

parts and spec.	quality levels and rel. factors								
integrated circuits (MIL-M-38510, MIL-STD-883)	S 1	B 2	B-0 4	B-1 6	B-2 13	C 16	C-1 26	D 35	D-1 70
discrete semiconductors (MIL-S-19500, MIL-STD-750)	JANTXV 1	JANTX 2	JAN 10	Lower 50	Plastic 100				
resistors (MIL-STD-199, MIL-STD-202)	S 1	R 3.3	P 10	M 33	NE 165	Lower 500			
capacitors (MIL-STD-199, MIL-STD-202)	S 1	R 3.3	P 10	M 33	L 50	NE 100	Lower 333		

The quality levels refer to devices processed and screened in accordance with supervised US-government specifications. European companies do not or do not exclusively manufacture according to this US-standards but to their own quality control systems. If the models of the MIL-HDBK are applied to these European products then it is necessary to know the quality level of these devices manufactured not in accordance with MIL-standards. To answer this question our Munich institute carried out some investigations and analyses /10/. It was found out that the quality of European commercial parts comes up to level C for IC's, to level JAN (an US-trademark) for discrete semiconductors and to level M ("established reliability" ER) for resistors and capacitors.

2.4 Learning Factor

With the introduction of a new product line difficulties may occur until conditions and controls have stabilized. In the first lot more weak devices may be fount than in the later ones. This fact is taken into consideration for IC's by a learning factor π_L which is 10 under any of the following conditions

- new device in initial production
- where major changes in design or process have occured
- where there has been an extended interruption in production or a change in line personnel.

The factor of 10 can be expected to apply for as much as six months of continuous production /6/.

2.5 Temperature Aspects

As the velocity of many chemical reactions increases with rising temperature the failure rate grows up with temperature, too. With the thermal activation energy E_a (eV), the Boltzmanns constant K (8,63·10⁻⁵ eV/K), and a proportional factor B the failure rate depends upon the temperature T (K) according to the Arrhenius equation as

$$\lambda = B\, e^{-\frac{E_a}{kT}} . \qquad (2.4)$$

Herein the activation energy is the energy which is required for a particular reaction to take place. Each failure mechanism has its own activation energy which is characteristic of that mechanism. It is 0,3 eV for oxide defects e.g. and 1,4 eV for ion migration. With a temperature rise from 25 to 100 °C the failure rate increases by a factor of 10 for E_a = 0,3 eV and by a factor of 100 000 for E_a = 1,4 eV.

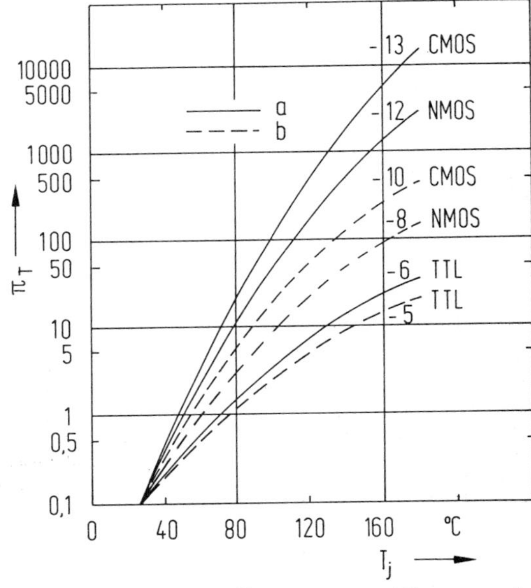

Fig. 2.2 Temperature Factor π_T with respect to Junction Temperature T_j /6/
a nonhermetic package, b hermetic package

The temperature dependence of the failure rate is a very serious effect (Fig. 2.2). In order to avoid these failures caused by temperature and to ensure a reliable operation

- a reduced power consumption
- a improved extensive heat sinking or
- an air conditioned operating area

can become necessary.

On the other hand the Arrhenius model is used for accelerated life testing. A high temperature test of a short duration is correlated to many hours of operation at lower temperatures. If λ_j is the failure rate at junction temperature T_j, then an acceleration factor F may be calculated as

$$F = \frac{\lambda_2}{\lambda_1} = e^{-\frac{E_a}{k}\left(\frac{1}{T_2} - \frac{1}{T_1}\right)} \quad . \qquad (2.5)$$

2.6 Electrical Stress

In addition to temperature, also electrical quantities such as voltage, current or power affect the part failure rate. In order to lengthen the useful life the operation of parts is recommended not with maximum acceptable ratings but with reduced stress conditions. Such a derating will improve the reliability and diminish the failure rate.

To give an example of this effect Fig. 2.3 shows the voltage factor π_V increasing with rising voltage. The failure rate of an IC is proportional to this factor. With a power supply of 15 V a CMOS IC may fail 10 times as much as if it is operated at 5 V.

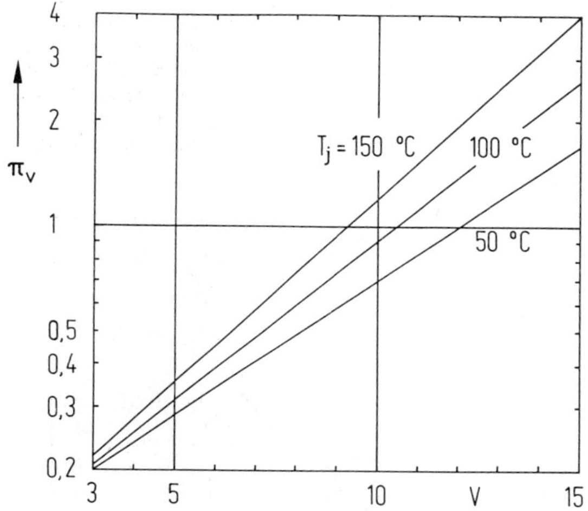

Fig. 2.3 Voltage Stress Factor π_V of CMOS IC's /6/

A serious problem is the electrostatic discharge (ESD) damage, which some reliability engineers have termed "the new contaminant of the age of microelectronics". It occurs in both MOS and bipolar devices. It can cause an immediate or delayed damage to circuits. In many cases no damage can be measured until months have passed and the equipment has failed in the field. Remember that a person walking across a carpet can seperate a charge generating a voltage up to 1000 to 6000 V. In fact, a voltage as low as 100 V is reported to be capable of damaging semiconductor devices. Most of the customer induced failures by voltage overstress (Fig. 2.1) are assumed to have occured by ESD. This happened inspite of input protective devices (e.g. diodes) on the IC to prevent the discharge from rupturing the gate oxide. Other provisions to protect against ESD are

- a proper floor finish
- grounding of furniture
- application of special sprays and
- control of humidity in the work area.

Another kind of damage is caused by current stress. The main effect is electromigration of metallization patterns. The rate of migration is proportional to current density and the failure rate increases approximately with the square of current density.

Derating is an effective procedure for improving reliability, but it has its limitations. The semiconductors must not be derated to such low power that more parts for signal processing are needed. Under these circumstances the reliablity would be reduced due to the larger number of components. Furthermore the signal to noise ratio must not fall below a given level in order to avoid disturbancies by electromagnetic interferences. And last not least mechanical parts such as relay contacts or moving shafts need a periodic operation to remain in good working condition.

2.7 Environmental Stress

The parts' reliability is affected by environmental stresses, too. Unsafe conditions are

- high temperatures
- temperature changes
- water condensations
- agressive gases in the ambient air and
- shocks, vibrations and accelerations.

The MIL HDBK summarizes all these effects by an environmental factor defined for various working areas. The industrial equipment use is encompassed by the areas

- G_B, Ground, Benign: Nonmobile, laboratory environment readily accesible to maintenance; includes laboratory instruments and test equipment, medical electronic equipment, business and scientific computer complexes.

- G_F, Ground, Fixed: Conditions less than ideal such as installation in permanent racks with adequate cooling air and possible installation in unheated buildings; includes permanent installation of air traffic control, radar and communications facilities.

- G_M, Ground, Mobile: Equipment installed on wheeled or tracked vehicles; includes mobile communication equipment.

The different parts such as diodes, transistors, bipolar or MOS IC's, optoelectronic devices have their specific environmental factors. To give an order of magnitude the failure rate in the ground, mobile area is about 10 times higher than in the ground, benign application. An IC installed within the control room of an electrical power station will be more reliable than the same one used for brake control of a car.

2.8 Summary

As discussed in the foregoing sections the part failure rate has not a given value but depends upon both the manufacturers quality system and the customers operation conditions. Especially the thermal, electrical and environmental stresses have to be considered. Fig. 2.4 shows the ranges which the failure rates are found in. The industrial application G_F is marked by a circle.

In the last years the complexity of IC's has been increased, the price has been decreased and nevertheless the reliablity has been enlarged. Thus it is recommendable to use IC's instead of discrete semiconductors not only with respect to the price but also with respect to a small failure rate. If the 1982 edition of the MIL HDBK is compared to the 1979 publication lower failure rates are found. They have been reduced e.g. by the following factors:

Si-diodes	5
Si-NPN-transistors	7
TTL-IC, 50 gates, P1	10
TTL-IC, 5000 gates, P1	50

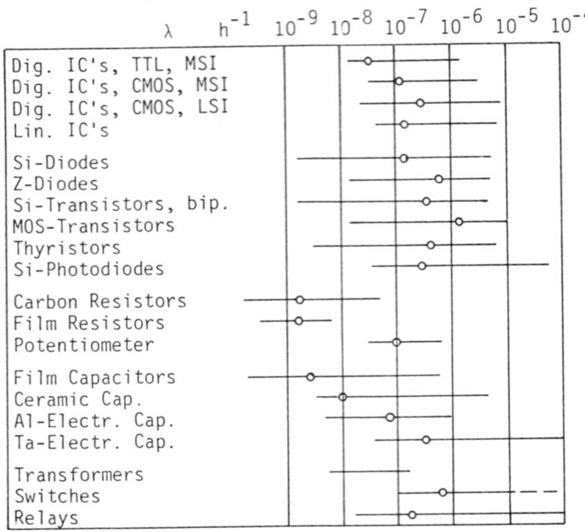

Fig. 2.4 Failure Rates

The failure rate data have been obtained from field
use of passed systems. As those data are used for
reliablity predictions of future systems an uncertainty exists until the new parts become tried and
true. With improved reliability of the latest devices it can be expected that the anticipated
"theoretical" failure rates give an upper limit and
a "conservative" estimate of the real behaviour.
The prediction is the more accurate the higher the
degree of similarity is between the new and the
operated devices both in hardware design and in the
anticipated environments.

3 EQUIPMENT FAILURE RATES

The equipment designer will take into account the
failure rate of the individual parts. He will apply
them to compose modules and the modules will result
in an equipment performing its required function.
One of the properties of the equipment is its reliability. The objectives of this chapter are to
deal with some characteristic quantities determining the reliablity and availability.

3.1 Failure Rates from Field Experience

Data collected from the field use of operating
equipment are assumed to be the most proved and
substancial. As with the parts equation (2.1) may
be used to find an equipment failure rate.

The failure rate is rather high at the beginning of
the production process of a new designed appliance
and decreases the more the better the production
process has been understood (Fig. 3.1). There is a
"learning factor" as with the IC's. Another similarity is found with respect of the environmental
stresses /11/. Resistance thermometers e.g. fail in
a chemical plant twice as much as in an electrical
power station. The failure rate is enlarged by ag-
gressive or corrosive gases and liquids. With mobile application it is increased by a factor of
three compared with the installation in permanent
racks.

Such failure rates based on experience are not
available for new designed appliances. In these
cases it can be extrapolated from used equipment to
the newly designed, if similar parts and modules in
a similar environment are used. If that is not
possible then theoretical analysis become necessary.

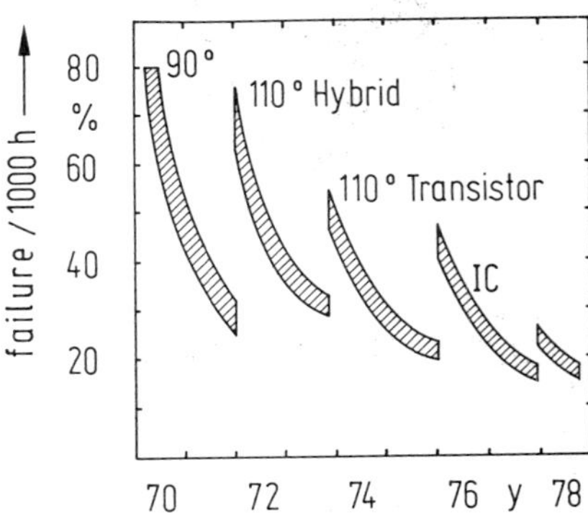

Fig. 3.1 Manufacturing of TV-sets; Learning Effect

3.2 Parts Counting Method

Let be E_i the event, that the part i with failure
rate λ_i is performing its required function. The
probability $p(E_i)$ that the event E_i occurs is with
exponentially distributed lifetimes

$$p(E_i) = e^{-\lambda_i t}.$$

The equipment may be composed by n parts in total,
i = 1, 2, ... n, and each part shall be necessary
for the equipment function. Then the event E (without subscript)

    event E: the equipment is performing its re-
        quired function

is the Boolean conjunction, the intersection of all
the events E_i

$$E = E_1 \wedge E_2 \wedge ... \wedge E_n. \tag{3.1}$$

The event E only occurs if all events E_i occur. The
probability $p(E)$ then results in

$$p(E) = p(E_1) \cdot p(E_2) \cdot ... p(E_n)$$
$$= e^{-\lambda_1 t} \cdot e^{-\lambda_2 t} \cdot ... e^{-\lambda_n t} = e^{-\Sigma \lambda_i t} = e^{-\lambda t}, \tag{3.2}$$

$$\text{where } \lambda = \sum_{i=1}^{n} \lambda_i \tag{3.3}$$

represents the anticipated failure rate λ of the
equipment.

Equation (3.3) is often used as it is easy to
handle. Like in book-keeping the part failure rates
only have to be summed up and an equipment failure
rate is achieved. Yet the problem arises that the
number calculated by Eq. (3.3) is not a confidental
statement. The assumption "all parts are necessary
that the equipment performs its required function"
does not always apply. In most cases an appliance
will fulfill its mission though one of the parts
(or some) has failed. It is even extremely difficult to design an equipment in such a manner that
it will fail with each part fault.

In this situation it is very important to know
whether or not a given fault affects the equipment
function. To answer this question a failure effect
analysis has to be carried out.

3.3 Failure effect Analysis FEA

The failure effect analysis is a systematic procedure for evaluating the failure consequences.

Failure Modes. At the beginning of such an analysis the part failure modes have to be identified. In linear circuits catastrophic failures are assumed. They are simulated by

- interrupting each of the part pins and
- shortening all adjacent pins.

By this way 6 failure effects are to investigate e.g. with a transistor of 3 pins. These are the interruption of base, collector and emitter and the connections of base-collector, collector-emitter and emitter-base. An IC in a package with N pins will have 2N different failure possibilities.

In digital circuits binary signals are processed. Here the failure mode of stationary stuck-failures is defined. In this case a signal can no longer follow any necessary change of its state and will remain "stuck" either at the LOW- or at the HIGH-level. Both such stuck failures are simulated at each pin of a digital circuit. In addition the shortening of all adjacent pins is accomplished as with the linear circuits.

Implementation. Fig. 3.2 shows a computer controlled test circuit for analysing a printed board loaded with discrete semiconductors and IC's. The test circuit needs a power supply, signal generator, digital meter, oscilloscope and a recorder. Before the start of an analysis it has to be made sure that the board performs its required function. Then the first failure is built in and its effect is reported. Sometimes the board is damaged by the investigated failures. In these cases the board must get repaired before the next failures can be simulated and their consequences can be evaluated. Up to 1000 or more failures have to be analysed with a single printed board.

The great advantage of the failure effect analysis is that it can be systematically carried out and that its completeness can be shown. As it is performed in accordance with the part list of the tested module each of the parts with each of the above defined failures is considered.

Grouping the Effects and Summarizing the Information. The gain of the FEA depends upon whether it succeeds in grouping the failure effects. The procedure shall be shown with Table 3.1 as an example. An isolation amplifier was analysed and the reported failure effects were classified into the five groups a to e. It is found that about half of the 555 simulated failures do not affect the amplifier output signal.

Now a failure detecting mechanism is assumed to be monitoring the amplifier output. That could be made e.g. by comparing the output signals, if 3 equal amplifiers are used. In this situation only large deviations between the signals can be discovered. It is possible that the larger errors of groups c and e are revealed, the others remain undetected.

In the next step some information about the application of the analysed equipment is needed. Let the amplifier be used as part of a temperature or pressure monitoring system which shall shut down the process if the input signal exceeds a given level. Here the amplifier is able to fail into two different directions:

- If the output current is too low then a shutdown is not executed in the required time. The safety is violated by the "unsafe" failures of groups d and e.
- If the output current is too high then a shutdown is operated though the process does not need this action. Thus the effects of groups b and c can be classified as "safe" and of course the true signals of group a, too.

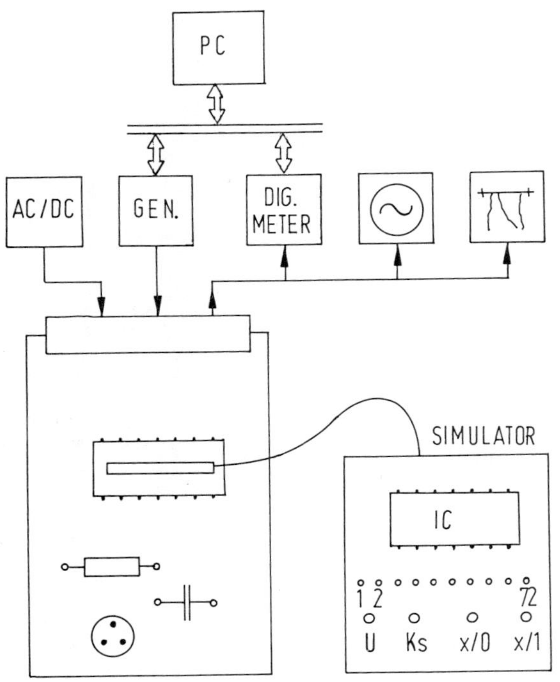

Fig. 3.2 Computer Aided Failure Effect Analysis /5/

TABLE 3.1   Failure Effect Analysis of an Isolation Amplifier
(input -10 ... 0 ... +10 mV; output -20 ... 0 ... +20 mA)

group	effect	number of failures	detectable	undetectable	safe	unsafe	$\Sigma\lambda_i$ 10^{-6} h^{-1}
	output current has						
a	- not changed	231		x	o		11,8
b	- increased by not more than 7,5 %	8		x	o		1,0
c	- increased by more than 7,5 %	53	x		o		5,7
d	- decreased by not more than 7,5 %	30		x		o	2,7
e	- decreased by more than 7,5 %	233	x			o	20,1
	sum	555					41,3

RISC-B

Safety Related Reliability and Failure Probability.
It is not difficult to convert the part failures
into the corresponding failure rates λ_i and to sum
up obtaining the group failure rates λ_j. With them
the total failure rate λ of an equipment can be
written as the sum of four specific failure rates
λ_{ij} (Table 3.2)

$$\lambda = \lambda_{11} + \lambda_{12} + \lambda_{21} + \lambda_{22} \tag{3.4}$$

with

λ_{11} rate of safe, detectable failures (group c)
λ_{12} rate of safe, undetectable failures (group a,b)
λ_{21} rate of unsafe, detectable failures (group e)
λ_{22} rate of unsafe, undetectable failures (group d)

TABLE 3.2   Breaking down of the Failure Rate

failures are	detectable	undetectable
safe	λ_{11}	λ_{12}
unsafe	λ_{21}	λ_{22}

The different failures are allowed to be considered
as not mutually exclusive events E_{ij} which can
happen at the same time. The equipment will perform
its required function in respect to the safety
- abbreviated as event E_s -, if both the events E_{21}
(no unsafe, detectable failure) and E_{22} (no unsafe,
undetectable failure) occur,

$$E_s = E_{21} \wedge E_{22}. \tag{3.5}$$

The corresponding probabilities are

$$p(E_s) = R_s(t) = e^{-\lambda_{21}t} \cdot e^{-\lambda_{22}t} = e^{-(\lambda_{21}+\lambda_{22})t}. \tag{3.6}$$

$\overline{E}_s$ is the event of an unsafe failure either detectable or undetectable

$$\overline{E}_s = \overline{E}_{21} \vee \overline{E}_{22} \tag{3.7}$$

with the probability

$$p(\overline{E}_s) = F_s(t) = F_{21} + F_{22} - F_{21}F_{22}$$

$$= 1 - e^{(\lambda_{21}+\lambda_{22})t} = 1 - R_s(t). \tag{3.8}$$

With the numbers of Table 3.1 the rate λ_{22} of an
unsafe, undetectable fault is only about 5 % of the
total failure rate and the corresponding failure
probability F_{22} in respect to the safety is appro-
priately low.

3.4 Consideration of an Repair Process;
    Availability and Unavailability

Maintenance Function and Repair Rate. Appliances
are considered having a failure detecting mechanism. As soon as a failed equipment is monitored a
repair process will be started. That will take some
time. Some repairs can be realized fairly fast,
others will take more time. As experience shows the
time to repair can be handled as exponentially distributed (Fig. 3.3). With a repair rate μ it is
possible to define a maintenance function M

$$M = \frac{\text{number of repaired appliances}}{\text{number of failed appliances}} = 1 - e^{-\mu t}. \tag{3.9}$$

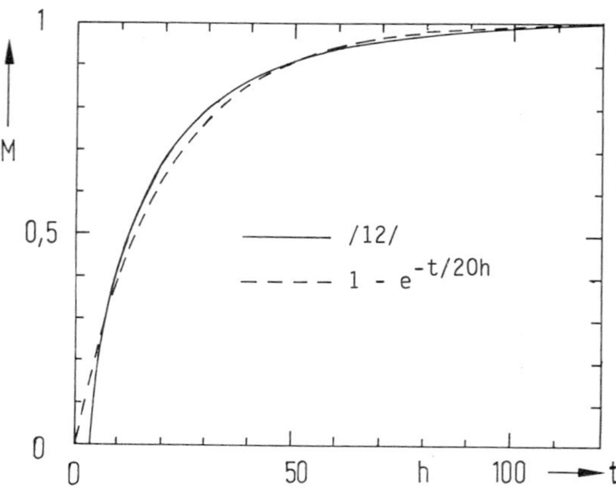

Fig. 3.3 Repair Time of Diesel Engines /12/

Including repair alternating uptimes (the equipment
is performing its function) and downtimes (the
equipment is not performing its function) will
appear. If the uptimes are averaged the mean time
between failure MTBF will result. With exponentially distributed lifetimes this is equal to the
inverse of the failure rate λ just like the MTTF,

$$MTBF = \frac{1}{\lambda}. \tag{3.10}$$

The mean of downtimes is the meantime to repair
MTTR which is determined by the repair rate μ.

$$MTTR = \frac{1}{\mu}. \tag{3.11}$$

Availability and Unavailability. A repair process
can formally be treated by the Markov chain of Fig.
3.4. State 1 is the successfull or working state;
state 2 is the failed one. The equipment passes
with the failure rate λ from state 1 to state 2.
There a repair process is started and after restauration with the repair rate μ the equipment is restored to the "as new" condition. If the probability of being in the state i is called p_i then the
differential equations results

$$\dot{p}_1 = -\lambda p_1 + \mu p_2 \tag{3.12}$$
$$\dot{p}_2 = \lambda p_1 - \mu p_2. \tag{3.13}$$

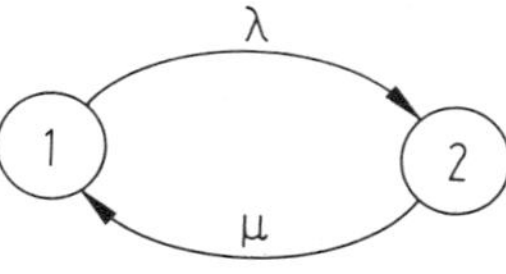

Fig. 3.4 Markov Chain of an Equipment with Repair;
        λ failure rate, μ repair rate

With the boundary condition $p_1(t) + p_2(t) = 1$ the
solutions are found as

$$p_1(t) = \frac{\mu}{\lambda+\mu} + \frac{\lambda}{\lambda+\mu} e^{-(\lambda+\mu)t} \tag{3.14}$$

$$p_2(t) = 1 - p_1(t) = \frac{\lambda}{\lambda+\mu} - \frac{\lambda}{\lambda+\mu} e^{-(\lambda+\mu)t}. \tag{3.15}$$

Fig. 3.5 shows the occupation probability p_2 of the
failed state with respect to time. Its course is
quite different from that of the failure probability. Whereas an equipment without repair will surely

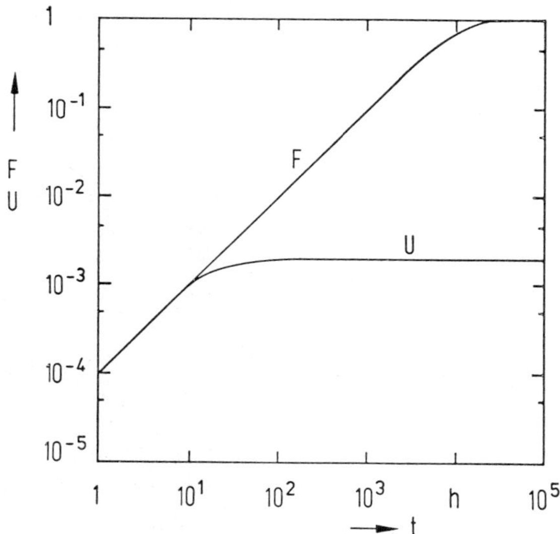

Fig. 3.5 Failure probability F and Unavailability U
         of an Equipment with Repair

failure rate $\lambda = 10^{-4}$ h^{-1};

repair rate $\mu = 5 \cdot 10^{-2}$ h^{-1}

have failed after some time $[F(t \to \infty) = 1]$, the
equipment with restauration will be found in the
failed state at most with the probability p_2

$$p_2(t \to \infty) = \frac{\lambda}{\lambda + \mu} \qquad (3.16)$$

$$\approx \frac{\lambda}{\mu} \quad \text{if } \lambda \ll \mu. \qquad (3.17)$$

It has to be distincted an irreversible process
without repair and a reversible one with renewal.
The terms <u>reliability</u> and <u>failure probability</u> belong to the first. The second is characterized by
the occupation probabilities p_1 and p_2. $p_1(t)$ is
defined as <u>availability A(t)</u> and its complement is
the <u>unavailability U(t)</u>. Reliability and availability respectively failure probability and unavailability are defined in the same manner. R(t) and
A(t) e.g. express the probability that an item will
perform its required function.

3.5 Consideration of the Failure Detection Time

<u>Failure Detection Function and Failure Detection
Rate.</u> Now the time needed for detecting the failures is taken into consideration. The facts will be
similar to those concerning the repair: Some failures will be discovered very fast after occuring,
others will be revealed only after some time. The
failure detection times were found to be exponentially distributed (Fig. 3.6). Thus a failure
detection function d(t) can be defined as

$$d(t) = \frac{\text{number of failures detected}}{\text{number of failures occured}} = 1 - e^{-\varepsilon t}. \qquad (3.20)$$

In this equation   represents the failure detecting
rate. Its reciprocal is the mean failure detection
time MFDT,

$$\text{MFDT} = \frac{1}{\varepsilon} . \qquad (3.21)$$

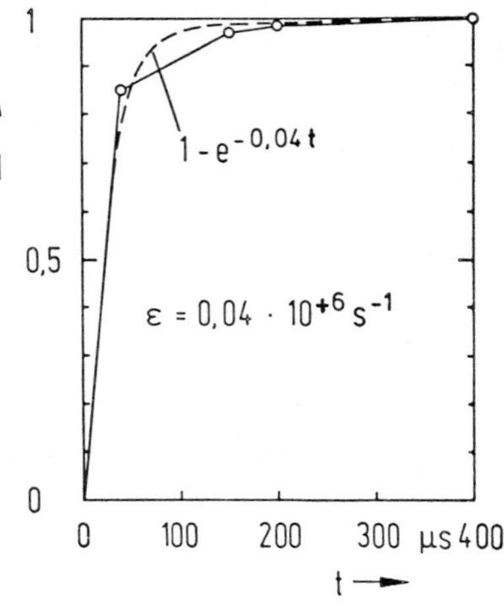

Fig. 3.6 Failure Detection Function d of a process
         computer /13/

<u>Availability and Unavailability.</u> Considering the
time needed for failure detection the Markov chain
of Fig. 3.4 can be expanded to that of Fig. 3.7
with the states

    State 1: equipment is performing its function
    State 2: equipment is not performing its func-
             tion;the failure is not yet detected
    State 3: equipment is not performing its func-
             tion; the failure is detected and the
             repair is started.

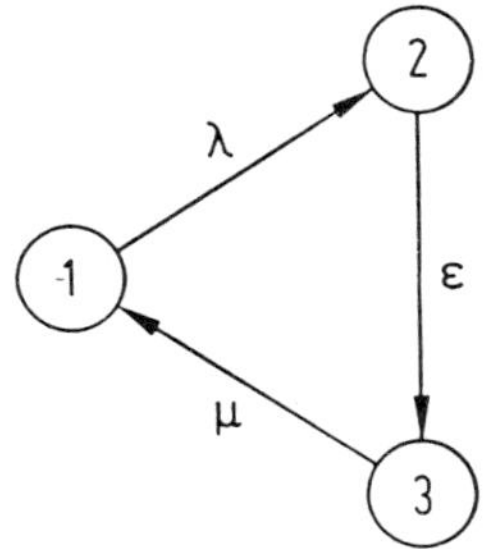

Fig. 3.7 Markov Chain of an Equipment with Failure
         Detection and Repair
         λ failure rate, ε failure detecting rate,
         μ repair rate

With the failure rate λ the equipment passes from
state 1 to state 2, with the failure detecting
rate ε from state 2 to state 3 and with the repair
rate μ the equipment returns to the "as new" state
1. The corresponding differential equations are

$$\dot{p}_1 = -\lambda p_1 + \mu p_3 \qquad (3.22)$$

$$\dot{p}_2 = \lambda p_1 - \varepsilon p_2 \qquad (3.23)$$

$$\dot{p}_3 = \varepsilon p_2 - \mu p_3, \qquad (3.24)$$

with the asymptotic solutions for $p_1 + p_2 + p_3 = 1$

$$p_1(t \to \infty) = \frac{\text{MTBF}}{\text{MTBF} + \text{MFDT} + \text{MTTR}} \qquad (3.25)$$

$$p_2(t \to \infty) = \frac{\text{MFDT}}{\text{MTBF} + \text{MFDT} + \text{MTTR}} \qquad (3.26)$$

$$p_3(t \to \infty) = \frac{\text{MTTR}}{\text{MTBF} + \text{MFDT} + \text{MTTR}} \qquad (3.27)$$

The required equipment function is available in
state 1,

12   E. Schrüfer

$$A(t) = p_1(t) \tag{3.28}$$

and not available in state 2 and state 3,

$$U(t) = p_2(t) + p_3(t). \tag{3.29}$$

<u>Asymptotic Infinitely Fast Repair.</u> Sometimes the failed equipment is replaced by a standby unit as soon as the failure was detected. Under those circumstances nearly no repair times occur and the repairrate μ becomes asymptotic infinite, $\mu \to \infty$. The three states of Fig. 3.7 are reduced to the two of Fig. 3.8. This Markov chain is identical to that of Fig. 3.3, if the failure detection rate ε is used instead of the repair rate μ. According to Eq. 3.16 the asymptotic unavailability U of the equipment becomes

$$U(t \to \infty) = \frac{\lambda}{\lambda + \varepsilon} \tag{3.30}$$

$$\approx \frac{\lambda}{\varepsilon} \quad \text{if } \lambda \ll \varepsilon. \tag{3.31}$$

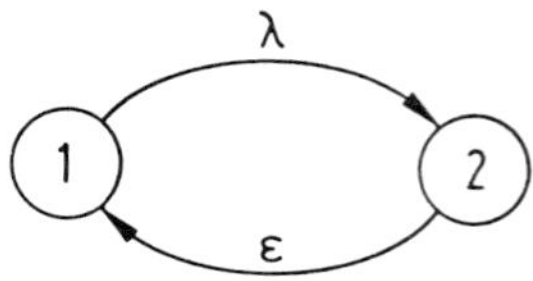

3.8 Markov Chain of an Equipment with Infinitely Fast Repair (MTTR = 0)

<u>Safety related availability and unavailability.</u> Only the unsafe failures with the rates λ_{21} and λ_{22} are related to the safety. If detectable failures occur then the equipment will be restored. No deficiencies exist against the undiscovered faults. With the events

$\overline{E}_s$: equipment does not perform its required safety function

$\overline{E}_1$: equipment has an unsafe detectable failure

$\overline{E}_2$: equipment has an unsafe undetectable failure

the equipment will not perform its required safety function if any of the two compatible events $\overline{E}_1$ and $\overline{E}_2$ or both will occure,

$$\overline{E}_s = \overline{E}_1 \vee \overline{E}_2 . \tag{3.32}$$

With the following probabilities of the events $\overline{E}_i$

$$p(\overline{E}_1, t \to \infty) = p_{21} = \frac{\lambda_{21}}{\varepsilon} \tag{3.33}$$

$$p(\overline{E}_2, t) = p_{22} = 1 - e^{-\lambda_{22}t} \tag{3.34}$$

$$\approx \lambda_{22}t \quad \text{if } \lambda_{22}t \ll 1 \tag{3.35}$$

the total safety related unavailability $p(\overline{E}_s) = U_s(t)$ comes out as

$$U_s(t) = p(\overline{E}_1) + p(\overline{E}_2) - p(\overline{E}_1) \cdot p(\overline{E}_2)$$

$$= \frac{\lambda_{21}}{\varepsilon} + (1 - e^{-\lambda_{22}t}) - \frac{\lambda_{21}}{\varepsilon}(1 - e^{-\lambda_{22}t}) \tag{3.36}$$

$$\frac{\lambda_{21}}{\varepsilon} + \lambda_{22}t \quad \text{if } \lambda_{22}t \ll 1$$
$$\lambda_{21} \ll \varepsilon. \tag{3.37}$$

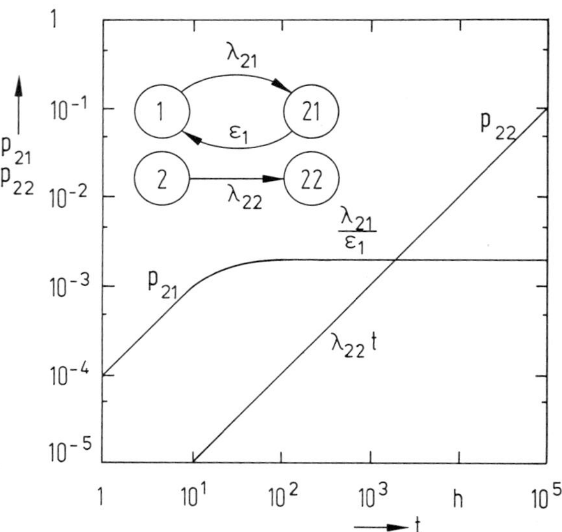

Fig. 3.9 Safety related Unavailability of an Equipment /5/;
rate of unsafe detectable failures
$\lambda_{21} = 10^{-4}$ h^{-1},
failure detecting rate $\varepsilon = 5 \cdot 10^{-2}$ h^{-1},
rate of unsafe undetectable failures
$\lambda_{22} = 10^{-6}$ h^{-1},
repair rate $\mu \to \infty$

The terms of this unavailability $U_s(t)$ with respect to time are shown in Fig. 3.9.

In the example the rate of undetectable unsafe failures is only 1 % of the rate of detectable ones. The failure detection rate corresponds to a MFDT of 20 h. For short times the more numerous detectable faults affect the unavailability more than the undetectable ones. Yet for longer times the influence of the undetectable failures grows until the safety related unavailability is only determined by those.

3.6 Characteristic Quantities of the Failure Detection

As the failure detection is one of the most important features determining the availability and the possibility to get fault tolerant systems some parameters shall be discussed and some useful formulas shall be given in the following sections.

<u>Efficiency.</u> If an test mechanism for discovering the failures has been designed then its efficiency has to be evaluated. This can be achieved by making use of a failure effect analysis. In this case the failures are built into the equipment monitored by the failure detecting mechanism to be checked. Then its failure detecting efficiency, its coverage factor can be found according to Eq. (3.20) as the ratio of the number of failures detected to the total numbers of failures built in

$$d = \frac{\text{number of failures detected}}{\text{number of failures built in}} . \tag{3.38}$$

<u>Total Efficiency of Several Failure Detecting Circuits.</u> Often several failure detecting means are applied. Each coverage factor is generally less than 100 % and no test mechanism will find every failure. Some failures may be revealed by one or more detecting routines; others will not be discovered at all. In order to evaluate the total efficiency of all the different test mechanisms the following not mutually exclusive events E_i and their corresponding probabilities are assumed:

event E_i: the failure detecting mechanism i discovers the failure
$$p(E_i) = d_i$$

event $\overline{E}_i$: the failure detecting mechanism i does not discover the failure
$$p(\overline{E}_i) = \overline{d}_i = 1 - d_i.$$

The bulk efficiency d_{tot}, the probability that anyone of the n applied test mechanisms, i = 1 ... n, will discover the failure is given by

$$d_{tot} = p(E_1 \vee E_2 \vee \ldots E_n) = 1 - \prod_{i=1}^{n} p_i(\overline{E}_i) =$$

$$= 1 - \prod_{i=1}^{n} \overline{d}_1 \qquad (3.39)$$

Portion of Undetected Failures. The complement of (3.39) represents the probability that a failure will not be discovered

$$\overline{d}_{tot} = p(\overline{E}_1 \vee \overline{E}_2 \vee \ldots \overline{E}_n) = 1 - d_{tot} = \prod_{i=1}^{n} \overline{d}_1 \qquad (3.40)$$

Profit of an Additional Test Mechanism. If several failure detecting circuits are applied the gain of an additional expense is of interest. It is useful, if it reveals failures not discovered by the other monitoring circuits. By defining the events $\overline{E}_1$ and $\overline{E}_2$

event $\overline{E}_1$: test mechanism #1 does not discover the failure

event E_2: test mechanism #2 discovers the failure

the additional efficiency $d_2^{\star}$ of the failure detecting mechanism #2, - the profit of this mechanism - can be found as the probability that both events will occur

$$d_2^{\star} = p(\overline{E}_1 \wedge E_2) = \overline{d}_1 d_2. \qquad (3.41)$$

In general the additional efficiency of the test circuit #n is given by

$$d_n^{\star} = p(\overline{E}_1 \wedge \ldots \overline{E}_{n-1} \wedge E_n) = \overline{d}_1 \ldots \overline{d}_{n-1} \cdot d_n. \qquad (3.42)$$

Weighted Mean Failure Detection Time. Each of the n different failure detection circuits has its own specific $MFDT_i$. From them the resulting $MFDT_{res}$ can be calculated as the weighted mean by

$$MFDT_{res} = \frac{d_1 MFDT_1 + d_2^{\star} MFDT_2 + \ldots + d_n^{\star} MFDT_n}{d_1 + d_2^{\star} + \ldots d_n^{\star}} \qquad (3.43)$$

and the resulting failure detection rate ε_{res} is achieved as the reciprocal of $MFDT_{res}$.

Example. The application of the above formulas is demonstrated by Table 3.3. There a system of 3 parallel computers is considered which are performing the same tasks. As failure detecting ciruits are used

- the comparison of the output signals
- a test of the input/output-interface
- a test of the central processing unit
- a test of the memory
- a main inspection and maintenance.

Each of these test mechanisms has an failure detection efficiency insufficient for high reliable applications. But an acceptable total efficiency $d_{tot} = 1 - 0,000048$ is achieved by the combination of all test circuits.

3.7 Fault of the Failure Detecting Circuit

At last the fault of the failure detecting circuit itself shall be considered. This situation will become unsafe if

- the monitoring circuit fails first and
- the equipment to be monitored fails second.

Hereby all failures remain undetected and a restauration cannot be carried out any longer. In order to evaluate the probability of this situation the following events are defined

event C: The failure detecting circuit (failure rate λ_k) has failed.

event D: The equipment has an unsafe detectable failure (failure rate λ_{21}) which can not be discovered.

event F: Both the events C and D occur: $F = C \wedge D$.

event G: Event C occured first given event F occured.

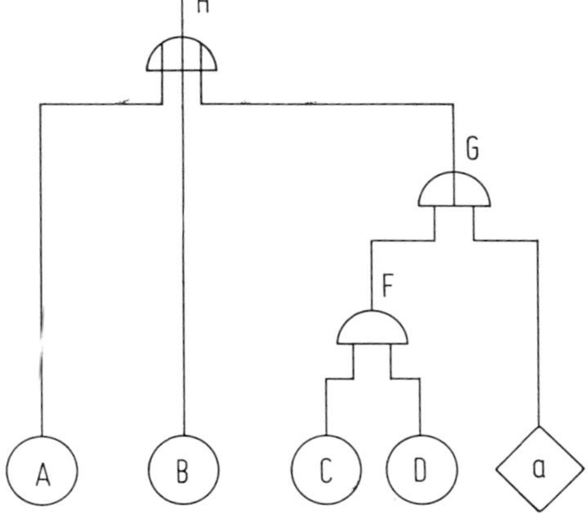

Fig. 3.10 Fault tree of a Monitored Equipment /5/

TABLE 3.3 Failure Detection in a triple-computer-system /5/

i	failure detection by	d_i	$\overline{d}_i$	$\overline{d}_{tot}$	$d_i^{\star}$	$MFDT_i$	$MFDT_{res}$
1	comparison of the results	0,7	0,3	0,3		1 s	
2	test of input/output-interface	0,6	0,4	0,12	0,18	1 s	1 s
3	test of central processing unit	0,9	0,1	0,012	0,108	3 min	20,6 s
4	memory test	0,6	0,4	0,0048	0,0072	1 h	46,5 s
5	main inspection	0,99	0,01	0,000048	0,004752	5000 h	23,8 h

After an operating time $t = 10\ 000$ h and with $\lambda_k = 10^{-7}$ h^{-1} and $\lambda_{21} = 10^{-6}$ h^{-1} the corresponding probabilities result as

$$p(C) = \lambda_k t = 10^{-7} \cdot 10^4 = 10^{-3}$$
$$p(D) = \lambda_{21} t = 10^{-6} \cdot 10^4 = 10^{-2}$$
$$p(F) = p(C) \cdot p(D) = 10^{-5}$$
$$p(G) = 0,5 \cdot p(F) = 5 \cdot 10^{-6}.$$

Now this probability $p(G)$ shall be compared with two other failure possibilities

event A: The equipment has an unsafe undetectable failure ($\lambda_{22} = 10^{-8}$ h^{-1})
$$p(A) = \lambda_{22} t = 10^{-8} \cdot 10^4 = 10^{-4}.$$

event B: The equipment has an unsafe detectable failure ($\lambda_{21} = 10^{-6}$ h^{-1}). The failure detecting circuit with the detecting rate $\varepsilon = 5 \cdot 10^{-2}$ h^{-1} has discovered the failure and the renewal is started. The asymptotic probability, being the worst case probability of the event B is
$$p(B) = \lambda_{21}/\varepsilon = 10^{-6}/5 \cdot 10^{-2} = 2 \cdot 10^{-5}.$$

The equipment fails, - the event H - if at least one of the events A, B or G has occured (Fig. 3.10)

$$H = A \vee B \vee G.$$

The corresponding probability $p(H)$ approximately results in

$$p(H) \approx p(A) + p(B) + p(G)$$
$$\approx 1 \cdot 10^{-4} + 0,2 \cdot 10^{-4} + 0,05 \cdot 10^{-4}$$
$$\approx 1,25 \cdot 10^{-4}. \tag{3.44}$$

The most important contribution to the safety related unavailability $U_s = p(H)$ is supplied by the undetectable equipment failures. That is the same behaviour as in Fig. 3.9. The probability of failing of both the equipment and the monitoring device is sufficiently low with the given failure rates. A further monitoring of the failure detecting circuit obviously is not necessary.

4 RELIABILITY AND AVAILABILITY OF SYSTEMS

Equipments are used to form systems. The equipments can fail either in random or in common mode. In order not to injure the ability of the system to perform its required function preventative measures become necessary. Thus the principle of redundancy is applied to get tolerant towards random failures, whereas other defences such as diversity and physical and electrical separation are necessary to serve common mode failures. As an efficient tool for evaluating the availability of redundant and diverse designed systems the fault tree analysis often is used.

4.1 Protection against Random Failures by Redundancy

4.1.1 Performance Criteria

This section starts with listing some performance criteria which will be helpful to compare the different systems. The quantities needed are

a) the reliability $R(t)$
b) the derivation of the reliability in respect to time for $t = 0$
c) the MTBF $= \overline{t}$
d) the failure probability $F(t)$
e) the approximated failure probability $F_a(t)$ for $\lambda t \ll 1$.

For a single, not redundant element with an exponentially distributed lifetime the criteria mentioned above are:

a) $R(t) = e^{-\lambda t}$

b) $\left. \dfrac{dR}{dt} \right|_{t=0} = -\lambda e^{-\lambda t} \Big|_{t=0} = -\lambda$

c) $\text{MTBF} = \overline{t} = \displaystyle\int_{t=0}^{\infty} R(t)dt = \dfrac{1}{\lambda}$

d) $F(t) = 1 - R(t) = 1 - e^{-\lambda t}$

e) $F_a(t) = 1 - (1 - \lambda t) = \lambda t \quad$ if $\lambda t \ll 1$

The same quantities are used in Table 4.1 to characterize the performance of various redundant systems.

TABLE 4.1 Reliability Criteria of Redundant Systems with Identical Units (Failure Rate λ)

	criteria	no redundancy single unit	active red. (1 out of n)- system	active red. (2 out of 3)- system	standby red. (1 out of 2)- system $\lambda_k = 0$
a	reliability $R(t)$	$e^{-\lambda t}$	$1 - (1-e^{-\lambda t})^n$	$3e^{-2\lambda t} - 2e^{-3\lambda t}$	$(1+\lambda t)e^{-\lambda t}$
b	$\left. \dfrac{dR}{dt} \right|_{t=0}$	$-\lambda$	0	0	0
c	$\text{MTBF} = \overline{t}$	$\dfrac{1}{\lambda}$	$\dfrac{1}{\lambda} + \dfrac{1}{2\lambda} + \cdots \dfrac{1}{n\lambda}$	$\dfrac{5}{6}\dfrac{1}{\lambda}$	$2\dfrac{1}{\lambda}$
d	fail. prob. $F(t)$	$1 - e^{-\lambda t}$	$(1 - e^{-\lambda t})^n$	$1-3e^{-2\lambda t}+2e^{-3\lambda t}$	$1-(1+\lambda t)e^{-\lambda t}$
e	approx. f. p. $F_a(t)$	λt	$(\lambda t)^n$	$3\lambda^2 t^2$	$(\lambda t)^2$

4.1.2 Active Redundancy without Repair

The active redundancy will use parallel units all operated simultaneously. The (1 out of n)-system will work as long as at least 1 of the n installed units is working. The (m out of n)-voting-system, in which the (majority) m of the n installed units determines the system function has some advantages.

Some reliability functions are shown in Fig. 4.1 and 4.2. The curves shall not be discussed in detail. It only shall be ventilated whether or not the MTBF is a good criterion to compare the systems.

After the MTBF the reliability has decreased down to about 40 %; or in other words, 60 % of all units have failed. As there is no technical process to be controlled with 60 % of all equipment out of operation such a situation happens quite unlikely. The MTBF will never be reached.

It is obvious that in practice the interesting reliability must be larger than 90 % and that therefore the mission time must be less than the MTBF. The behaviour of a system within short times compared with the MTBF can be seen, if the failure probability function is presented in a logarithmic scale. If this is made, then the failure probabilities of different systems are found to differ by some orders of magnitude. In the region where the exponential function $\exp(-\lambda t)$ may be approximated

by $(1-\lambda t)$, the failure probability of the single element is increased with λt, that of the (1 out of n)-system with $(\lambda t)^n$. As λt is a very small fractional number, the failure probability of the parallel system is much less than that of the single element. The gain of the active redundancy is caused by the increase of reliability within a given period of time after starting and not by the enlargement of the MTBF. For this reason the approximated failure function $F_a(t)$ or the gradient dR/dt of the reliability function is more suitable to describe the performance of redundant systems.

This statement is approved by discussing the (2 out of 3)-system of Fig. 4.2. Its MTBF_{2v3} is shorter than the MTBF_1 of the single element,

$$\mathrm{MTBF}_{2v3} = \frac{5}{6}\,\mathrm{MTBF}_1.$$

Would the MTBF be the most important quality, then it would be nonsense to install 3 units instead of 1. The advantage of the (2 out of 3)-system is to have the better reliability for short mission times. But its failure probability is always higher than that of a (1 out of 2)-system. This is the price to be paid for the fact, that the (2 out of 3)-system works bimodally. It will perform its required functions not only if a single unit fails to give a signal when called upon to do so, but also if a single unit gives a signal when not called upon to do so.

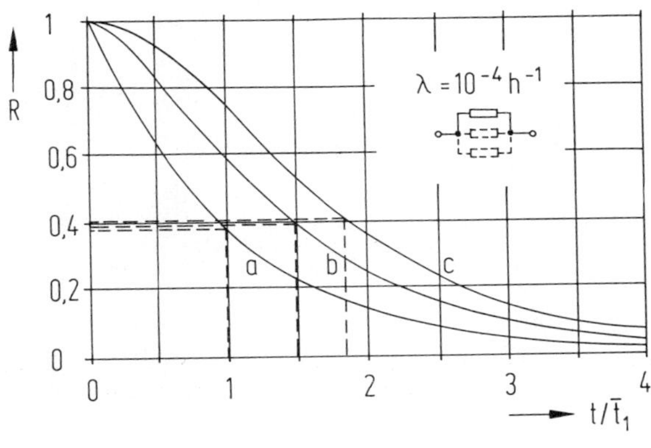

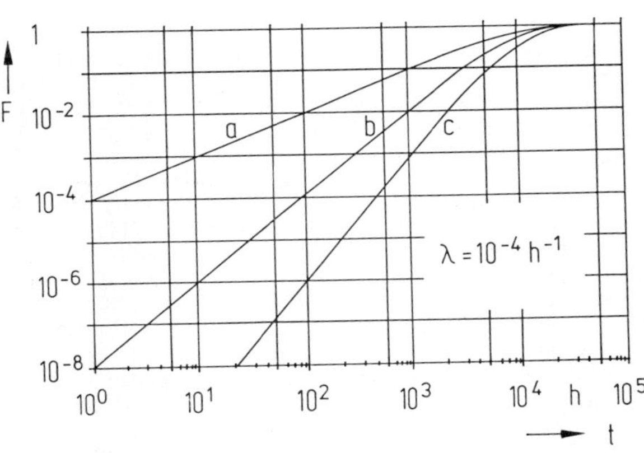

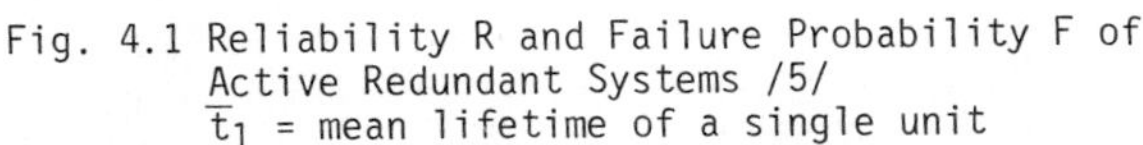

Fig. 4.1 Reliability R and Failure Probability F of Active Redundant Systems /5/
$\overline{t}_1$ = mean lifetime of a single unit

   a) not redundant single unit
   b) (1 out of 2)-system
   c) (1 out of 3)-system

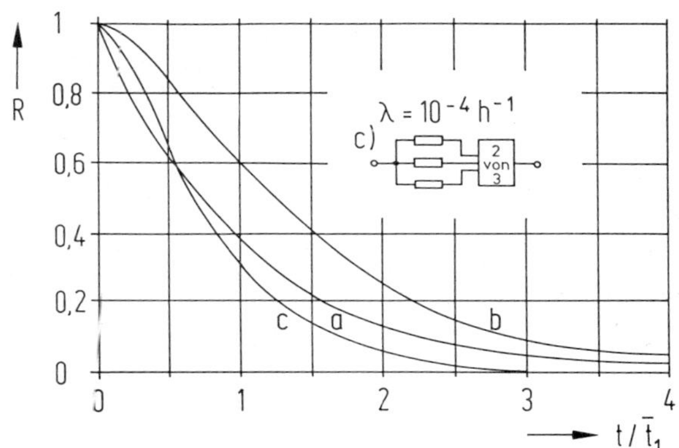

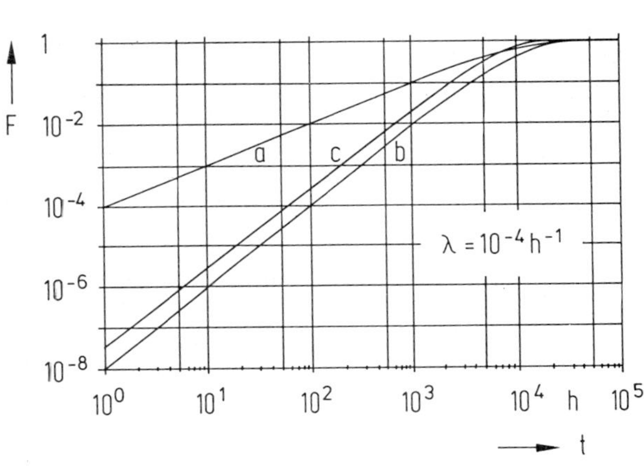

Fig. 4.2 Reliability R and Failure Probability F of Active Redundant Systems /5/
$\overline{t}_1$ = mean lifetime of a single unit

   a) not redundant single unit
   b) (1 out of 2)-system
   c) (2 out of 3)-system

4.1.3 Standby Redundancy without Repair

A standby redundant system in the simplest manner (Fig. 4.3) consists of two units - one of them being operated and the other serving as a spare - and of an additional master element which has

- to decide whether or not the operated unit has failed and
- to switch to the standby unit if needed.

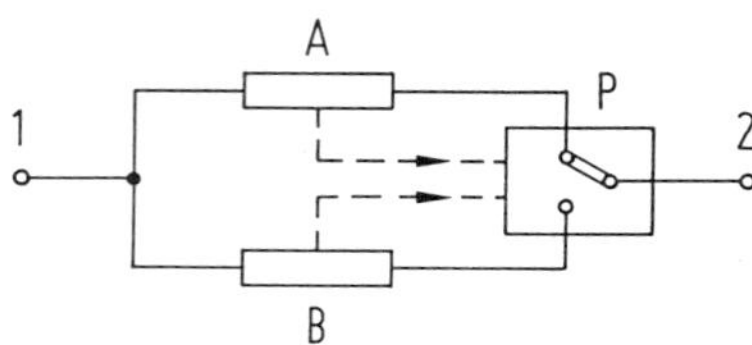

Fig. 4.3 (1 out of 2)-System with Standby Redundancy

The reliability prediction of such an standby system is mostly based upon the following - not realistic - assumptions:

- the standby unit cannot fail before it is operated
- the master element discovers all failures of the operated unit and
- the master element cannot fail to switch ($\lambda_k = 0$).

Only on these idealized suppositions the reliability of the standby-system is higher than that of the corresponding active redundant system (Fig. 4.4). But this advantage is lost, if a failure rate $\lambda_k \neq 0$ of the master element is assumed. If its failure rate is only of the same magnitude as the failure rate of the operated unit, than the reliability of the standby system is less than that of the active redundant one.

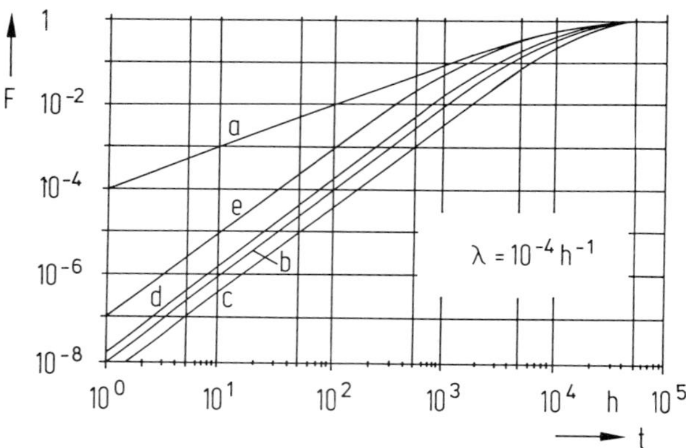

Fig. 4.4 Failure Probability F of Systems /5/
  a) not redundant single unit
  b) (1 out of 2)-system active redundancy
  c) (1 out of 2)-system standby redundancy,
       $\lambda_k = 0$
  d) (1 out of 2)-system standby redundancy,
       $\lambda_k = 10^{-4}$ h^{-1}
  e) (1 out of 2)-system standby redundancy,
       $\lambda_k = 10^{-3}$ h^{-1}

Another point of view has to be mentioned. With the standby system only the operated unit is connected to the process to be controlled. Only the operated unit receives input signals and generates output signals. If the standby unit is switched on, it has a lack of information about the process. Under these circumstances it hardly will be possible to get a smooth transition from the operated to the standby unit. For this reason the standby redundancy is not applied with units with exponentially distributed lifetimes, but only with those having wear out failures.

4.1.4 Redundant Systems with Repair

Maintenance and repair are the most effective means of improving the reliability and availability. This shall be shown by a (1 out of 2)-system as an example. Fig. 4.5 shows the Markov chains of both an active and a standby system. They only differ in the transition rate from state 1 to state 2. As at the active redundancy in state 1 both the installed units are working the corresponding failure rate is twice as high as at the standby redundancy. In both chains the state 2 may be left into two directions, either by repair of the failed unit and return to the operating state 1 or by failure of the second unit and arrival in the failed state 3. As an irreversible process is assumed the failed system is not repaired and state 3 is an absorbing one.

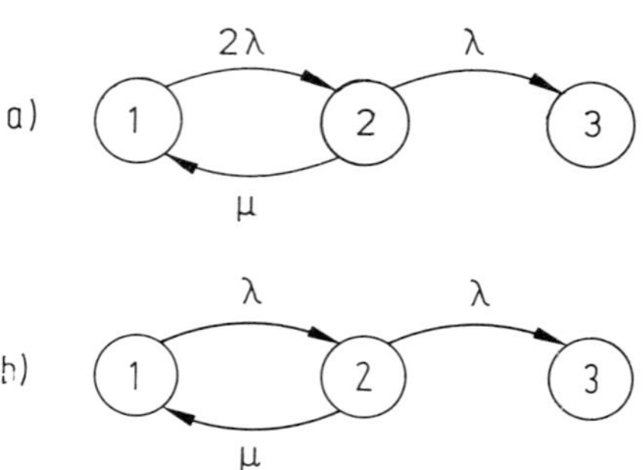

Fig. 4.5 Irreversible Markov Chains
         of (1 out of 2)-Systems /5/
         a) active redundancy
            state 1: both units are operating
            state 2: one unit has failed; the re-
                     pair has been started
            state 3: both units have failed; the
                     system has failed
         b) standby redundancy
            state 1: unit A is operating
            state 2: unit A has failed; the repair
                     has been started,
                     unit B is operating
            state 3: unit B has failed before unit
                     A has been repaired;
                     the system has failed

The unavailabilities of both systems are found by solving the corresponding differential equations. The results are shown in Fig. 4.6. The availability of the repaired systems is near at unity for all times and cannot be distincted from the upper line of the diagram. The difference between the active and standby redundancy only can be expressed by the logarithmic scale of the failure probability diagram. For short mission times the course of the unavailability is not affected by repair. If there are no failures, then no repair is needed. With unavailabilities higher than 10^{-6} the repair process gets effective. The curve b (no repair) and the curves c and d (with repair) begin to seperate. The standby system is a bit more available than the active one. But both systems will fail after some time ($U(t\to\infty)=1$) as there is no possibility to leave the absorbing state 3.

If in state 3 the repair will be allowed then the process becomes reversible. The asymptotic unavailability of the active system will reach the value

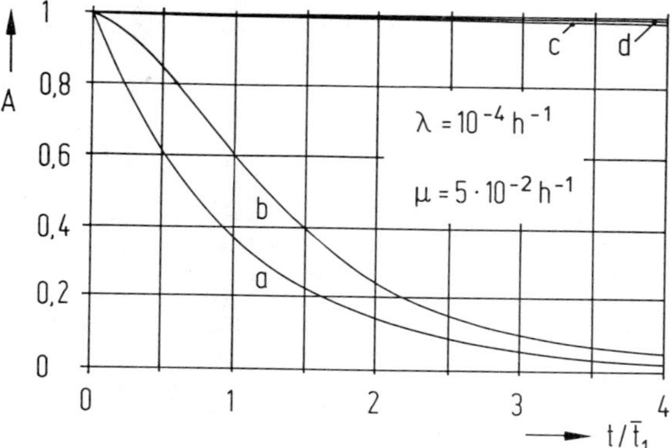

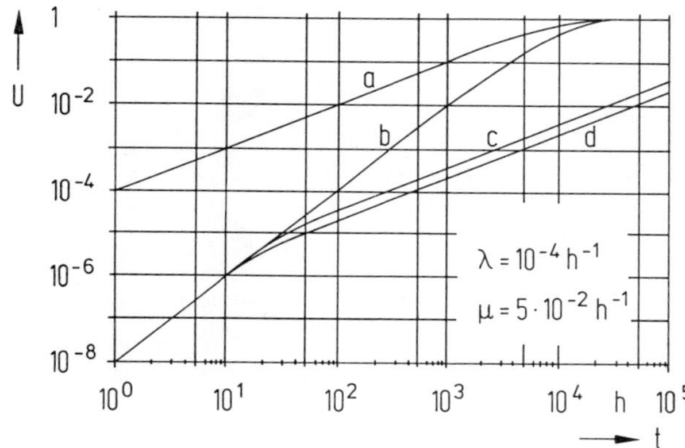

Fig. 4.6 Availability A and Unavailability U of
         different Systems /5/
         a) not redundant single unit
         b) (1 out of 2)-system with active redun-
            dancy without repair
         c) (1 out of 2)-system with active redun-
            dancy and repair
         d) (1 out of 2)-system with standby redun-
            dancy and repair

$$U_{1v2}(t \to \infty) = \left(\frac{1/\mu}{1/\lambda + 1/\mu}\right)^2 = \left(\frac{MTTR}{MTBF + MTTR}\right)^2 \quad (4.1)$$

and that of the standby system is about half as
much.

Besides the active and standby redundancies discussed above there are a lot of other structures partly proposed and partly realized. With the "dynamic"
redundancy the necessary hardware expense is
minimized. Such a structure e.g. is applied in
fault tolerant computer systems. With a failure the
computer will no longer be able to fulfill all his
tasks. By means of a "graceful degradation" only
the most important tasks are carried out or the
tasks are carried out in a lower frequency until
after error diagnosis, repair of failed modules and
reconfiguration the fully operating mode is reached
again /14/.

4.2 Protection against Common Mode Failures by Separation and Diversity

The foregoing discussed redundancy is a preventative measure against random failures. Always there
was assumed that not all parallel units fail on
account of the same cause at the same time. Some
units were considered to remain still operating.
Unfortunately the principle of redundancy is not
effective against common mode failures which will
damage all parallel units simultaneously. In order

to indicate the different efficiency of random and
common mode failures the probability calculation is
used.

4.2.1 Conditional Probability

Two parallel operated active redundant units are
assumed with the events

    event $\overline{E}_1$: unit 1 failed

    event $\overline{E}_2$: unit 2 failed

    event $\overline{E}_s = \overline{E}_1 \wedge \overline{E}_2$: both units failed;
                         (1 out of 2)-system failed.

The probability that the two events will both occur
is the product of the probability that one of the
events will occur and the conditional probability
that the other will occur given that the first
event occured:

$$p(\overline{E}_s) = p(\overline{E}_1 \wedge \overline{E}_2) = p(\overline{E}_1) \cdot p(\overline{E}_2|\overline{E}_1) \quad . \quad (4.2)$$

There are three different situations to be distincted:

a) The events $\overline{E}_1$ and $\overline{E}_2$ are independent. Then follows

$$p(\overline{E}_2|\overline{E}_1) = p(\overline{E}_2) \quad (4.3)$$

$$p(\overline{E}_s) = p(\overline{E}_1) \cdot p(\overline{E}_2). \quad (4.4)$$

All remarks in section 4.1 are based upon this
assumption of "independence".

b) The events $\overline{E}_1$ and $\overline{E}_2$ have a common cause. They
occur commonly; they are tightly coupled with

$$p(\overline{E}_2|\overline{E}_1) = 1 \quad (4.5)$$

$$p(\overline{E}_s) = p(\overline{E}_1) \cdot 1 = p(\overline{E}_1). \quad (4.6)$$

All parallel channels will fail simultaneously. The
redundancy is not effective and cannot improve the
reliability.

c) The events $\overline{E}_1$ and $\overline{E}_2$ are dependent, but not
strongly coupled. Here is

$$p(\overline{E}_2) < p(\overline{E}_2|\overline{E}_1) < 1 \text{ and} \quad (4.7)$$

$$p(\overline{E}_1) \cdot p(\overline{E}_2) < p(\overline{E}_s) < p(\overline{E}_1). \quad (4.8)$$

As situation b injures the reliability more than
all big efforts must be taken in order to avoid
such common mode failures.

4.2.2 Causative Factors of Common Mode Failures

Especially human errors and unacceptable environmental stresses are causative factors of common
mode failures. Some possibilities shall be listed:

a) Design Deficiency.

- Misunderstanding of process variable behaviour, if the dynamic response of the instrumentation e.g. is not fast enough to follow the speed of
the process.

- Inadequate design of equipment; some times ago
in a plant filters were used to reduce electromagnetic interferences. These filters have cut off the
input signals. Due to this limitation the setpoint
of the power control could never be reached and a
vessel was destroyed by superheating.

- Unrecognized interdependence between moduls,
units or subsystems which were assumed to be independent. Such problems may happen within the logical network of a complex protection system or within the software of a process computer.

- Unrecognized electrical or mechanical dependence on a common element such as the power supply or the earth line.

b) Operational and Maintenance Errors.

- Miscalibration, due to outdated instructions e.g. or to not accurately working instruments.

- Inadequate or improper testing, if it is made by unskilled operators, perhaps with wrong instruments. In the testing mode the instrumentation is separated from the process and connected with test signalgenerators. If after testing an instrument is not switched back to the process variable, it is not available for process control.

- Careless in maintenance. By the use of improper oils or lubricates e.g. the part surfaces can be damaged and mobile parts can get fixed together.

- Misinterpretation of (ambigous) process signals can cause wrong human actions.

c) Environmental Impacts.

- Temperature; with a failure of the common air condition system the temperature can get too high or too low. Sometimes 5 redundant diesel generators could not be started as their cooling water was frozen due to the fault of the building heating.

- humidity

- vibration

- electrical interference, lightning-stroke

- external catastrophes as e.g. storm, fire, flood, aircraft crash, earthquake.

All these situations can make redundancy ineffective. Nevertheless high reliable systems have to withstand these common mode failures. Thus preventative measures become necessary to reach this goal.

4.2.3 Preventative Measures against Common Mode Failures

Basic Requirements. The engineers who are designing, licensing, constructing and operating the systems must work carefully in order to avoid common mode failures. It's useful to applicate

- parts with established reliability
- standardized modules
- tested and proved equipments
- failure detecting mechanisms and
- periodic operational tests.

A design is recommended to be as simple as possible and clearly structured. In addition high reliable systems are to be separated and to be diversified physically and electrically.

Physical and Electrical Separation. The redundant units are to be installed in different areas and the connecting cables have to run on different ways. The signals must be electrically isolated by means of optoelectric or magnetic decoupling devices if entering or leaving the redundant areas. This design shall guarantee that failures are limited to only one redundant area. It is helpful peculiarly against environmental and catastrophic stresses.

Diversity. A system is diversified if different means of performing a required function do exist. Other physical quantities or principles or techniques are used to solve the same task. For monitoring a steam vessel e.g. its pressure or its temperature can be chosen. The logical network for actuating a shut down can be realized by relais, electronic gates or computers. The actuators can be powered electrically, hydraulically or pneumatically. If such different means are supplied, then it is quite unlikely that the same design, operation or maintenance error will affect all the diversified units.

High Reliable Software. A region which is extremely susceptible to common mode failures is the software of computers. While the hardware can be doubled or tripled in order to withstand the random failures this possibility is not recommended in respect of the software. Even if different teams would develop the programs, these are based upon only a single specification and the same errors could be produced. Would the programs be written in different languages to get a diversity as far as possible, then the operation and maintenance of the system would be complicated with the consequence of a decrease in availability. The best way is assumed to work very carefully and to ensure a lot of recommendations /15/. In these cases in which the software is verificated this procedure can be considered as a diverse measure to programing.

4.3 Fault Tree Analysis

Principle. If high reliable systems are designed with redundancy and diversity the problem of evaluating the system's unavailability arises. This can be achieved by means of a fault tree analysis.

The fault tree analysis is a technique by which failures that can contribute to an undesired event are organized deductively and represented pictorially. The tree is developed top down. The most undesired event, the top event must be defined at first. Then those events leading to the top are considered whereby the logical relations are expressed by the symbols of the Boolean algebra. This procedure is repeated step by step until the basic events are reached. Thus a treelike strukture is resulting and with known probabilites of the basic events the likelihood of the top event may be evaluated.

Example. The generation of a fault tree shall be demonstrated by means of the vessel of Fig. 4.7. It is monitored by 2 pressure and 1 temperature channels. If pressure or temperature are higher than the setpoint of the limit switch, then the relais contacts R1, R2 and R3 open and the current through the coil of a solenoid valve is interrupted. The valve is deenergized and shuts down the vessel by interrupting the fuel flow. The unavailability of this monitoring system shall be predicted:

The top event A is defined "current not interrupted". It will occur if both the pressure AND the temperature measurement will fail. The pressure is controlled by two redundant channels. These can fail (event B) either by random OR common mode (event G) failures. Owing to random failures the active redundant monitoring system is not available if both the contacts R1 AND R2 have failed (event D). Any of the instrumentation channels i does not operate successfully (event F_i), if one of the following basic events occurs (OR):

event H_i: transmitter i supplies an unsafe output signal

event I_i: limit switch i is not available due to detectable failures

event K_i: limit switch i has failed undetected.

The probabilities of these events - considering the given failure, repair and failure detection rates after an operating time t = 10 000 h - are

$$p(H_i) = 1,25 \cdot 10^{-4} \qquad \text{(Eq. 3.44)}$$

$$p(I_i) = \lambda_{21}/\varepsilon = 2 \cdot 10^{-6} \qquad \text{(Eq. 3.33)}$$

$$p(K_i) = \lambda_{22}t = 1 \cdot 10^{-4}. \qquad \text{(Eq. 3.35)}$$

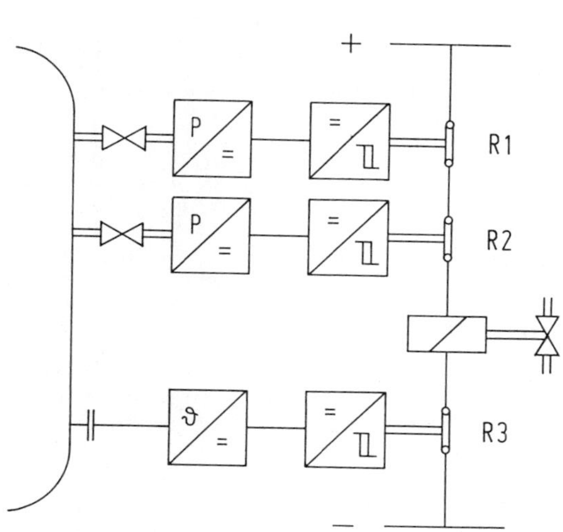

Fig. 4.7 Vessel Monitored by 2 Pressure and 1 Temperature Channel

The probability that any of those events will occur results as

$$p(F_i) = p(H_i \lor I_i \lor K_i)$$
$$\approx p(H_i) + p(I_i) + p(K_i) \approx 2{,}2 \cdot 10^{-4}.$$

The probability that both pressure channels fail is

$$p(D) = p(F_1 \land F_2) = p(F_1)p(F_2) = 4{,}8 \cdot 10^{-8}.$$

Now the common mode failures are to be considered. They are not known, otherwise they would have been corrected. Thus they only can be assessed. Their probability is assumed to be 1 % of the probability of the not detected transmitter faults; that gives

$$p(G) = 1 \cdot 10^{-6}.$$

With this number the common mode failures are dominant and the probability that both pressure channels fail either by random or by common mode failures is obtained as

$$p(B) = p(D \lor G) \approx p(D) + p(G)$$
$$\approx 4{,}8 \cdot 10^{-8} + 1 \cdot 10^{-6} = 1{,}048 \cdot 10^{-6}.$$

There would be no advantage, in tripling the pressure channels. In this case the probability p(D) still would be lower, $p(D) \sim 10 \cdot 10^{-12}$, but would be overridden again by $p(G) = 1 \cdot 10^{-6}$.

However the temperature channel is effective with an unavailability of

$$p(C) = 2 \cdot 10^{-4}.$$

In combination with the pressure measurement the failure probability of the whole monitoring system becomes sufficiently low

$$p(A) = p(B \land C)$$
$$= p(B)\, p(C) = 1 \cdot 10^{-6} \cdot 2{,}2 \cdot 10^{-4} = 2{,}2 \cdot 10^{-10}.$$

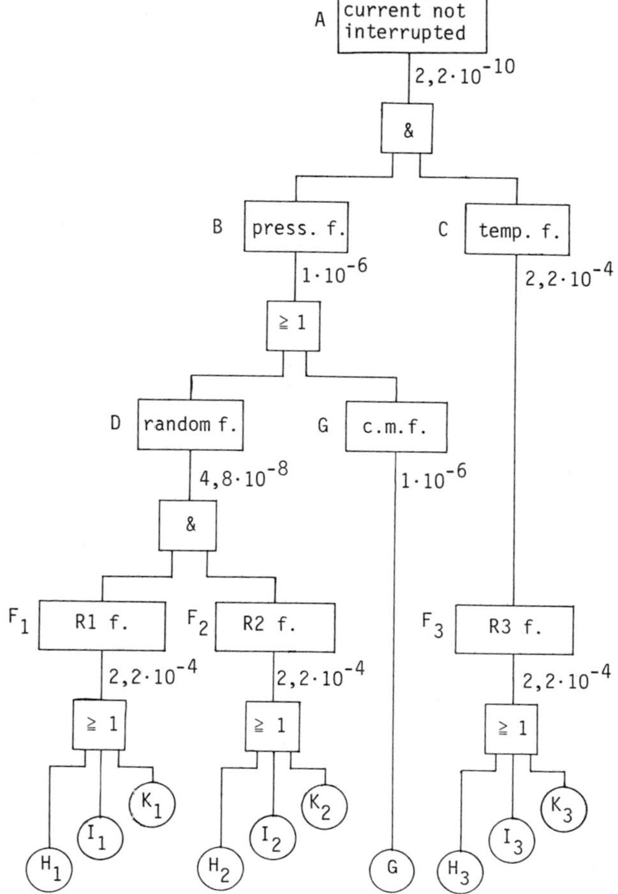

Fig. 4.8 Fault Tree
       of the Monitoring System Fig. 4.7 /5/
$$p(H_i) = 1{,}2 \cdot 10^{-4}; \quad p(I_i) = 2 \cdot 10^{-6};$$
$$p(K_i) = 1 \cdot 10^{-4}; \quad P(G) = 1 \cdot 10^{-6}$$

In the fault tree no direct line leads from a basic to the top event. To reach the top, events must be combined together in an intersection. Thus the single failure criterion is fulfilled by the monitoring system. It will not fail due to only a single event.

The fault tree analysis yields the wanted probability of the top event. Furthermore the weak points within the system are revealed, so that they can be improved. And last not least the analysis allows to compare competitive systems in order to select the better one based upon a quantitative decision.

While the fault tree analysis works top down, the failure effect analysis operates down up. Both are applicable for equipments as well as for systems. The completeness of the failure effect analysis can be proved, that of the fault tree analysis is dependent upon the experience of the analysing engineers. But it offers the possibility to consider only those events which affect the most undesired event.

CONCLUSION

Reliable systems are not obtained by chance but only by careful outline. Reliability is designed in, defects are tested out. The design is intrinsic not only to the parts' but also to the system's reliability.

The reliability concept starts with "picking good parts and using them right". The last is proved by a stress analysis. At the equipment level an efficient failure detection in combination with restauration is the basis of fault tolerant systems. By application of redundancy and diversity and by a possibly necessary redesign the reliability can be improved until the desired level is achieved. The availability of high reliable systems is better than that of the only moderately reliable components. It can be approved by analytical and modeling techniques.

REFERENCES

/1/ Green, A.E.; Bourne, A.J.: Reliability Technology, John Wiley & Sons Ltd., London 1977.

/2/ IEEE Std. 352-1975: IEEE Guide for General Principles of Reliability Analysis of Nuclear Power Generating Station Protecting Systems.

IEEE Spectrum, Reliability, October 1981, Vol. 18, No. 10, S. 33 - 103.

/3/ Dryden, M.H.: Design for Reliability, Microelectronics and Reliability, Vol. 15, S. 399 - 436, Pergamon Press, 1976.

/4/ Weygang, A.: Fehleranalyse an integrierten Halbleiterschaltungen, Elektronik 1979 Heft 12, S. 55 - 61.

/5/ Schrüfer, E.: Zuverlässigkeit von Meß- und Automatisierungseinrichtungen, C. Hanser Verlag München 1984.

/6/ US Department of Defense, Washington, Military Handbook 217, Edition D 1982.

/7/ The Reliability Handbook 1982; National Semiconductor Corporation, Santa Clara, Cal.

/8/ European Semiconductor Hi-Reliability Programs; MOTOROLA Semiconductors.

/9/ Birolini A.: Qualität und Zuverlässigkeit technischer Systeme, Springer Verlag Berlin 1985.

/10/ Schrüfer, E.: Ermittlung von Bauelementausfallraten aus Gerätebetriebsstatistiken, in Meß- und Automatisierungstechnik (INTERKAMA-Kongreß 1980), Springer Verlag Berlin 1980, S. 288 - 299.

/11/ Bartels, G.: Geräteausfälle in Abhängigkeit von den Einsatzbedingungen, Regelungstechnische Praxis 1983, Heft 9, S. 352 - 355.

/12/ Rasmussen, N.C.: Reactor Study - An Assessment of Accident Risks in US Commercial Nuclear Power Plants, United States Nuclear Regulatory Commission, WASH-1400 (NUREG-75/014), 1975.

/13/ Tasar, V.: Analysis of Fault Detection Coverage of Self-Test Software Program, 8th Symp. on Fault Tolerant Computing FTCS-8 1978, S. 65 - 71.

/14/ Färber, G.; Demmelmeier, F.: Taskspecific assignment of redundancy in the fault-tolerant multicomputersystem FUTURE, Preprints EURO-MICRO Symposium 83, 13. - 16. September 1983, Madird. In Mircrocomputers: Developments in industry, business and education. North-Holland, Amsterdam, 1983, S. 245 - 255.

/15/ IEC TC 45 Subcommittee 45A: Software for Computers in the Safety System for Nuclear Power Stations (Draft 1984)

European Workshop on Industrial Computer Systems Committee EWICS-TC7
Development of safety related software Nr. 268 (1981)

Guidelines for verification and validation of safety related software Nr. 333 (1983)

Techniques for verification and validation of safety related software Nr. 400 (1985)

System requirements specification for safety related systems Nr. 444 (1985)

RELIABILITY OF PROCESS CONTROL SOFTWARE

M. L. Shooman

Department of Electrical Engineering, Polytechnic University, Brooklyn, New York, USA

Abstract. One of the most urgent problems in the design and development of modern digital systems is the reliability of computer software. Many system failures occur due to the excitation of residual errors in the software when a particular combination of inputs occurs. These residual errors endure into operation because of the impossibility of testing more than a fraction of the hugh number of input combinations. The random occurrence of the inputs and the unknown location of the errors make this a probabilistic problem.

This paper introduces some of the simpler software reliability models which are used in the field. These result in a reliability function which predicts the probability of failure free operation for a specified time period and the mean time between system failures due to software errors. Evidence is presented showing that the models work well when used with complete and accurately gathered data. More advanced models are referenced. The appropriate data needed to support the models are discussed along with suggestions for the creation of a software reliability data base.

Keywords. Probability; reliability theory; software.

INTRODUCTION

The Probabilistic Basis of Software Errors

Many people resist the concept of software reliability, reasoning erroneously that hardware reliability is a probabilistic problem because it wears out, but software does not fail since it does not wear out. Software fails in a different mode than hardware, due to residual software design errors which are excited for some particular combination of initial and input conditions. Software design errors persist in operation because for practical systems it is impossible to test more than a small subset of all the possible input combinations during development. The goal is to minimize the number of residual errors, their frequency of occurrence, and their effect on system performance. The probabilistic aspects of the problem are due to the random nature of the inputs, the unknown mechanisms of human failure which create errors in the development process, and the randomness of the testing process used in detecting errors.

Software Failures in Instrumentation and Control

Virtually every modern instrument and control system contains one or more microprocessors or microcomputers as is the case in almost all military, space, and industrial systems. Many such systems are called embedded computer systems because failure of the computer means system failure.

The importance of human safety and the economic consequences of failure in process control are in many ways similar to those of manned space flight. The major difference is that the space program is revolutionary whereas process control is largely an evolutionary process, thus process control can benefit by using mature technologies and methodologies which have been tested on military and space systems.

The Problem

The fundamental problem in the area of software reliability is complexity. The cost of computer hardware has decreased rapidly due to advances in integrated circuit technology. With the advent of microprocessors, high density memory chips, and a myriad of powerful and inexpensive LSI and VLSI integrated circuits tremendous computing power is available with modest weight, volume, and power requirements, and at low cost. This has led planners of modern electronic systems to propose tasks which could only have been dreamed of a decade or so ago.

With all the great achievements of modern technology, there have been numerous problems when a complex system or product is conceived without concomitant engineering, management, and manufacturing methods to carry out the plan. Problems have occurred in the U.S. automotive industry, the nuclear power industry, and with consumer products. These areas must meet increasing safety, reliability and environmental requirements in the face of increasing complexity and cost minimization objectives. In many cases safety and reliability has suffered in the face of these conflicting requirements and complexity.

In the software area, the complexity of the tasks which the software must perform has grown faster than the technology for designing, testing, and managing software development. Furthermore, software costs are primarily labor intensive, rather then technological dependent, and man-hours spent on software development is roughly proportional to the size of the program measured in lines of code. Thus, as software complexity has increased over the years the man-hours for a typical project have increased, as have labor costs due to increased size and

21

inflation. By contrast, the advances in integrated circuit technology have resulted in a relative decrease in hardware costs. The net result is that an increasingly larger portion of computer system _development_ costs are due to software. See Fig. 1. Of course we have only spoken of development costs, and for a product which requires large volume production, the impact of development costs on the per unit production cost may become small.

Definition of Software Reliability

Just as in the case of hardware, or human, or systems reliability, software reliability really has two different meanings. The first meaning is the collection of all the techniques which can be used to design and test the software so that it is relatively error free. Some people call these the techniques which apply to the design of _reliable_ _software_. The second meaning, which we will stress in this paper, is the probabilistic definition of software reliability. The definition of software reliability which we will use is given below (Shooman, 1983a, p 312):

> _Software_ _reliability_ is the _probability_ that a given _software_ _system_ _operates_ for some time period without software _error_, on the machine for which it was designed given that it is used within _design_ _limits_.

The key words in the above definition are underlined. Software reliability is a probability, and therefore obeys all the laws of probability theory. We must also define successful operation of the software, a seemingly obvious but in practice a nontrivial task. For example, suppose a computer program has been developed to calculate the value of three time functions $a(t)$, $b(t)$, and $c(t)$ for various values of time t. We would normally expect the output to consist of four columns: t, $a(t)$, $b(t)$, and $c(t)$. If one of the columns were missing or if there were wrong numbers anywhere, we would certainly call it a system failure. But how would we classify a misaligned heading or extra blanks between columns. Clearly it would be a nuisance, but would it be a system failure?. One way of dealing with problems of this type is to classify errors as critical, major, and minor (Shooman, 1983b). This of course implies that we might consider three different reliabilities

$$R_C(t) =$$

Pr{no critical error in interval 0 to t}   (1)

$$R_{CM}(t) =$$

Pr{no critical or major error in
    interval 0 to t}   (2)

$$R_{CMM}(t) =$$

Pr{no critical or major or minor
    error in interval 0 to t}   (3)

It is also necessary to count only system failures due to software problems as software failures. Thus, we have to investigate all system failures carefully and classify the causes as software, hardware, operator, or unknown. In some cases this is very difficult. For example, suppose signals from two different radars are to be converted from analog to digital form, compared, and analyzed by a computer program. If the timing between the two radar signals drifts a bit, there could be large differences between the converted programs. Is this an error in the hardware or in the design of the software comparison algorithm?

Early History

It is difficult to state when software reliability, or more precisely, probabilistic modeling of software errors began. However, it is clear that by the late 1960's software designers and theoreticians had begun to think a lot about reliable software, and were groping with the concept of software reliability with little results (Shooman, 1983a, 1984b). Perhaps the earliest meetings which directly addressed these issues were the two Conferences on Software Engineering sponsored by the NATO Science Committee, which were held in Garmisch, Germany from October 7-11 1968, and in Rome, Italy from October 27-31 1969. The conference proceedings appear in (Naur, 1976).

The earliest efforts at software reliability modeling were a Markov birth-death model (Hudson, 1967) and the fitting of a growth function to cumulative error removal curves (Ellingson, 1967). The earliest software reliability models published in the open literature were independently developed by Jelinski (1971) and Shooman (1971a, 1971b).

Approaches to Software Reliability Modeling

Shortly we will discuss some of the leading software reliability models as well as reference sources for further study of the literature. This section will briefly describe the approach which most of the models have taken.

At the outset we should point out that the quantity of real interest is System Reliability. System reliability depends upon proper performance of the hardware, the software, and in some cases the system operator(s). In fact if we broaden the definition of the term _error_ in Eq. 3 (Eq. 1, Eq. 2) to include all system errors regardless of the cause we have defined system reliability. However, it is much easier to deal with system reliability, R_{SY}, if we can easily decompose it into separate software reliability, R_S, hardware reliability, R_H, and operator reliability, R_O. One case where such a decomposition is easy is where the classes of hardware, software, and operator errors are mutually exclusive. In such a case (Shooman, 1983a, p. 351-353), the system reliability is merely a product of the other three reliabilities.

$$R_{SY} = R_S \times R_H \times R_O \qquad (4)$$

We will assume that this is the case and will concentrate on modeling R_{SY}.

Availability is generally used (along with reliability) to evaluate the goodness of a system which has repair capabilities. The definition of software availability is similar to reliability, however it allows failures as long as they are followed by repairs. Thus, it is essentially a measure of the probability that the system is up at any instant of time, specifically

> _Software_ _availability_ is the _probability_ that a given _software_ _system_ _is_ _operating_ _successfully_, according to specifications at a given point in _time_.

Availability models are generally Markov type probability models which define a set of up states and down states of the system. A system of coupled first order differential equations is written, one for each state, and then solved for the individual probabilities of being in any state. Since the state probabilities are mutually exclusive, the availability is the sum of the up state probabilities.

A further distinction among the various software reliability models is whether the effect of an error (say a critical error) in one portion of the program is treated the same as in other portions of the program. In essence we are saying an error is an error regardless of where it occurs. Such models are generally referred to as Macro models. The software is treated as if it is a "black box" containing E_T total errors. In general the reason for such a simplification is that it is difficult to make a probabilistic model of a program's structure, a Micro model, and even more difficult to obtain data which is required to apply such a model. Most of the models which we will discuss are Macro models.

At this point we are now able to discuss what is the basis of most of the Macro models which we will focus on. All these models assume that the rate of failure is proportional to the number of errors present in the software.

Thus, embedded in the software reliability modeling process there is generally a software error model. These software error models generally depend on the error data collected on software which is undergoing development testing. We divide the development cycle into a specifications phase, a design phase, a coding phase, a code module (code unit) test phase, a module integration test phase, an acceptance phase, and an operational phase. In theory, we could begin collecting error data at the specification phase based upon inconsistencies in the specifications, missing specifications and wrong specifications. Most such errors would be found by careful reading of the specifications, specification reviews, attempts by the design group to implement the specifications, and analysis by computer programs of specification inconsistencies. Most projects are poorly organized at such a stage and no careful count of errors is made. For most program developments, careful control of the project configuration and error record keeping does not start until the beginning of the integration testing phase. Consequently, most error models are begun at that point. A typical set of error data for four projects is shown in Fig. 2.

The cumulative number of errors removed is denoted by $E_r(\tau)$, where τ is the number of months since the beginning of integration testing. Note that we are using a different time variable so as to avoid confusion with operational time t, (cf. Eq. 1-3). Sometimes it is convenient to try and compare error data for various programs. Clearly, it is unreasonable to directly compare error data for large and small programs, thus one often plots the normalized number of errors, $E_r(\tau)/I_T$, where I_T is the number of object instructions (compiled machine language statements). Experience has shown that the ratio $E_r(\tau)/I_T$ is approximatly equal to .01 for many programs (Shooman, 1983a, p. 323-329). Sometimes it is more instructive to study the error removal rate $r(\tau)$ or its normalized counterpart.

$$r(\tau) \quad = \quad dE_r(\tau)/d\tau \qquad (5)$$

$$r(\tau)/I_T \quad = \quad (\ dE_r(\tau)/d\tau \ )/I_T \qquad (6)$$

A set of normalized error removal rate curves corresponding to the cumulative curves shown in Fig. 2 is given in Fig. 3. If one postulates that the error removal rate is equal to the error detection rate, and that the detection rate is

proportional to the number of remaining errors an exponential error removal model ensues (Shooman, 1983a, p. 333-335). This intuitively appealing model is not substantiated, since only the data for Application program B of Fig. 3 seems to fit such a conclusion. However, we have been dealing with months of test time rather then the more useful independent variable, man-months. Unfortunately, it is very difficult to obtain accurate man-month data in most projects.

Given that an error model has been made based on error removal data, we postulate that the software failure rate is proportional to the number of remaining errors according to some formula, containing empirical constants. Generally, the way we evaluate these empirical constants is to use data on times to failure, number of failures, and number of test hours recorded during software testing as a means of determining the constants of the failure rate model. Once a failure rate model (hazard function, z(t)) has been determined one merely substitutes into the well known formulas for reliability, R(t), and mean time to failure, MTTF (or MTBF) (Shooman, 1986).

$$R(t) \ = \ \exp(- \int_0^t z(x) \ dx) \qquad (7)$$

$$MTTF \ = \ \int_0^\infty R(t) \ dt \qquad (8)$$

Thus, the usefulness and accuracy of these models depend not only on good error data but on the aforementioned test data. The best (most realistic, most accurate) test data is that taken once the system in question is deployed and has begun operation. However, this is so late in the project that the only possible actions are: to celebrate the success if the numerical results of the model are favorable (adequate MTTF or low probability of failure for a given operating period); or to contemplate the means for reliability improvement and their costs if the numerical results are unfavorable. Clearly, the earlier in the development cycle we can test, the earlier we can predict the operational reliability, and the more time we have to take action to improve the software in response to a gloomy prediction. Unfortunately, the earlier back we go in the test cycle, the less realistic, the less documented, and the more scattered are the test results. At present most analysts use either integration test data or simulator test data taken during the integration test phase.

Design of Reliable Software

The design techniques used to design reliable software are all aimed at providing structure, order, and logic to what is otherwise a chaotic procedure of interchange among many workers. The process begins with formal specifications for the project written in prose along with algorithms, lists of variables, constants, etc. Sometimes a formal specification language is used. The specifications are subjected to a design review. A preliminary design is developed based on these specifications which is expressed in terms of design representations such as : flow charts, pseudo code, program design language, (PDL), HIPO diagrams, descriptions of data structures, tables of variables, etc. The design is reviewed. The program modules are coded in a modern high level language (with very few small linked assembly language subroutines if absolutely necessary) using structured, modular, defensive techniques (Shooman, 1983a, Ch. 2). The code modules are tested thoroughly and then they are combined into a complete program in steps during the integration testing phase. Design reviews are followed by simulation testing, acceptance testing, field testing and deployment. Software quality assurance has the responsibility to sit in on all design reviews, review all documents, review test plans, and constantly check for completeness, thoroughness, and accuracy of all documents, and review any reliability predictions and measurements.

Progress to Date

Before we begin a more detailed treatment of the leading software reliability models, it is useful to summarize very briefly the state-of-the-art.

Most reliability researchers and modelers have used their models and those of their colleagues on a number of sets of data. The data varied from one extreme where they supervised the collection and understood any weaknesses, to the other extreme where a system operator was quizzed for two days to recover, classify, and document the narrative contained in two notebooks of how a computer system behaved during final integration tests conducted at an alpha test site (eg. the first field-like test conducted by the developer). In all cases the results were very useful and quite adequate. In the former case they, predicted the field measured MTTF within about 30%. In the latter case they predicted that another 6-12 months of integration testing would be required to achieve a tolerable, but lower than specified MTTF for the software. The result was that management decided to abandon the software development studied and place all their resources behind another parallel and better planned, but much later initiated development of the system software. Thus, the modelers feel that most of their models work adequately and that they should be widely applied. In essence they say, try them they work.

There is a diversity of opinion among software engineers. Some of them have studied these successful efforts and have tried some of the models themselves. In general this group is supportive of using the available software reliability models. Their opinion is balanced by those who have read some of the literature, have not tried the models, and are cautious. This latter viewpoint is buttressed by some software designers who do not like and distrust probabilistic models, and claim the only way to make reliable software is to write the code without any errors in the first place. Experienced managers realize that there is nothing such as error free code, however, they are somewhat confused and wary of the controversy.

One of the real problems in moving ahead to settle this controversy is the lack of an adequate data base to use in testing the models, demonstrating their accuracy, applicability, and differences; and for use in early predictions of software reliability.

LEADING MODELS

Introduction

In this section we discuss, in summary fashion, some of the many different software reliability models which have been introduced in the literature. Several comprehensive surveys have been conducted documenting the numerous models in the literature, and the reader is referred to these papers for a more comprehensive treatment (Shooman, 1983a, Chap. 5; RADC, 1979; Littlewood, 1980; Musa, 1982; Farr, 1983).

The models which we shall discuss all have much in common. All relate the failure rate of the software to the number of remaining errors. The probabilistic basis of each is dependent on changing combinations and sequences of input data driving the program down the same or different paths with varying input parameters. A system failure occurs when an error is encountered along some path for the given input values.

The Author's Model

The Author's model begins by assuming a simple model for software errors. A curve of the cumulative error data for supervisory system A of Fig. 3 is shown in Fig. 4. Except for the normalization factor, Fig. 4 shows the same behavior as the curves of Fig. 2; all build up initially with an increasing slope, followed by a decreasing slope, and finally the curves appear to become asymptotic. If we assume that all detected errors are immediately and accurately corrected, then $E_c(\tau) = E_d(\tau)$. If the number of generated errors during the debugging process is zero, then we have a simple error-balance equation; i.e., what remains is what we started with, E_T, minus what was removed:

$$E_r(\tau) = E_T - E_c(\tau) \qquad (9)$$

(A model which includes error generation is described in (Shooman, 1983a, Chap. 5)).

If we rewrite Eq. (9) in terms of normalized quantities, i.e., divide all terms by the total number of object instructions, we obtain

$$\frac{E_r(\tau)}{I_T} = \frac{E_T}{I_T} = \frac{E_c(\tau)}{I_T} \qquad (10)$$

The above error model is now incorporated in a failure rate model by writing an expression for the probability that a bug is encountered in the time Δt after t successful hours of operation: $z(t)\Delta t$. We make the assumption that this probability is proportional to the fractional number of remaining bugs $E_r(\tau)/I_T$. Musa has substantiated this assumption experimentally (Musa, 1979).

Thus we obtain

$$P(t < t_f \leq t + \Delta t \,|\, t_f > t) = z(t)\Delta t = KE_r(\tau)\Delta t \quad (11)$$

$$z(t) = KE_r(\tau) = \frac{K}{I_T}[E_T - E_c(\tau)] \qquad (12)$$

where $t_f \equiv$ operating time to failure (occurrence of a software error)

$P(t < t_f \leq t + \Delta t \,|\, t_f > t)$

$\equiv$ probability of failure in interval Δt, given no previous failure.

K = an arbitrary constant[1]

-----

[1] In earlier work (Dickson, 1972) an attempt was made to achieve a semimicro model by splitting K into two factors: K, an arbitrary constant, and r_p, the instruction-processing rate. This elaboration is not included here, since, to date, insufficient data have been obtained to define or calculate r_p.

Note that in Eq. 12 two time variables appear: first there is t, the operating time in hours of the system, and second there is τ the debugging time in months (or more generally, the debugging resource variable). Once the assumptions in Eq. (12) have been made, the reliability and mean time to failure functions follow directly from classical reliability theory.

By combining Eqs. (12) and (7) and assuming that K and $E_r(\tau)$ are independent of operating time t, we obtain for the reliability function

$$R(t) = E^{-[\frac{KE_r(\tau)t}{I_T}]} = e^{-\frac{K}{I_T}[E_T - E_c(\tau)]t} \tag{13}$$

For a fixed value of τ, $KE_r(\tau)$ is a constant which we call γ for convenience.

$$R(t) = e^{-\gamma t} \tag{14}$$

Basically the above equation states that the probability of successful operation without software errors is an exponential function of operating time. When the system is first turned on, t = 0 and R(0) = 1. As operating time increases, the reliability monotonically decreases as shown in Fig. 5. We depict typical reliability functions for three values of debugging time, $\tau_0 < \tau_1 < \tau_2$. From this curve we may make various predictions about the system reliability. For example, looking along the vertical line t = $1/\gamma$, we may state:

1. If we spend τ_0 hours of debugging, then $R(1/\gamma) = 0.35$.

2. If we spend τ_1 hours of debugging, then $R(1/\gamma) = 0.50$.

3. If we spend τ_2 hours of debugging, then $R(1/\gamma) = 0.75$.

The constants of the model must be determined as discussed below to complete the model.

A simple way to summarize the results of the reliability model is to compute the mean time to (software) failure MTTF by substituting Eq. (13) into Eq. (8), yielding

$$\mathrm{MTTF} = \frac{I_T}{KE_r(\tau)} = \frac{I_T}{K[E_T - E_c(\tau)]} \tag{15}$$

In order to interpret Eq. (15), one must assume a model for $E_c(\tau)$. For simplicity, let $r(\tau)$ be modeled by a constant rate of error correction r_0 (Dickson, 1972), then $E_c(\tau) = r_0\tau$ and solution of Eq. (15) yields

$$\mathrm{MTTF} = \frac{I_T}{K(E_T - r_0\tau)} = \frac{1}{\beta(1 - \alpha\tau)} \tag{16}$$

For convenience we define normalized parameters

$$\beta = \frac{E_T}{I_T}K \quad \text{and} \quad \alpha = \frac{r_0}{E_T}$$

In Fig. 6, $\beta \times$ MTTF is plotted versus $\alpha\tau$. We see that the most improvement in MTTF occurs during the last quarter of the debugging. This is an extremely important point. If we have no model to guide us, we may expend our resources, patience, and credibility during the debugging process short of achieving our goals. (In practice one uses previous experience or extrapolation of the data already taken on the project to predict future values of $r(\tau)$ later in the development cycle.)

RISC-C

One can qualitatively explain the shape of the MTTF curve given in Fig. 6. Suppose $E_T = 1000$ and during the first quarter of the debugging period we remove 250 errors. The difference in the MTTF displayed by the program with 750 residual errors is only 33 percent greater than that at the start (1000 errors), since MTTF is proportional to the reciprocal of the number of remaining errors. Once we reach the last phase of debugging with say 100 errors left, the MTTF has increased tenfold. Removal of another 50 errors provides another factor of 2 gain for a total increase of 20. Thus, removal of the same number of errors early in the integration process has much less effect on MTTF than when the removal takes place near the end.

Further examination of Eq. (16) reveals that when $\tau = 1/\alpha$, the MTTF $\to \infty$. This is mathematically correct but not very satisfying, since practically, zero errors is unreachable, and our probabilistic model assumptions become weak when the residual number of errors $\to 0$. If we assume one remaining error, we might consider I_T/K to be a sort of upper bound on the MTTF.

The parameters of the above model, K and E_T, must be evaluated from test data. The earliest stage at which an entire system can be functionally tested is during system integration using the system simulator (functional test) program. If this test is performed at the beginning of system integration, the result will be a succession of very short runs and immediate crashes. The model will predict a poor reliability, but help us estimate how rapidly this will improve as more errors are removed.

The test data which must be recorded for each run of the system test program is: how long the test ran, whether an error occurred, if the error is a software error, and the time of failure.

There are n total runs, r of these represent failure, and n − r represent success. The n − r successful runs represent $T_1, T_2, \ldots, T_{n-r}$ hours of success, and the r unsuccessful runs represent $t_1, t_2, \ldots, t_r$ successful run hours before failure. The total number of successful run hours H is given by

$$H = \sum_{i=1}^{n-r} T_i + \sum_{i=1}^{r} t_i \tag{17}$$

Assuming that the failure rate is constant, we denote it by λ and compute it as the number of failures per hour:

$$\text{Failure rate} = \lambda = \frac{r}{H} \tag{18}$$

The MTTF for a constant failure rate is the reciprocal of the failure rate (Shooman, 1986):

$$\mathrm{MTTF} = \frac{1}{\lambda} = \frac{H}{r} \tag{19}$$

We assume a known program size and careful collection of error data, thus I_T and $E_c(\tau)$ are known values and only the constant K and E_T remain to be determined. These two unknowns, K and E_T, can be evaluated by running a functional test after two different debugging times $\tau_1 < \tau_2$ chosen so that $E_c(\tau_1) < E_c(\tau_2)$. We then equate the MTTF expressions given by Eqs. (15) and (19) at times τ_1, and τ_2:

$$\frac{H_1}{r_1} = \frac{I_T}{K[E_T - E_c(\tau_1)]} \tag{20}$$

$$\frac{H_2}{r_2} = \frac{I_T}{K[E_T - E_c(\tau_2)]} \tag{21}$$

Simultaneous solution of the above two equations allow us to solve for our constants.

In statistical terms, the evaluation of model parameters is called parameter estimation. Specifically, equating expressions as was done in Eqs. (20) and (21) is a form of the moment method of parameter estimation. Another estimation technique, maximum-likelihood estimation (MLE), can be used to develop a different estimate (Shooman, 1983a, p. 370-372; Shooman, 1973; Lloyd 1977).

A simple and very useful method of estimation is that of least-squares. We begin our estimation procedure by rewriting the failure-rate equation, Eq. (12). The failure rate is assumed to be a constant once we stop debugging; however, it decreases from one test interval τ_i to the next. We write $z(t) = \lambda_i$, and rearranging Eq. (12) we obtain

$$\frac{E_c(\tau_i)}{I_T} = \frac{E_T}{I_T} - \frac{1}{K}\lambda_i \tag{22}$$

Thus, a plot of $E_c(\tau_i)/I_T$ versus λ_i yields a straight line with an intercept of E_T/I_T and a slope of $-1/K$. Thus, one can plot the experimental values of $E_c(\tau_i)/I_T$ obtained from recorded error data versus the λ_i values obtained from Eq. (18). One can then choose the best straight line by eye or use least-squares estimation. An example of the use of least-squares fitting is given in Fig. 7.

Jelinski-Moranda Model

Jelinski (1971) proposed a software hazard function of the form

$$z(t) = \phi[N - (i - 1)] \tag{23}$$

where ϕ = constant of proportionality
N = total number of errors present
i = number of errors found by debugging time τ_i

Comparison of Eq. (23) with Eq. (12) shows them to be equivalent for

$$E_T = N \tag{24}$$

$$\frac{K}{I_T} = \phi \tag{25}$$

$$E_c(\tau) = i - 1 \tag{26}$$

Schick-Wolverton Model

Schick (1978) modifies Jelinski's (1971) model and assumes that the failure rate is proportional to the number of remaining errors and increases with operating time t:

$$z(t) = \phi[N - (i -)]t \tag{27}$$

One rationale for postulating an increase in z(t) would be if operation were viewed as a succession of different trials which gradually closes in on the remaining errors (sampling without replacement). This author disagrees with this assumption. On the contrary, he feels that z(t) should decrease with t during development testing since latter errors are the subtle ones which take a long while to encounter in operation. Similarly, in most cases of large, intricate, well-tested real-time systems the hazard will remain constant once the initial field debugging of a new release is finished. The small number of subsequent "patches" generated between releases should not be significant. Failure will be caused by rare combinations of input data and path traversals, with the time between failures governed by an exponential distribution, yielding a constant hazard.

Musa Model

Musa (1975) has developed a model similar to those previously described. He defines τ to be the execution time or central processor time used in testing the program (rather than months of calendar development time; see Fig. 8), and τ' to be the execution time of the program after release. The reliability model which he obtains is

$$R(\tau') = \exp(-\tau'/T) \tag{28}$$

where T is the mean time to failure, which is defined by the expression

$$T = T_0 \exp(\frac{C_\tau}{M_0 T_0}) \tag{29}$$

The parameters in the Eq. (29) are

T_0 = MTTF at start of test ($\tau = 0$)
C = ratio of equivalent operating time to test time
M_0 = number of failures which must occur to reveal all errors ($M_0 = E_T$)
τ' = $C\tau$

If we wish to compare Musa's model with he one developed in Sec. 2.2, we must account for the differences in the various time and notational definitions. We begin by setting t equal to τ'; $M_0 = E_T$; and set $E_c(\tau) = r_0\tau'$ yielding

$$T_0 = \frac{I_T}{KM_0}$$

$$MTTF = \frac{T_0}{1 - \dfrac{r_0\tau'}{M_0}} \tag{30}$$

If we compare the series expressions of e^x and $(1 - y)^{-1}$ we see that

$$e^x = 1 + x + \frac{x^2}{2!} + ...$$

$$\frac{1}{1 - y} = 1 + y + y^2 + ...$$

The first two terms are identical and the third term differs only by a factor of 2. Thus, Eqs. (29) and (30) will behave similarly over the range where

$$\frac{1}{T_0} \approx r_0$$

One advantage of the Musa model is that it allows us to use actual test data rather than simulator data. The disadvantage is that the input stream is less representative of operation and in fact the constant C in Musa's model is introduced to compensate for this fact. Of course the introduction of C results in an additional parameter which must be estimated from past experience and historical data.

The Littlewood Model

A somewhat different approach is taken by Littlewood (1974). They use Bayesian mathematics to formulate a software reliability model. The assumptions are: (a) An exponential failure rate model, (b) a uniform prior distribution, and (c) a set of n times between software failure, $t_1, t_2, \ldots, t_n$. The model results in complex expressions which require computer evaluations. More recent work is described in Littlewood (1980) and Goel (1978).

Availability Models

As was previously discussed, in addition to reliability, availability is an important measure of performance of any system. Many computer systems do not have the rigid requirements that a real-time system has for continuous operation. Time-shared systems and batch systems can, and often do, go down without causing severe problems to the user. The real item of importance here is how frequently the system goes down and for how long it stays down.

The mathematical technique generally used to model system availability is a Markov model (Shooman, 1983a).

SOFTWARE RELIABILITY PREDICTION

Hardware vs. Software Reliability Prediction

It is natural to draw parallels between hardware and software reliability prediction. As much as we may criticize hardware reliability prediction results, by and large they do provide guidance for selection among alternate designs, management of the system development, early warning of trouble spots in the design, planning for field maintenance, and prediction of the operational reliability of the system. This is achieved by decomposing the system into a structural model composed of the various system elements. One then consults failure rate manuals or other data sources for the element failure rates taking into account the environment, and substitutes the failure rates into the structural model with the aid of pencil-paper-calculator or a computer reliability modeling program. The end result is reliability functions and mean times to failure for the system and the major subsystems.

In the case of software reliability, the macro reliability models serve the same role as the structural model. However, the macro models do not contain any structural information. The micro models, as yet only partially developed, provide more of a system structure. If we attempt to carry the hardware-software analogy further, there is an additional problem, the concept of a hardware component is well defined, whereas, the definition of software modules is a largely unexplored field. More work must be done on defining what we mean by a particular type of program module before we can begin collecting error and failure rate data at the module level.

The following information is needed to establish a software reliability data base for the system software.

1. A paragraph or two describing the nature and scope of the project, the organizations involved, the time frame, the outcome, field performance, and any predecessor and successor projects. A knowledgeable point of contact within each major organization should be provided.

2. A brief high level (1/2 page) summary of the requirements and specifications should be stated.

3. The program language, compiler, operating system, major tools used, and target machine should be specified.

4. The size of the program (source and object, comments and executable instructions), the development approach, key milestones, and the number of man-hours should be stated.

5. The number of errors discovered and corrected should be recorded each week (or month) during integration testing, and if possible during module testing.

6. The number of tests, level, severity, running time to failure, and running time without failure should be carefully recorded, for development testing and for simulation testing.

7. Any other pertinent qualitative or quantitative information should be recorded.

Given a set of such data, one can make reliability predictions using any of the models discussed above. Much of the data collected to date is either fragmentary or missing key facts which are necessary for software reliability modeling.

Experience in Applying Models

It is difficult for a researcher to collect software reliability and error data. If he does so for a small experimental project, all including the researcher will criticize that this small "antiseptic" project is not the real world. If the researcher goes to a program manager, he seldom achieves cooperation for data gathering. Although the manager may be sympathetic, he receives no credit for data gathering, and he generally feels that somehow the error gathering will slow down his progress. (The author has shown in a few instances that even if no error gathering is planned initially on a project, it will take only 1-2% of the programmer's time to add such data gathering during integration test.) The only way such an arrangement will work is if the researcher has the sponsorship and cooperation of the project manager's boss, or has sufficient funds to pay for an extra programmer to "compensate" for any lost time. Obviously this seldom happens, although researcher's are the ones most able to specify what data should be collected and analyzed.

There are several isolated sets of software reliability data in the literature, however, only two major sources exist, one for 20 data sets, and one for 10 data sets. Musa (1975) collected error and integration test data for 20 different software projects at Bell Laboratories in the early 1970's. In the case of four of these projects, the software MTBF was estimated at release and then compared with operational field data. (See Table 1.) The average deviation between prediction and field performance was 24%. Shooman (1983a, p. 368) applied his model to Musa's four projects (those with associated field data) and recorded the model parameters in tabular form. In addition, a set of data collected by Miyamoto (1975) was analyized and the resulting mean time between software errors appears in Fig. 9. Note that both Fig. 9 and 8 both confirm the rapidly rising MTTF vs. development time shape predicted in Fig. 6.

Shooman's (1983b) data includes simulation testing of the Space Shuttle data processing software. This data was used to predict the software failure rate, which was taken as the mean occurrence rate of a Poission probability distribution. The Poission model predicted the number of errors expected, during the first shuttle mission, which is shown in Table 2. Note the good agreement between prediction and practice, especially when all the severity levels are combined -- enlarging the sample size.

THE PROMISE OF THE FUTURE

Introduction

Software reliability is an established dicipline, albeit a controversial one. What is needed is additional research, data gathering, and wider use and support for the existing techniques.

Traditionally, the military has sponsored research in the reliability field. Some companies supported modest in house research in these areas, however, these were largely firms with large amounts of government business. Thus, although there are some recent changes, the majority of reliability research is government sponsored. What is lacking is a well coordinated, purposeful, and adequately funded research effort.

During the early 1970's, Rome Air Development Center (RADC) established a sizeable research program in the areas of software measurement, reliability, and engineering. Many of the models discussed above were developed under this support. By the end of last decade, the program funding was cut and many projects were phased out. What is lacking is a well coordinated, purposeful, and adequately funded research effort and a properly planned and directed data collection effort. Some felt that in the 1980's, the Department of Defense would now provide direction and funding within the "Software for Testable and Reliable Systems," STARS, program. The sizeable funding planned has been considerably cut and the development of software development tools in the Ada and other languages is being stressed. At present it is likely that little research in software metrics, (the Computer Scientist's name for performance and reliability models and measurements), will be supported under this program. Progress will be slow in this area unless sufficient funding is obtained.

Creating an Adequate Data Base

As stated above, this is a prime objective if the field of software reliability modeling is to advance. In order to accomplish this objective, one must satisfy the following conditions:

1. Institutional support which provides an air of authenticity and insures wide distribution and comment.

2. The services of talented hardworking individuals to insure that the data is properly recorded, analyized, and documented, and that several models are fitted to the data and evaluated.

3. Necessary support via volunteers, military contract support, or unsponsored company research.

4. The cooperation of software managers to provide data on contemporary or retrospective software projects.

Predicting Early in the Life Cycle

The purpose of reliability prediction is to serve as a guide for the remainder of the development cycle. Clearly, the earlier this occurs the more benefit the project can derive from the predictions. Any predictions during the early design stages can only be done by relying on adaption of data on similar previous projects, ie. the necessity of the data manual. At present, the existing software reliability models work during the integration test phase or during field deployment. It would be a great improvement if analysts could find a technique for applying software reliability prediction during the module (unit) test phase.

Micro-Models

As discussed above, the micro-models as a class have not been sufficiently exploited (Shooman, 1983a; Lloyd, 1977). They have the potential to link the reliability with a model of the software structure. Furthermore, if one can identify paths in the program with module sequences executed, they may also serve as a means of modeling the module test phase.

Hardware-Software Models

Almost all complex systems seem to be computer controlled nowadays. This means that there is a great chance that the hardware and software errors will be intertwined. In the case of dependent hardware-software errors, our modeling skills are sorely taxed. The inclusion of BIT further complicates this problem. This author and other are presently doing research on this problem (Shooman, 1984a, 1986).

ACKNOWLEDGMENT

This work was supported in part by the US Naval Air Systems Command Contract No. N00019-83-C-0330 and in part by the US Office of Naval Research Contract No. N00014-75-C-0858.

REFERENCES

Dickson, J., J. Hesse, A. Kientz, and M. L. Shooman (1972). Quantitative analysis of software reliability. Proc. Annual IEEE Reliability and Maintainability Symp.

Ellingson, O.E. (1967). Internal memoranda on predicting the remaining number of program errors. System Development Corp.

Farr, W.H. (1983). Survey of software reliability modeling and estimation. Naval Surface Weapons Center, NSWC TR 82-171.

Goel, A.L., and K. Okumoto (1978). Bayesian software prediction models. Rome Air Development Center Report No. RADC TR-78-155.

Hudson, G.R. (1967). Program errors as a birth-and-death process. System Development Corp. Report SP-3011.

Jelinski, Z., and P.B. Moranda (1971). Software reliability research. McDonnell Douglas Astronautics Co., Huntington Beach, CA., MADC Paper WD1808. Conf. Statistical Methods for the Evaluation of Computer Systems Performance, Brown University. Also Printed in Statistical Computer Performance Evaluation, (1972) W. Freiberger (Ed.), Academic Press, NY, 465-484.

Littlewood, B., and J.L. Verrall (1974). A Bayesian reliability model with a stochastically monotone failure rate. _IEEE Trans. Reliability_, _R-23_, 108-164.

Littlewood, B. (1980). Theories of software reliability: how good are they and how can they be improved?. _IEEE Trans. Software Eng._, _SE-6_, 5, 489-500.

Lloyd, D.K, and M. Lipow (1962). _Reliability: Management, Methods, and Mathematics_. Prentice Hall, Englewood Cliffs, NJ. 2nd edition published by the authors, (1977), 201 Calle Miramar, Redondo Beach, CA.

Miyamoto, I. (1975). Software reliability in on-line real time environment. _IEEE Int. Conf. Reliable Software_, CHO 940-7CSR, 194.

Musa, J.D. (1975). A theory of software reliability and its application. _IEEE Trans. Software. Eng._, _SE-1_, 3, 312-327.

Musa, J.D. (1979). Validity of execution-time theory of software reliability. _IEEE Trans. Reliability_, _R-28_, 3, 181-191.

Musa, J.D., and K. Okumoto (1982). Software reliability models: concepts, classification, comparisons, and practice. _Proc. Electronic Systems Effectiveness and Life Cycle Costing Conf._, NATO Advanced Study Series, Springer Verlag, Heidelberg, 1983, 385-424.

Naur, P., and others (Ed.) (1976). _Software Engineering Concepts and Techniques_. Petrocelli/Charter, NY.

Rome Air Development Center (1979). Quantitative software models. Data and Analysis Center for Software, NY. Report No. SRR-1.

Schick, G.J., and R.W. Wolverton (1978). Analysis of competing software reliability models. _IEEE Trans. Software Eng._, _SE-4_, 2, 104-120.

Shooman, M.L. (1971a). Preliminary concepts of software reliability. Supplementary Notes, Course EE686 Reliability Analysis II, Dept. of Electrical Engineering, Polytechnic University, Brooklyn, NY.

Shooman, M.L. (1971b). Probabilistic models for software reliability prediction. Conf. on Statastical Methods for the Evaluation of Computer Systems Performance, Brown University. Also Printed in _Statistical Computer Performance Evaluation_, W. Freiberger, (Ed.), Academic Press, NY, 485-502.

Shooman, M.L. (1973). Operational testing and software reliability estimation during program development. _IEEE Symp. Computer Software Reliability_, NY. Catalog No. 73 CHO741-9 CSR, 51-57.

Shooman, M.L. (1983a). _Software Engineering: Design, Reliability, Management_. McGraw-Hill, NY.

Shooman, M.L., and G. Richeson (1983b). Reliability of shuttle mission control center software. _Proc. Annual IEEE Reliability and Maintainability Symp._, 125-135.

Shooman, M.L. (1984a). Research on combined hardware software reliability models. Polytechnic University, Brooklyn, NY Report No. POLY 84-003, Vol. 1, Overview and Summary.

Shooman, M.L. (1984b). Software reliability - historical perspective. _IEEE Trans. on Reliability_.

Shooman, M.L. (1986). _Probabilistic Reliability: An Engineering Approach_. Reprint Edition, Kreiger, Melbourne, FL. Original edition (1968), McGraw-Hill, NY.

TABLE 1   Comparison of Measured and Predicted MTBF, h

	Project 1	Project 2	Project 3	Project 4
Measured (during use period	14.6	31.4	30.3	9.2
Predicted (at end of test period) using maximum likelihood point estimate	19.1	35.2	24.4	12.3
50% confidence range	13.5-28.8	> 19.8	> 12.9	6.4-23.6
New or modified instructions*	19,500	6600	11,600	9000
Total program size	21,700	27,700	23,400	33,500
Number of programmers	9	5	6	7
Project length, months	12	11	12	10

Source: Musa (1975)
*Size given is the number of assembly or machine language instructions.

TABLE 2   Comparison of Predicted and Observed Number of Software Errors (Shooman, 1983b)

Errors	Observed	Expected Number	95% Confidence Interval
Critical+ Major+ Minor	17	11	17<
Critical+ Major	7	5	9.8<
Critical	0	2	4.4<

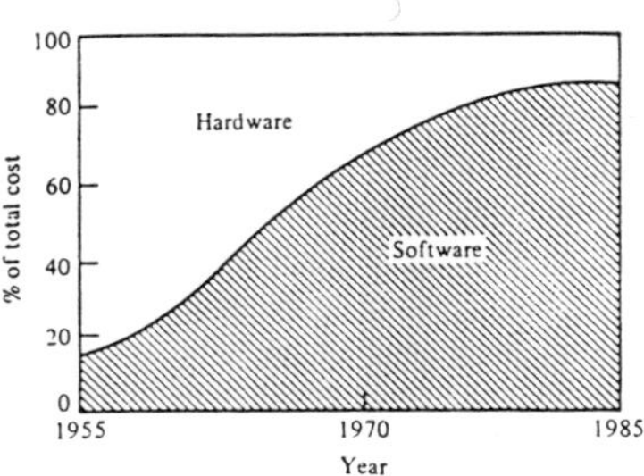

Fig. 1 Hardware and software cost trends (Shooman, 1983a, p. 11).

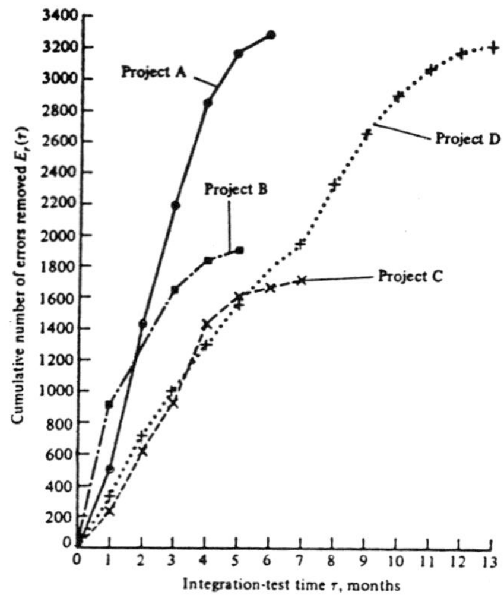

Fig. 2   Cumulative curves of the number of
         errors removed versus test time for
         four projects involving applications
         programs (Shooman, 1983a, p. 322).

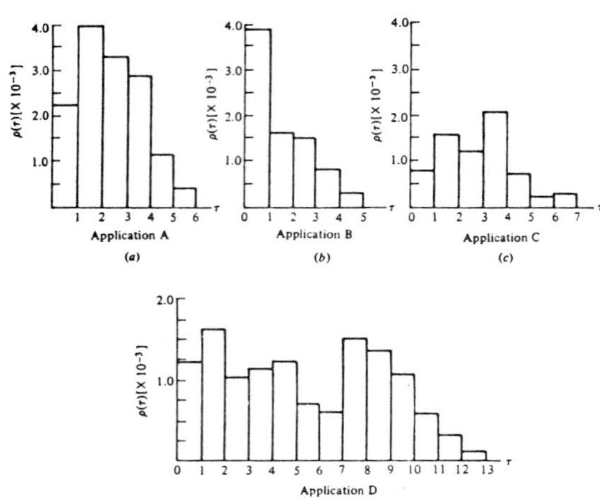

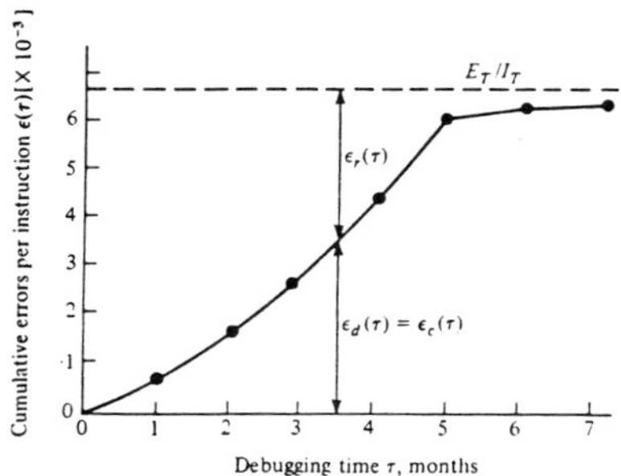

Fig. 3   Normalized error rate versus debugging
         time for four applications programs.
         (Shooman, 1983a, p. 331).
         Note: $\rho(\tau) = r(\tau)/I_T$.

Fig. 4   Cumulative error curve for a supervisory
         system A given in Shooman (1983a, p. 332).
         Note: The Greek letters represent
         normalized quantities, i.e.
         $\varepsilon_i(\tau) = E_i(\tau)/I_T$.

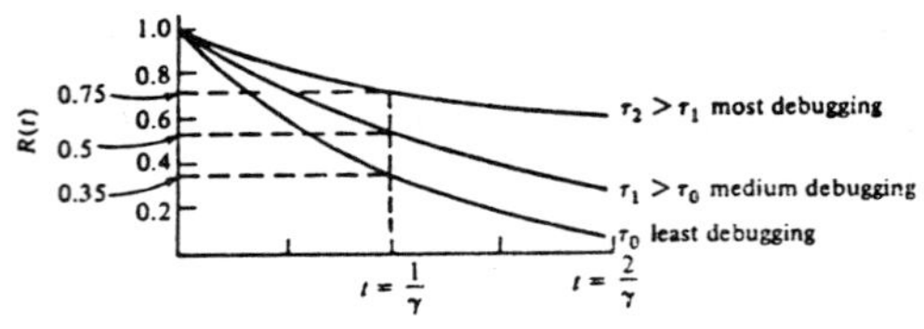

Fig. 5   Variation of reliability function (R)t
         with debugging time τ. (Shooman, 1983a,
         p. 357).

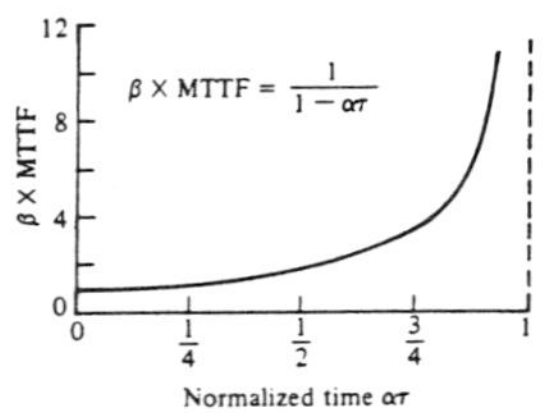

Fig. 6   Comparison of MTTF with debug time for
         the constant-error-rate model (Shooman,
         1983a, p. 357).

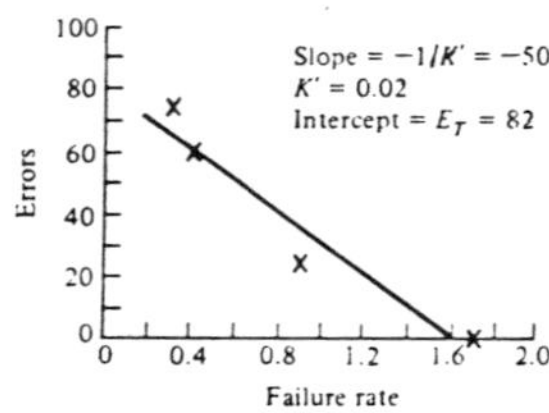

Fig. 7   Least-squares fits of data to Eq. 22
         (using TRS-80 computer and Radio Shack
         Statistics Package No. 26-1703)
         (Shooman, 1983a, p. 373).

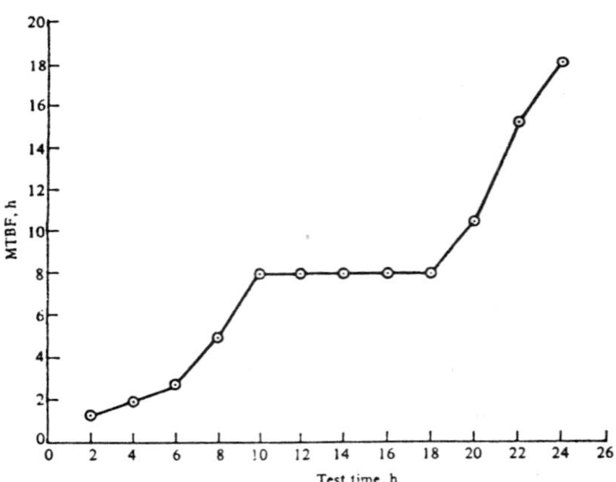

Fig. 8   MTBF versus test time for project 1
         (replotted from the data of Fig. 3.,
         Musa, 1975) (Shooman, 1983a, p. 360).

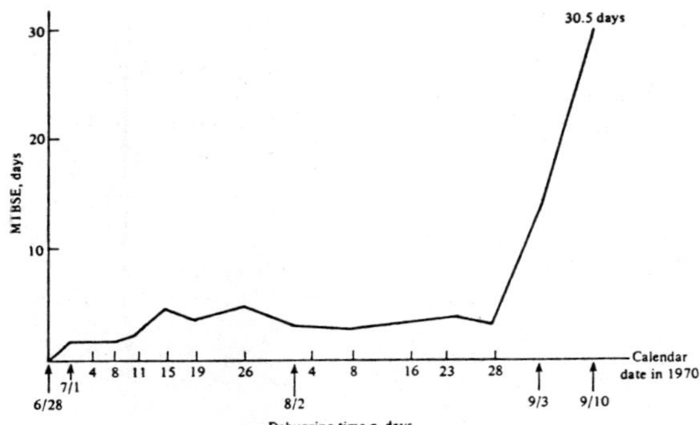

Fig. 9 Growth curve of software reliability
        (mean time between software errors).
        (From Miyamoto, 1975, Fig. 6).
        (Shooman, 1983a, p. 359).

MATHEMATICAL TOOL FOR SYSTEMS
RELIABILITY ANALYSIS

M. Elbert

Honeywell Information Systems, 245 Merrimack Street, Lawrence, MA 01843, USA

Abstract. The subject of this paper is the mathematical tools and methods for systems reliability analysis.

Although there are possibly as many approaches and models to system reliability analysis as there are analysts, the methods presented hereby, if applied informally, do provide general methods for reliability evaluation, prediction and optimization of physical systems (that can be mechanical, electronic, electrical).

Keywords. Mathematical tools, methods, reliability, systems.

INTRODUCTION

The importance of improved reliability, availability and maintainability to different industries have been well documented.

Reliability engineering is a broad discipline of which modeling techniques and mathematical tools constitute an important part.

It is well known, for example, that an improper reliability model can often lead to substantially lower predicted reliability and increase the system overdesign.

Before a detailed discussion about mathematical tools, just a few words about two trivial concepts which are considered as the basis in utilization and development of mathematical tools and models at present time. It helps to discuss in detail the objective of the paper.

1. The concept that there is an optimal reliability for systems at minimum cost dominates the market at the present time for different industries. It is considered that reliability engineering requires both engineering (technical) and management decision activities. Such an approach creates reliability life cost/benefit and optimization concepts. Reliability analysis and optimization are considered as two different sides of the same coin, each of which cannot exist without the other.

2. The development of methods, techniques and tools which can be successfully used to solve practical engineering needs and problems are very strong requirements and a basic direction in the development of reliability engineering at the present time.

Status of the paper restricts opportunities for deep and broad discussion of all aspects of this very deep and broad engineering area.

This paper is restricted by discussion of the methods and mathematical tools which are used to solve the most topical and practical engineering reliability problems at the present time and which have been confirmed by the successful result of their utilization. Moreover, the methods and techniques will be divided in two groups, in particular, traditional and well recognized mathematical concepts and new or non-traditional concepts which were

developed and utilized to specific reliability engineering tasks. The techniques are presented together with examples illustrating the application of techniques.

Among well recognized mathematical concepts two topical methodologies are presented: reliability analysis and optimization tool for the system during design stage of the development of the product; system reliability growth management and prediction models. As an example of new concepts two methodologies are described, particularly, maintainability analysis without actual maintainability data and reliability improvement methodology.

TRADITIONAL MATHEMATICAL TOOL

Design reliability analysis and optimization.

The important requirement of reliability engineering during design stage is knowledge about the level of system reliability for each design approach, model and action, and selection of the final solution among numerous design models and alternatives. Every design change has a potential for enhancing or degrading system reliability, as well as system cost and performance.

This requires both engineering (technical) and management decision activities.

The basic part of the engineering activities are to evaluate reliability and develop technical solution to reach reliability requirements.

The management activities are to provide the optimum reliability, that is to expose those areas of the system where improvements are meaningful.

To solve this problem reliability evaluation and optimization methodology has been developed (M. Elbert, B. B, V. N, 1985).

Several important considerations and mathematical concepts of the methodology are discussed below briefly.

The method of analysis.

An iterative method of reliability analysis and

optimization is used in the methodology. This technique gives rise to related problems of choosing part and design method of improvement and decides the effect of these actions upon comparative analysis reliability and cost/performance trends. Thus, the method relies on calculations of changes in system reliability based on the difference between assessments of reliability with and without consideration of the improvement.

Reliability evaluation method.

The Markov Model is selected and recommended as the reliability evaluation method for conceptual design stage. The selection Markov Model is based on a number of advantages compared with other traditional methods.

Reliability evaluation criteria.

Each level of computer modeling usually requires different reliability criteria for adequate description and comparison of alternative designs. Several and multiple criteria are required on the highest level of modeling, that is the system level. The most frequently used single criteria – failure rate or MTBF is poor or at least partially invalid for evaluating massively redundant and complex systems. Moreover, the most effective target for improvement is not always the one with the highest failure rate.

Combination of standard criteria, which include MTBF, MDT, steady state availability and probability reliability function is utilized in the methodology to evaluate the reliability level for the entire system.

Optimization model and cost/performance analysis.

Basically in the cost model the cost of the system is weighted against the cost of service and down time and the design alternatives are optimized for user ownership over life time of a system.

However, since adding redundancy to a system often affects performance, the cost of the system should be modified by a performance scale factor.

Thus, criteria for developing an optimum solution is the total cost per unit of performance and reliability parameters. The minimal cost/performance of user ownership over life time of the system is considered as the decision function.

The application of the methodology.

This methodology was used for systematic analysis and optimization of different design alternatives of computers, including conventional systems, high up time systems and fault tolerant systems (M. Elbert, B. B, V. N, 1985).

Several global design goals and objectives were set before reliability analysis. Among them are the following:

1. The objective of the study was to formulate a hardware and software plan which describes how to develop a future system.

2. It is well known, that in order to satisfy given reliability requirements the failure can be attacked by two basic approaches: fault preventive (avoidance) and fault tolerance. It is considered at the beginning of analysis that both approaches have "equal rights" to be used for all three analyzed design alternatives.

More than thirty different design alternatives for various system configurations were developed either duplicated (redundant) or improved in their design (resilient) to improve reliability and to reach reliability requirements.

As a basic system which was used as a basis for comparison of design alternatives, the traditional system with series configuration, no redundancy or resiliency and not repairable on line was selected and called Model 1. The Model 1 is used as a comparative and reference basis for design changes and analysis.

The basic hardware system configuration of Model 1 and list of component MTBF criticality are represented in Table 1. The components are prioritized on their roles and contribution on system reliability.

The field reliability data and data from MIL-STD-217D were basis of failure rate information. The Bayes' Theorem was used to combine this information.

Information in Table 1 was used among others to develop and prioritize the design changes to improve system reliability.

As a result of MDT analysis three different values of MDT for each optimum replaceable unit are considered in this study, particularly, 1 hour, 4 hours, 24 hours.

The evaluation of reliability and availability parameters for each alternative and of the effect on their betterment design options for each model were developed utilizing Markov Model. Standard block-diagrams, transition diagrams and matrix, and calculations were developed. All Markov Model concepts, definitions and mathematical formulations well developed and widely utilized during the last two decades are not discussed hereby.

Some results of availability analysis are presented in Table 2 and Figure 1.

The data from Table 1 and 2 were used to perform cost/performance/benefit analysis and optimization.

The system optimal decision depends on many factors. The major ones were: reliability and availability level, performance level, initial system cost, cost of down time and loss of data. Such components as cost of site preparation, supplies, discount rates were also considered in analysis. System costs and performance factors developed in analysis are shown in Table 2.

Results of cost/benefit analysis are illustrated on Figure 2. All curves follow the typical parabola shape of standard cost/benefit analysis. The lowest point on each curve identifies the "best choice" of this system design for certain combinations of variables and coefficients.

It should be emphasized that the initial cost of the system modified by a performance scale factor results in different "best choice" compared with models without utilization of the performance scale factor.

The decision table is presented in the Figure 2.

The results of analysis were used to make numerous important conclusions.

One of them is developing the best design model. Another is developing the group of equations which provide general design optimal decision (M. Elbert, B. B, V. N, 1985).

It should be emphasized that if at the beginning of design reliability analysis there are special reliability requirements for the system similar analysis should be developed for design alternatives where reliability level reached requirements.

A few words about integrity. Integrity which is
the ability to detect failures when they occur
is important to provide a fault tolerant performance.
The analysis showed that integrity for various
electronic optimal replaceable units ranges from
low of 56% to a high of 96%. One of the design
alternatives is that the unchecked components can
be designed using increased quality parts. Based
on analysis, assuming a 3 to 4 fold increase in
the reliability for undetected failures was substantiated using parts with higher quality factor.
However, it leads to associated increase in price
for the system of about 4-5%.

Reliability growth management concept

It is common for new products to be less reliable
during early development than in the program, when
improvements have been incorporated as a result
of failures observed and corrected.

Emphasis on reliability performance prior to the
final demonstration could substantially increase
the chance of meeting reliability requirements.
This can be accomplished by the utilization of
a reliability growth management concept (MIL-HDBK-
189, 1981). A comprehensive approach to reliability
growth management consists of planning, evaluating
and controlling the growth process throughout the
development program.

For planning, evaluating and controlling reliability
three types of reliability growth curves are considered. These are idealized, planned, and tracking
curves.

The idealized, planned and tracking curves are
concepts and methods for realistically setting
reliability objectives and assessing what has been
achieved against (interim) goals and the requirements.

The reliability growth requires a special mathematical model.

Different mathematical mdoels for reliability growth
are utilized in literature.

The mathematical model from (MIL-HDBK-189, 1981)
gives a realistic method to describe a reliability
growth of the entire system.

The approaches and mathematical model are simple.
This model assumes that the cumulative failure
rate versus cumulative time is linear on log-log
scale.

However, the utilization of this model is not
restricted only by quantifying and planning of
reliability growth during the development stage
of the program. This model can be successfully
used to make a failure rate prediction for the
entire system.

Reliability prediction.

It is known that the reliability prediction can
be realized in two ways, particularly, based on
bottom-up and top-down approaches.

In bottom-up approach the failure rate of components
is defined as a function of the certain model and
assumptions. After this reliability component
data are combined in subsystems and systems levels.

However, more often than not predictions based
on bottom-up approach do not match the real result.
Nominally identical equipments may have widely
different failure rates in service.

This statement can be proved by example of electronic
device failure rate prediction.

Well known sources for electronic device failure
rate prediction based on the bottom-up approach
in the USA are (MIL-HDBK-217, 1979) and (British
Handbook of Reliability Data for Electronic
Components, 1984) in which stress analysis is a
basis for failure rate modeling.

However, the failure rate prediction can rarely
be made with high accuracy or confidence based
on these standards (J. Spencer, 1986).

Moreover, in many practical cases not enough data
is available on the components information which
is necessary to develop prediction.

Therefore, the reliability prediction can be made
for some cases only on the basis of an overall
analysis for the entire system with future allocation
of this data between subsystems and components.
Moreover, development and utilization of both of
top-down and bottom-up approaches will create a
basis for comparative analysis and will increase
the accuracy and confidence of the failure rate
prediction.

Thus, a reliability prediction for a system based
on comparison analysis of both approaches is
recommended.

(MIL-HBDK-189, 1981) is recommended as a basis
for development of the top-down failure rate
prediction. This model which was described before
provides a means of viewing the reliability prediction in an integrated manner for the entire system.

NON-TRADITIONAL MATHEMATICAL TOOL

Structural method for analyzing maintainability.

It is well known that traditional maintainability
analysis is based on actual data. However, more
often than not designers and analysts lack sufficient
data or the available data has a low confidence
level. Insufficient data is especially common
for a newly designed system without operational
experience. Reliability and maintainability analysis
is limited by this problem.

To solve this problem for maintainability analysis,
special structural method of calculating maintainability of a system, in a non-traditional way,
that is, without actual maintainability data, has
been developed (M. Elbert, 1983). The Method provides
opportunity to analyze maintainability of the system
without actual MTTR.

The basic philosophy and mathematical principle
behind this method of calculating maintainability
is Graph Theory. Graph Theory gives the opportunity
for any physical systems to link the level of
maintainability of the system with special quantitative parameters associated with the overall system
design.

The flowcharting sequence for each operation of
repair or maintenance for a system may be represented
by a finite directed graph with a rooted vertex.
This is accomplished by associating the individual
components of the system with graph vertices and
repair and maintenance process (interrelation between
respective components) with graph arcs.

Components of the system are represented by graph
vertices which are drawn as circles. The repair
sequence, disconnections and removals of individual
components are denoted by arcs as straight lines
with arrows connecting vertices.

To use this technique, the only information required
is the construction scheme which is used as the
basis of the breakdown sequence of repair or

maintenance for systems and their components. The
maintainability of the system is evaluated by the
special characteristic $\Delta\phi$ (See Fig. 3) on the basis
of created graph.

The method is simple. It is necessary to perform
just a few calculations and actions. Example of
utilizating the method is presented in Figure 3.

Reliability improvement methodology.

One of the most difficult problems faced by a lot
of industry is the large number of design, manufacturing, and operating factors that simultaneously
affect a system's ultimate reliability in complex
ways. Usually the level of reliability of the
system is due to a complex combination of simultaneously acting factors. Separation and determination
of effects and contributions of individual factors
on level reliability of the system is the key point
for reliability analysis, prediction and improvement.

Traditional methods like random balance plans,
design of experiment and multiple regression analysis
are relatively complex. They require special data
systems, computer power and take a long time.

The engineering method, entitled "The Method of
Basic Parameters" was developed for analyzing
dependencies between reliability parameters and
factors to solve this problem (M. Elbert, 1984).

A few features of this method follow. In this
method only input and output characteristics are
considered, particularly, input is different factors,
output is certain levels of reliability. It is
considered that each factor on each level contributes
a certain proper portion to system reliability
and each level of system reliability is associated
with a corresponding level of main factors.

The mathematical model behind the Method of Basic
Parameters is the principle of describing the
function of several variables in the form of multiplication of functions of one variable. Based
on this principle, the mathematical model of the
method for analyzing relationship between reliability
and factors, is presented by the equation in this
figure. Calculations are produced in a table form
by interactive procedure techniques. The final
result of the analysis is the set of curves and
mathematical expressions which separately define
reliability as a function of different factors.
The main step of calculations is the moving of
curves to satisfy the basic equation.

The method can be applied effectively to analyze
and evaluate the complex effect of different factors
on system reliability. Even in cases of many complex
dependencies between factors, the analysis is simple
and the result is unambiguous. Only one simple
equation is used (See Fig. 4). This method is
especially effective when significant uncertainty
exists due to small data base.

Example of utilization of this method is presented
in Figure 4.

CONCLUSION

The experience described in this paper concerns
a reliability improvement aimed at providing a
mathematical solid basis for management decision
concerning system reliability analysis, prediction
and optimization.

REFERENCES

Elbert, M. (1983). Structural method for analyzing
     maintainability. International RAM Conference
     for Electric Power Industry, pp. 205-209.

Elbert, M. (1984). Availability improvement
     methodology, Reliability Conference for
     Electric Power Industry, pp. 117-122.
Elbert, M., B. Bodendorf, V. Negi (1985). Design
     reliability analysis and optimization for
     computer systems. Vol. 1, IECON'85, pp.
     118-124.
Handbook of Reliability Data for Electronic
     Components used in Telecommunications Systems,
     (1984). British Telecom., pp. 1-39.
MIL-HDBK-189, (1981). Reliability Growth
     Management, USAF Rome, pp. 1-148.
MIL-HDBK-217, (1979). Reliability Prediction for
     Electronic Systems, USAF Rome, pp. 1-279.
Spencer, J. L. (1986). The highs and lows of
     reliability predictions, Proceedings Annual
     Reliability and Maintainability Symposium,
     pp. 156-165.

TABLE 1. Failure rate analysis. Component criticality list

DESCRIPTION	FAILURE RATE CONTRIBUTION FOR COMPONENT, PERCENTAGE	QTY	FAILURE RATE CONTRIBUTION FOR SUBSYSTEM, PERCENTAGE
HARD DISK	14.60	2	22.16
PRINTER	20.59	1	15.63
MEMORY	9.70	2	14.73
DUAL PROCESSOR	14.61	1	11.09
COMM. (RS422)	4.87	2	7.39
CACHE	8.90	1	6.75
LAN	6.21	1	4.71
VBSDC	5.50	1	4.17
COMM. (HDLC)	5.07	1	3.85
UPC	4.66	1	3.54
SMF	3.86	1	2.93
POWER	0.95	3	2.13
AXIAL DC FAN	0.24	3	0.55
AXIAL DC FAN	0.24	2	0.37
TOTAL FOR SYSTEM	100.00		100.00

TABLE 2. RAM and Cost/Performance System Analysis

MODEL	RAM ANALYSIS					COST/PERFORMANCE ANALYSIS		
	AVAILABILITY FOR MDT HOURS			HOURS DOWN IN 1 YEAR 4 HRS MDT	% IMPROVE- MENT BY MODEL	SYSTEM COST (RELATIVE FIGURES)	PERFOR- MANCE FACTOR	COST/ PERFOR- MANCE ANALYSIS
	1	4	24					
MODEL 1	0.99918	0.99673	0.98070	28.65	–	1.00	1.0	1.000
MODEL 1A	0.99925	0.99702	0.98238	26.11	8.9	1.00	1.0	1.000
MODEL 2A	0.99957	0.99827	0.98966	15.10	42.2	1.33	1.0	1.330
MODEL 2B	0.99968	0.99873	0.99236	11.16	26.1	1.47	1.7	0.865
MODEL 2C	0.99970	0.99878	0.99270	10.66	4.5	1.53	1.7	0.900
MODEL 2D	0.99972	0.99888	0.99328	9.81	7.9	1.97	1.7	1.159
MODEL 2E	0.99973	0.99891	0.99347	9.53	2.9	2.00	1.7	1.176
MODEL 3	0.99978	0.99912	0.99468	7.67	19.5	1.82	1.7	1.071
MODEL 4	0.99999	0.99999	0.99969	0.08	99.0	2.39	1.7	1.406
MODEL 5	0.99999	0.99999	0.99989	0.02	66.7	3.05	3.0	1.017
MODEL 6	0.99999	0.99999	0.99992	0.01	23.1	3.34	3.0	1.113

M. Elbert

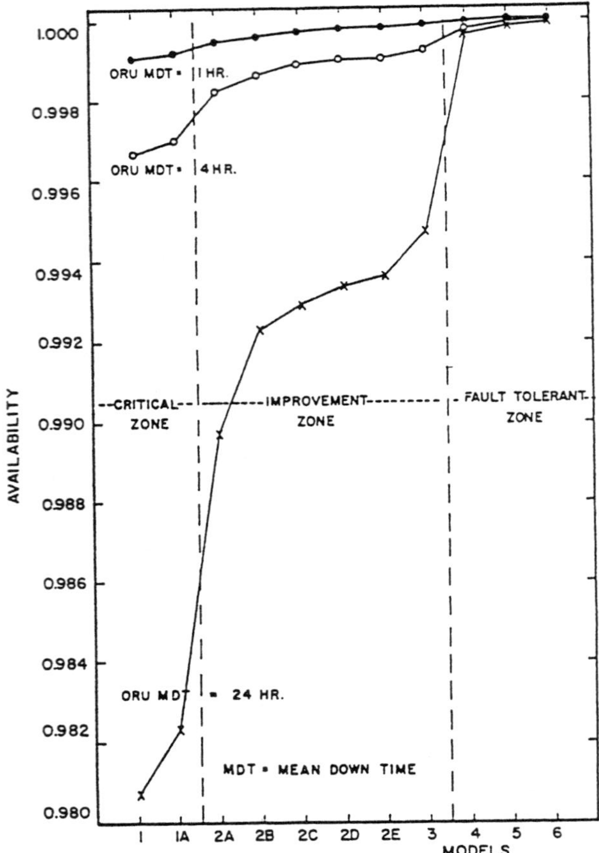

FIGURE 1. Availability Analysis

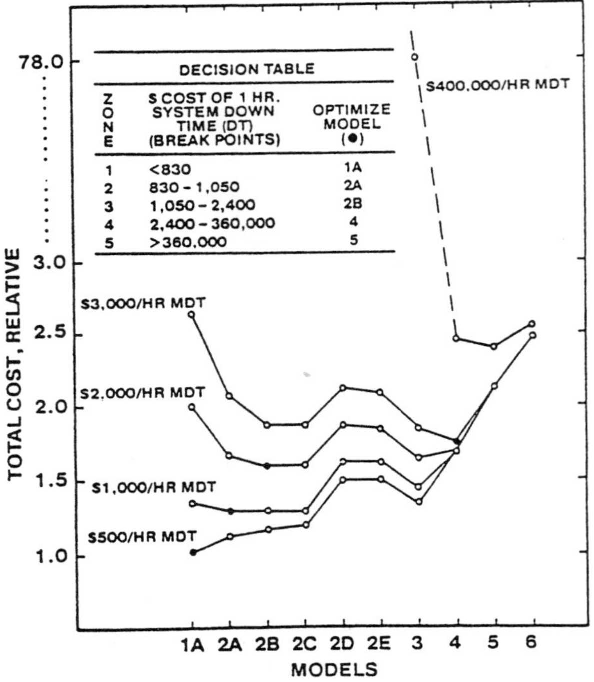

FIGURE 2. Cost/Benefit Analysis

MAINTAINABILITY FUNCTION

$$\Delta\phi = \phi_{d(x_o x_i)} - \phi_{m(x_i)}$$

where:

$\Delta\phi_{d(x_o x_i)}$ — summary paths from vertex x_o to vertex x_i

$\phi_{m(x_i)}$ — summary weight of graph vertices

Example of calculations of Maintainability Functions

Component	Design Construction	Maintainability Function
Graph of accessibility and disassembly of pump gland seal	Modular Construction, Fig. A	-1264 + 90 = -1174
	Non-modular Construction, Fig. B	-136 + 30 = -106
Graph of accessibility and disassembly of control transfer pump valve	Modular Construction, Fig. C	-246 + 54 = -192
	Non-modular Construction, Fig. D	-214 + 40 = -174

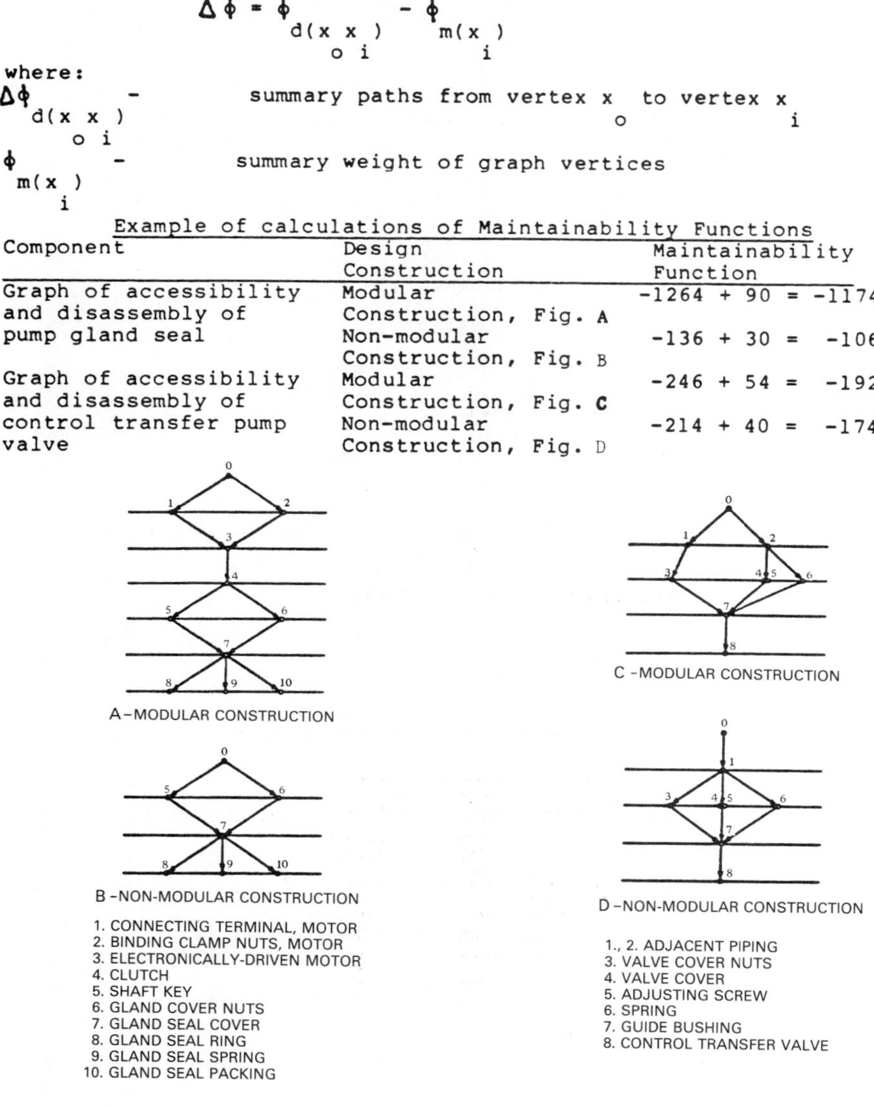

Figure 3. Example of utilization of the Structural Method

M. Elbert

Basic   Equation

$$Ai = Ao \prod_{j=1}^{n} \Delta f_{ij}$$

where:
Ai   — the availability for i level
Ao   — the availability in the starting point of analysis
Δf_{ij} — the correction coefficient for the j factor on i level

$$\Delta f_{ij} = \frac{\overset{o}{Aij}}{Ao}$$

$\overset{o}{Aij}$ — the current availability for the j factor on the i level

i   — the number levels/observations, i=1, 2....., m

j   — the number of factors, j=1, 2....., n

Availability (fraction) for Ship Set as a function of Major
factors using the Method of Basic Parameters

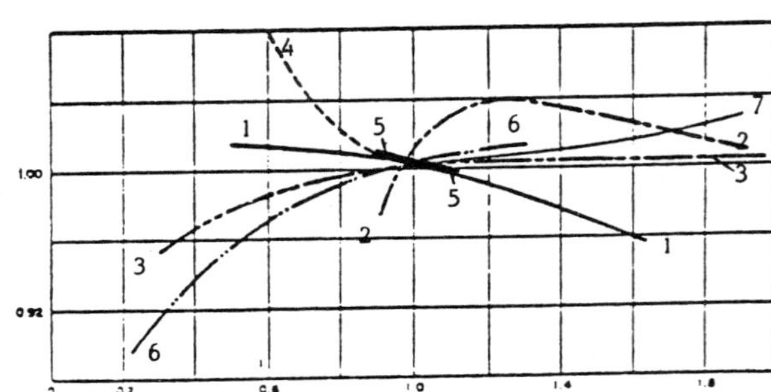

Legend: 1 — Age
            2 — Cycle of repair
            3 — Manhours of maintenance
            4 — Coefficient of utilization of useful time
            5 — Coefficient of utilization of speed
            6 — Repair productivity
            7 — Sequence of ships in the set

Figure 4. Example of utilization of the Method of
Basic Parameters for analyzing system availability

HUMAN RELIABILITY CONSIDERATIONS

A. Carnino

Electricité de France, Direction Générale, 32 rue de Monceau, 75384 Paris Cedex, France

Abstract

Since the technical aspects of systems have reached now a high reliability, the human
beings involved in operating and designing a nuclear plant have become a major con-
tributor to unreliability. Various improvements are made to improve the man-machine
interface. When quantification of human errors is needed, some techniques and models
do exist but still need validation on raw data. The data are still scarce and need to
be collected, especially in the field of cognitive operator behaviour. The trend
should now be to design the plants for men, and not adapting the men to the
processes.

Keywords: reliability; human factors; man machine systems; human error
quantification.

INTRODUCTION

How should be taken into account the potential human errors during the life of a plant :

Human factors considerations should be
adressed during the whole life
of an installation : design phase,
fabrication and construction phase,
licensing phase and operation phase.
Errors of design, of quality assurance
or control should also be analysed as
they may induce later operation errors.

Design phase :

For designing reliable equipments and
systems, rules have been established
such as :
- single failure criterion, active or
passive redundancies,
- anticipated transients studies and
accident prevention,
- design basis accidents etc ...
For human beings, only rules of simple
ergonomics are used in the design such
as accessibility, readibility, space... It means that for human beings
there is no equivalent to equiment and
system design rules. For instance for
designing safely and man-adapted, it
would be necessary to integrate in the
design how, by who, etc... the installation will be operated, how, by
who... it will be maintained, what is
the human redundancy required to accomplish certain tasks..., what is the
optimum level of repartition between
human and automation, how to apply the
"single failure" criterion or its
equivalent to the men etc... Safety
considerations have lead to analyzing
in the design phase anticipated transients, design basis accidents, where
human factors have received very little attention. Human errors in acci-

dent prevention and in accident management are not only errors of omission
or commission but also representation
and diagnosis errors. This kind of
errors are due to cognitive behaviour
and may lead to potentially more severe consequences if no means for
error discovery and situation recovery
are provided.

Fabrication and construction phase :

A lot of efforts have been devoted to
quality assurance and quality control.
Quality assurance manuals have been
developed, but their implementation
relies entirely on men. I must say I
have never seen any human factor studies integrated in these manuals. It
is also obvious that the quality controls performed entirely by men have
not received enough attention from
human factors specialists. This is an
area which in my view would deserve
more attention.

Licensing phase :

In this phase, if I consider essentially what is done for the nuclear
power plants, human reliability is
certainly now taken into account at
various levels :
- training and habilitation of plant
personnel,
- operating procedures (normal operation, transients, accidents),
- probabilistic evaluation of accident
sequences,
- operating experience analysis.
But discussions still take place on
the uncertainties associated with the
human reliability data, on the effects
of perfomance shaping factors on human
behaviour and on the diagnosis errors
and consequences.

RISC-D

<u>Operational phase :</u>
Most of the human factors studies have been devoted to the operational aspects : it is the most obvious interface between the man and the machine, which can be measured by the availability and the safety of the installations.

The interface includes :
- operating rules, procedures in any situation (normal, transient, accident...)
- quality assurance and control in operation, inspections
- control room design, lay-out and level of automation
- training and selection of operators
- use of operating experience to prevent human error re-occurrence
- communication problems, etc...

In order to improve the various aspects of the man-machine interface, many parameters have to be taken into account, such as :
- task
- performance shaping factors, both internal and external
- stress and work load
- work organisation and work situation
- cognitive behaviour
- cause-consequence analysis of potential human errors
- etc...

Since quantification of human errors is needed for probabilistic studies of systems or of accidents, it will lead to perform a lot of detailed analysis including all the previous aspects for a good comprehensive assessment.

2 - How to improve the man-machine interface
2.1 - Procedures
As procedures constitute the real interface between the operators and the process, they are of vital importance. It would be misleading to quantify human errors in a procedure badly designed or technically not well sounded. It should be noted that until recently procedures were written by engineers, mostly designers and no account was taken from the operators themselves.
Procedures should be developed not only for operation from the control room but also for testing, maintaining and repairing. They address normal operation as well as incident, accident situations.
All the different kinds of procedures should include considerations such as:
- good layout of the document to help the users to perform the task involved reliably and efficiently
- good task repartition, good control repartition, good communication between the members of the crew
- validation of the procedures before final use by the operators for checking completeness, clarity and consistency
- work organization, composition of the crew, equipment and means for executing the procedure, work environment, documentation needed
- job and task analysis in order to determine potential consequences of human errors, and to provide recovery whenever it is feasible...

2.2 - Control room layout and operator aids
As a lesson from the TMI accident, many of the control rooms haved been revised in the nuclear power plants in order to integrate more ergonomic considerations in the lay-out and in the information display and processing. Many aids have also been developed since.
The improvement of the lay-out of existing control is essentially based on ergonomic considerations, such as functional regrouping, better labeling, use of colors etc...
The safety related information which the operator needs for identifying any perturbed situation,for checking the availability of the safety functions and for determining the evolution of the installation, are located in the control room on various indicators and recording systems. The new system (Safety Parameter Display System or Safety Panel) processes and displays the information on CRTs. It displays the parameters, characteristics of the main safety functions, the functioning point and its location within the permitted domain, a hierarchy of the alarms, etc... An engineer-Shift Technical Advisor in USA or Safety Engineer in France-was adjoined to the usual crew in the control room in order to advise the shift supervisor and to diagnose accidents by an approach based on physical phenomena analysis.
This is a very brief list of some improvements of control rooms and I am sure that many others which were developed by other countries should be added to this list. But, on the other hand, it illustrates my introductory comment : these improvements are still a way to adapt the man to the machine -whilst the reverse would be desirable.

2.3 - The training :
Training of operators has also been modified on both aspects: the means for training and the programs. To full scope simulators, it was added function simulators, computer assisted training which, located near the control room, allows for the operators to refresh their knowledge on physical and functional behaviour of plant components and systems, and on procedures, especially during night shift and week-ends. The full scope simulators are enhanced by the simulation of accidents called "beyond design".
Some aspects of human factors considerations should also be included in training such as how to fight stress conditions, human relations in a crew, team behaviour etc...

3 - Data banks
For the needs of quantification which come from the probabilistic safety studies, data on human errors are necessary.

The best and most well known set of human reliability data is compiled in NUREG 1278*. The problem of data is still very difficult to solve : are the data contained in NUREG 1278 applicable to other countries, with a different cultural and educational background ? are the performance shaping factors all quantifiable ?...
Data can also be derived from simulator exercises or directly from training on simulators. But these data have to be interpretated with great care since they do not represent real life and do not take always into account all the performance shaping factors needed for quantification. Man is much more complex and difficult to predict than components.
An other solution for obtaining data is the use of expert judgment. Different techniques have been developed and applied to estimating human error data. I shall list them very briefly: Paired comparison method, Psychological scaling, SLIM-MAUD, etc.
The use of operating experience -that is incidents- is a source of data difficult to use for quantification. The reports of incidents need further detailed analysis for understanding and categorizing human errors. The number of opportunites for committing the error is not easy to reach and the number of reported errors may also be biased due to human reluctance to report and due to non safety related consequences.

The examples of data sources listed here show how difficult it is to get real good data. This is why large uncertainties still have to be associated with these data.
There is also a field of human errors where data are scarce : cognitive behaviour which leads to diagnosis errors or representation errors.
All these difficulties illustrate the fact that prior to any quantification and selection of data, a thorough analysis should be performed on the activities involved (especially when dealing with teams or crews which may have a behaviour different from individual's).

In general, quantification is concerned with the effects of human errors, but study of the causes and mechanisms of human malfuction permit qualitative analysis and prevention of similar errors.

The ideal situation would be to have precise knowkedge of the errors for each task (and the conditions and environment in which it is executed) -a situation which appears to be very far in the future when one considers the current situation regarding data banks on human reliability. It is preferable to make a qualitative analysis, then deduce preventive measures which will allow one to be sure that the probability or consequences of the errors considered have been sufficiently reduced so that they no longer constitute the limit to the safe operation of the installation.

The methods and models for quantifying human errors have developed since a few years. The most well known and most used is THERP (Technique for Human Error Rate Prediction), used in human reliability analysis.
Task analysis involves breaking down system-required human actions (or tasks) into small units of physical or mental performance (steps) as well as identifying to the extent possible likely human actions not required by the system but having the potential for degrading certain system functions. These small units are then fully described and analyzed in terms of the Performance Shaping Factors that affect each of and combinations of them. The performance models and theories that make up THERP are then applied to these steps. Possible human errors are identified and estimates of the probability of occurrence of each error are derived. The end product of an HRA for a PRA is a set of system success and failure probabilities that reflect the probable effects of human errors. These system-based probabilities are in a form such that they can be entered on a task performance or component availability basis on the system fault trees.
The SHARP project has been conducted by EPRI/NUS Corporation to develop a framework for more systematically including human interactions in PRA studies. Seven distinct analysis steps have been identified which define the minimum number of activities that both a PRA Systems Analyst well versed in human factors techniques and a Human Reliability Analyst well versed in PRA techniques should perform to produce a quality description of human interactions that enhances the study's credibility.
The process can provide the structure for identifying human interaction, stating key assumptions, focusing on key human interactions, describing key human interactions in detail, noting where models of cognitive behavior are needed, incorporating the influence human actions have on events modeled in

PRA studies, quantifying the impact and documenting the results. It can help the PRA analysts provide utility personnel with an improved understanding of how human interactions impact risk.
The Operator Action Tree (OAT) method was developed for quantification of errors related to decision and where time plays a major role. The method is based on models of actions linked to detection of the event, thinking and diagnosis, and required response. The curve of failure probability versus time for operators to detect and think gives the failure probability.
These three methods listed above are not the only ones that can be used. Many others could be described, each having its own application domain and its own limitations. Progress in this field is very rapid and will not be addressed here.

4 - An example of human error probability quantification:

Readjustment of bistable amplifiers after reactor trip (exercise proposed by Dr SWAIN):

Background:

In this exercise, a reactor trip has occurred. It has been determined that trip was caused by a miscalibration of all three bistable amplifiers monitoring the output of the level sensors in the pressurizer. The specified trip point for high-level trip is 93%, but the amplifiers were tripping at 88%. A calibration technician has been assigned to readjust the three amplifiers and is working as rapidly as possible to minimize the downtime for the plant. After the three amplifiers are adjusted, the technician must retest the coincidence circuits to verify that the reactor trip signal will occur when any two amplifiers trip.

Task description:

No written procedure will be used for this task. The technician will perform the following operations from memory.

1. Obtain the reference voltage source which simulates the imputs from the level sensors in the pressurizer.

2. Adjust the reference voltage to a value of .93 on the digital readout on the simulator.

3. Connect the simulator to amplifier * 1.

4. Apply voltage to amplifier * 2.

5. Adjust trip to proper voltage level (.93). The point at which the trip occurs is indicated on a digital readout on the simulator.

6. Repeat steps 3,4 and 5 for amplifiers * 2 and * 3.

7. Test the coincidence circuits.

The coincidence circuits will be tested in conjunction with an operator in the control room who will confirm the alarm trips when the calibration technician trips successive pairs of amplifiers. Under normal operating conditions, the probability of his failing to note the status change is negligible. The HEP for this two man task will therefore be disregarded, as will the recovery factor it affords. Upon completion of this test, the plant may be brought up to normal operation.

Problem to be worked:

Assume that the technician is experienced and that he is responding to a direct order from his supervisor. Determine the probability that the recalibration task will be carried out correctly. Prepare an event tree oulining the various steps and their associated HEPs. Indicate the most likely errors that the technician may make. The recalibration task is defined as steps 1 through 5 of the procedure. The coincidence circuit test is not considered to be part of this problem.

Performance shaping factors:

1. No written procedure is used

2. The technician is responding to an oral instruction from his shift supervisor

3. Because of the time pressure, assume a moderately high level of stress

4. The correct adjustment of all three amplifiers is completely dependent on the initial adjustment of the simulator—there are no alerting cues in this situation. The technician expects to find all the amplifiers out of adjustment by a considerable amount. Therefore, if he adjusted the simulator incorrectly, there would be no reason for him to become suspicious when each of the amplifiers required a large adjustment.

5. Assume that the simulator is designed so that it will not operate if it is incorrectly connected to the amplifiers. Disregard any errors in hook up, since the technician will have to correct them in order to perform the task.

6. Assume that the simulator has been calibrated properly.

In this exercise, the supervisor gives a direct order to the operator to perform a task which he then checks to see if it was performed. In this case, the HEP for the operator's unrecovered failure to initiate the task is negligible.

For our initial analysis we will assume that an experienced technician performs the tasks. The tasks to be performed are as follows (along with their unmodified HEPs):

Task	HEP	Table NUREG 1278
"A" adjust simulator (involoves on digital readout)	.001 (.0005 to .005)	20-5, (p. 20-11)
"B" adjust amplifiers	Negligible	20-5, (p. 11-15)

Adjustment of the amplifiers involves turning a control until a change in a status indication is obtained, such as a lamp turning on or off. Although the HEP for this task is usually negligible, we are assigning a value of .001 because of the stress level involved. This value is somewhat arbitrary; it is a reasonably small number that reflects our feeling that, under stress, even very well-practiced, easy tasks are subject to disruption. By the same reasoning, for the first task, we take the nominal HEP of .001 (.0005 to .005) as the HEP for adjustment of each of the three amplifiers.

Note that the adjustment of all three amplifiers is completely dependent on the adjustment of the simulator: there are no alerting cues in this situation. The technician expects to find all the amplifiers out of adjustment by a considerable amount. Therefore, if he adjusted the simulator incorrectly, there would be no reason for him to become suspicious when each of the amplifiers required a large adjustment.

The event tree for the recalibration consists of just two branches, as shown in diagram below. (The success probability for branch b is raised to the third power because there are three operationally similar amplifiers.

Using above modified HEPs, the failure probability is $1-(.998 \times .999^3) = .005$.

Regarding the most likely error that would be made, the error in setting up the simulator would be limited to misreading a digital display. One of the most common errors would be that of confusing an 8 for a 3, resulting in the trip point's being set for 98% instead of 93%

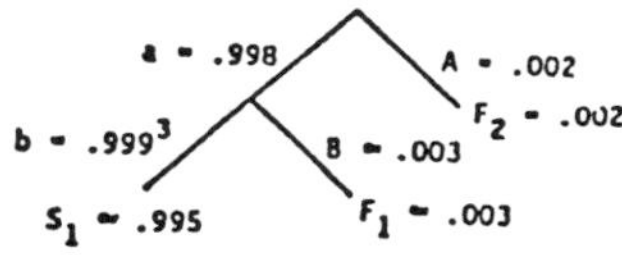

Error	HEP	Table
A = Adjust simulator incorrectly	.002 (.001 to .01)	20-5
B = Adjust at least 1 amplifier correctly	.001 (.0005 to .005) (per amplifier)	20-5

Event Tree of Readjustment of Three Bistable Amplifiers after Reactor Trip

In determining the potential operator's errors, I would personally add:

- selector of wrong cabinet: since the operator is under high stress he can open the wrong cabinet;

- wrong adjustment of simulator: this error could be of 2 types:
. omission of the setting of the reference voltage;
. setting the reference voltage to a wrong value by error of memory or by misreading.

- wrong adjustment of all 3 amplifiers: I personnally would think that the error to consider should be the wrong setting of all 3 amplifiers at the same time; the errors leading to 1 or even 2 amplifiers wrongly set should be discovered during the coincidence test;

- failure to restore multiposition selector: this is a failure not negligible from my own experience.

This example shows how important is the analysis on site by the walk/talk through which would give the information I need for taking into account the additionnal errors listed above.

5 - Conclusion :

In this paper I have tried to cover some of the aspects of human factors or human reliability that may be of interest to the participants. I certainly have not been exhaustive in describing all human factors activities and methods of quantification. There is not just one recipe to solve any human problem. But my personal view is that the value of any human reliability assessment relies on the quality of the analysis. Man cannot be compared to a component for which failure modes, mechanisms of failure, etc... can be well established and predicted.

Improving the man-machine interface or the work conditions in which men have to operate or maintain nuclear plants certainly lead to decrease both the probabilities and the consequences of human errors.

Quantification is certainly needed for decision making in the field of nuclear safety. But it should not be considered as a prediction of human behaviour.

The areas where studies and research are still needed are : cognitive behaviour, representation errors, team behaviour.

A lot of work is underway presently and should give very important results in one or two years. Do not forget that if man is still an operator in our installation, this is due to its faculty of being a "thinking person". I would like also to conclude by the fact that the machines should now be designed for men and not the reverse, so that they be man-adapted. The frequency of human errors could then be reduced to a level comparable to components and systems.

RELIABILITY ANALYSIS OF SYSTEMS
CONTAINING COMPLEX CONTROL LOOPS

M. Galluzzo* and P. K. Andow**

*Istituto di Ingeneria Chimica, Universita di Palermo, Palermo, Italy
**Department of Chemical Engineering, Loughborough University of Technology,
Loughborough, UK

Abstract.   Many Process Plants employ complex control and trip systems to improve
process safety. Reliability prediction requires determination of the failure
mechanisms followed by quantification of the failure rates. The fault tree technique
is often used for reliability prediction. Fault tree construction is a skilled task
and can be complex for systems containing multiple protective devices. The paper
compares 3 approaches to the problem of fault tree construction. The problems of
quantification and common mode failure is also considered.

Keywords.   Reliability theory; Control Equipment; Safety Analysis; Alarm Systems;
Fault Trees; Common mode.

INTRODUCTION

Modern process plants are complex. High reliability of protective systems is often required in
order to ensure safety. Safety analysis is needed
to assess system reliability. A full safety
analysis consists of 4 steps:

1.   Definition of Safety criteria

2.   Identification of hazards

3.   Discovery of failure mechanisms

4.   Quantification of failure mechanism

Step 1 is often pre-defined by legislation,
company standards or established practice. Step
2 may consist of use of a technique like the Hazard
and Operability Study (HAZOP) as described by
Lawley (1974). Step 3 is most commonly achieved
by use of a Fault Tree. A fault tree is simply a
diagram showing those basic failures (or failure
combinations) that give rise to a given final
event - such as an explosion, fire or toxic release.
A fault tree is therefore a representation of the
failure mechanisms. Step 4 requires quantification
of the basic failures and calculation to find the
system failure rate summed over all the failure
mechanisms discovered at Step 3. If a fault tree
is used at Step 3 then Step 4 is essentially
"putting numbers on the fault tree". This paper
concentrates on the problem of constructing a
fault tree for Step 3. In practice this step is
often the most difficult of the 4 to carry out
successfully. Some comments are also made
concerning Step 4.

FAULT TREE CONSTRUCTION

A fault tree is always constructed by working backwards from the undesired final event. For example,
if an explosion is the final event then the logical
precursor is the combination of an explosive
mixture and a source of ignition (i.e. an AND gate
is used to specify this combination). The
analyst would then concentrate on finding all the
possible causes of an explosive mixture (i.e. OR
gates are used to add alternatives) and all the

possible causes of a source of ignition. Lawley
(1980) gives an excellent example of fault tree
construction and quantification. Lawley uses a
method involving 3 stages:

1.   Construction of a "cause tree" that does not
include allowance for the actions of protective
systems.

2.   Addition of AND logic gates to the cause tree
to include provision for both automatic
protective systems and operator actions.

3.   Quantification of the fault tree by using the
basic event failure rates and probabilistic
calculations at AND and OR logic gates.

This is conceptually simple and is very effective
when applied carefully by such a skilled analyst
as Lawley. Less-experienced users of the technique
have often considered the use of computer-aided
methods of fault tree construction - particularly
for construction of large fault trees or those
that arise from complex control systems. (The
authors are not advocating that manual fault-tree
construction should be replaced by a computer aid
- but a computer aid might be useful for checking
a manual analysis or for updating an analysis
following system modification).

Lapp and Powers (1977, 1979) have devised a method
for computer-aided construction based on the use
of a signed digraph. A digraph, see Fig. 1,
consists of nodes (representing plant variables)
and edges (representing the links between one
variable and another). The magnitude of the edges
give the strength of interaction. The digraph is
an intermediate representation that helps the
analyst to understand the plant dynamics. The
analyst derives the required fault tree by examination of the digraph rather than directly from the
Piping and Instrumentation Diagram. A particular
feature of the Lapp and Powers methodology is that
the system behaviour is derived by recognising
feedback and feedforward loops in the digraph.
From a control engineering viewpoint this is a
natural way of understanding system behaviour -
and one that is particularly suited to systems
containing complex control loops. The basic idea
encapsulated in a fault tree is that disturbances

and failure events propagate through the plant. Lapp and Powers recognised that control loops are crucial to this process. In particular:

1.   A control loop can itself cause a disturbance (e.g. because of a failure or an incorrect setpoint).

2.   A control loop can smooth out some disturbances by means of its normal action.

3.   A control loop is incapable of controlling some very large disturbances

These features of control loop behaviour are represented in "loop operators" - Fig. 2 shows a simplified form of the negative feedback loop (NFBL) operator.

In outline the Lapp and Powers method consists of the following steps:

1.   Draw system digraph

2.   Identify loops on the digraph

3.   Identify failure state on digraph

4.   Trace backwards from failure state along edges towards basic causes

5.   When a loop is encountered apply the appropriate loop operator (such as the NFBL operator).

6.   Continue this process until all paths end in basic failures

This basic method can be applied to complex control systems. It can (and has been) automated by means of a computer program. There have been a number of criticisms of the results obtained by the method although in the authors opinion this is largely due to the details of the example systems rather than to the method itself. Allen (1984) gives a number of further examples of use of this method. A difficulty is that the fault trees produced by the method often contain extra logic gates and can be considerably simplified. In some cases the additional complexity is such that the tree cannot be easily "read" unless it is simplified. This is a serious weakness since one of the goals of fault tree construction is to provide a clear and concise representation of the failure mechanisms. By contrast the trees produced by Lawley's method are clear. In summary the Lapp and Powers method has a sound and attractive philosophical basis and is probably easier for the new user to apply but often produces rather untidy fault trees.

Shafaghi et al (1984) have proposed a further method of fault tree construction that is intended to use the same underlying philosophy as Lapp and Powers. This method is based on the observation that it is the interactions between control loops that characterises system behaviour rather than the simple variable interactions that Lapp and Powers use. From a philosophical viewpoint the method is therefore seen more strongly control-oriented. The method uses a digraph that has individual control loops as nodes. The edges then show how the loops interact. The method also distinguishes between internal and external disturbances see Tables 1 and 2. As with Lapp and Powers scheme the tables show that there is a recognition that some disturbances can only propagate through a control loop if there is a further failure in the loop itself. Shafaghi refers to these as "complementary conditions" (e.g. control valve stuck). Shafaghi (1984) gives an example containing 6 control loops with a detailed analysis of loop failure mechanisms. The fault trees produced by this method are clear and well-organised (although there is still some tendency to produce a small number of redundant

- although, it is noted, not correct - logic gates). The reason that the fault trees are easy to read is that Shafaghi's method closely reflects the way in which an experienced control engineer might work.

Fig. 3 shows the top of a tree produced by this method. Note that the faults are collected into categories having similar features.

In summary the method is based on a rigorous and sound philosophy which is easy for the user to apply but also produces a clear result.

The process of fault tree construction is the most critical stage of the fault tree technique - whichever construction method is used. In the introduction to this paper it was noted that the full safety analysis consisted of 4 steps and that fault tree construction (at the third step) is commonly used to discover the failure mechanisms. The fourth step (quantification of the failure mechanisms) is clearly important but must be based on a complete and correct fault tree.

QUANTIFICATION

Sometimes a safety study is carried out to determine a frequency of failure (e.g. how many times/yr will a feedwater system fail?). For other studies a probability is needed (e.g. what is the chance that a trip system will fail to operate on demand?). In both cases it is necessary to find the total failure rate by combining the contributions from different failure mechanisms. Fig. 4 shows the rules for combining probabilities at a gate. Fig. 5 shows the equivalent rules where a frequency is required. Lawley (1980) illustrates how a large tree can be quantified directly by application of these rules. The rules are applied at each logic gate and the combined value passed up the tree to the next gate until the "top event" is quantified.

Direct quantification is straightforward and helps to convey a clear understanding of the most significant failure modes. A problem can arise for complex systems involving a high degree of redundancy - such as those found in the nuclear and aerospace industries. The problem arises because some failures occur at several different places in the same fault tree. This may lead to the same physical failure being "counted twice" (or more) and hence the overall failure rate is overestimated. This is not to say that the same failure cannot occur in different failure mechanisms - but it should not be counted twice in one particular failure mechanism. For highly redundant systems this problem is avoided by using the method of "minimum cut sets". A minimum cut set is simply a combination of one or more failures that will cause the top event to occur - it is one mechanism of failure. Note that all of the events in at least one minimum cut set must occur in order for the top event to occur. Highly-redundant systems may yield huge numbers of minimum cut sets. Computer codes are available to find the minimum cut sets of a fault tree. These codes are widely used - particularly in the nuclear industry. Computer codes can also be used to quantify the top event from a knowledge of the minimum cut sets. Both of these processes can of course be carried out by hand for simple fault trees. Fig. 6 shows the overall cycle for using fault trees in a design environment where safety targets have to be met by design changes.

COMMON MODE

A further problem that often occurs is that of common-mode failure. The reason why we get an advantage from use of redundancy is shown in Fig. 4 - we <u>multiply</u> the probabilities at an AND gate representing coincident failure. This implies that the inputs to the AND gate are <u>independent</u>. In practice similar items of equipment nearly always contain one or more common-mode failure mechanisms. In the simplest case a pump might have a failure rate of 0.06/demand composed of 0.04/demand from internal failures and 0.02/demand from loss of power. If 2 pumps are connected to the same power supply then failure of that supply will cause simultaneous failure of both pumps - the failures are clearly <u>not</u> independent. Power supply failure is therefore a common-mode failure. Fig. 7 shows how failure to recognise common-mode leads to a serious <u>underestimation</u> of failure - in this case by nearly an order of magnitude. Common-mode failures need to be taken into account at the top of the relevant part of the tree, as in Fig. 7.

DISCUSSION

Many safety analyses are carried out as part of the design process. In most cases some target figure for acceptable frequency of failure is used to evaluate the design. The initial analysis will often highlight certain critical event sequences which dominate the calculations. As the design evolves the intention is to either eliminate these sequences or to reduce the frequency of occurrence to an acceptable level. This may be achieved by basic design changes or use of protective systems. Fault-tree analyses may be carried out in greater detail to identify the most unreliable or unsafe components in one of these critical sequences. In this way expenditure on safety improvements can be made most effectively. Following these changes the analyses will be modified and the cycle repeated until no unacceptable sequences remain. The final design fault-trees may be used as a documentation aid to demonstrate the plant safety to a regulatory authority.

In order to evaluate fault-trees and event-trees, it is necessary to have access to the appropriate failure data. Unfortunately this data is frequently not available either because the hardware is of a new type or its environment differs from that on which the failure data (if any) is based. Even when data is available it usually happens that the failures rates vary over <u>at least</u> an order of magnitude. At first sight this makes analysis very difficult. In practice it very often occurs that the system is very sensitive to failure of one or two key components. This becomes obvious if the tree is inspected and/or evaluated with approximate data. Even where no real data is available it is usually possible to make a conservative (i.e. pessimistic) estimate of failure rate. The fact that accurate data is not available for many of the components is then largely irrelevant. The safety analyst can then concentrate his efforts on estimating or unearthing the failure rates for the key components.

An important concept in the philosophy of safety and reliability is that few plants are inherently safe irrespective of design. It should be noted that maintenance procedures affect failure frequencies. The predicted failure rate of the plant will be too low if actual maintenance procedures are not as regular or as efficient as those assumed in the failure analyses. Operational procedures also affect the assumptions made in the analyses. The reliability of a passive component (e.g. a relief valve) depends on how often it is proof-tested. These factors help to illustrate the point that the operation and maintenance of

the plant affect the devisions taken in the design process. Viewed from the opposite angle, it is clear that we can improve the predicted safety and/or reliability of a plant by changing the <u>requirements</u> for the way the plant is operated and maintained. It is however very important to ensure that these requirements are realised in the working plant.

In practice it is common to make minor design change to improve safety, but it may be useful to enumerate some of the various ways in which changes may be made to improve a system:

a) Major changes to the process flowsheet. This includes different process routes and changes in inventory.

b) Detail changes to the process flowsheet. Typical examples include addition of isolation and/or one-way valves.

c) Use of more reliable components. This includes more reliable active components and more frequent testing of passive components.

d) Faster repair. This could imply holding higher stocks of spares and/or more maintenance men.

e) Use of redundancy. This may be in the form of multiple items of similar type (e.g. pumps in parallel) or diverse systems fulfilling the same function.

f) Use of instrumentation. This includes alarms to give the operator a more rapid indication of failure or trips to reduce the risk associated with failure.

g) Use of space. This includes modifications in plant layout to reduce consequential damage and the use of remote plant to protect the operators and/or the public.

The aim of this long list is to show that plant safety and reliability is a function of many parameters. If an initial study shows that a plant does not meet design criteria then there are usually several potential methods of improvement.

CONCLUSIONS

The fault tree technique can be used to effectively discover and quantify the various ways in which a system may fail. For systems without a high degree of redundancy the method illustrated by Lawley will yield good results. For complex systems with interacting control loops it may be better to use the method illustrated by Shafaghi. This method uses the Lapp and Powers basis of recognition of control loops but also produces a clear fault tree as an end product.

Fault tree quantification is straightforward provided that the tree does not contain large numbers of repeated events or common-mode failures. Unrecognised common-mode failures are particularly serious since they lead to underestimation of the failure rate. Great care should be taken to identify common-mode failures due to utilities, design errors, manufacturing problems, plant layout, maintenance practices etc.

REFERENCES

Allen, D.J. (1984) Digraphs and Fault Trees, <u>Ind. Eng. Chem. Fund.</u>, <u>23</u>, 175-180.
Lapp, S.A., and Powers, G.J. (1977). Computer-Aided Synthesis of Fault Trees, <u>IEEE Trans. Reliability</u>, <u>R-26</u>, 2-13.
Lapp, S.A., and Powers, G.J. (1979). Update of Lapp-Powers Fault-Tree Synthesis Algorithm

IEEE Trans. Reliability, R-28, 12-14.
Lawley, H.G. (1974). Operability Studies and Hazard
    Analysis. Chem. Eng. Prog., 70(4), 45-57.
Lawley, H.G. (1980). Safety Technology in the
    Chemical Industry: A problem in Hazard Analysis
    with Solution. Reliability Engineering, 1,
    89-113.
Shafaghi, A., Andow, P.K., and Lees, F.P. (1984).
    Fault Tree Synthesis Based on Control Loop
    Structure, Chem. Eng. Res. Design, 62, 101-110.
Shafaghi, A. (1984). Modelling of Process Plants
    for Risk and Reliability Analysis. Paper
    presented at AIChE 1984 Annual Meeting, San
    Francisco, California, U.S.A.

TABLE 1. Internal Disturbance types.

Internal Disturbance	Complementary Condition
Control or Process Element Failure (EF)	--
Control Element Invariant Failure (SK)	Controllable Disturbance (CD)
Element Susceptibility (ES)	Failure-Inducing Disturbance (FD)

TABLE 2. External Disturbance types.

External Disturbance	Complementary Condition
Uncontrollable Disturbance (UD)	--
Controllable Disturbance (CD)	Element Invariant Failure (SK)
Failure-Inducing Disturbance (FD)	Element Susceptibility Failure (FS)

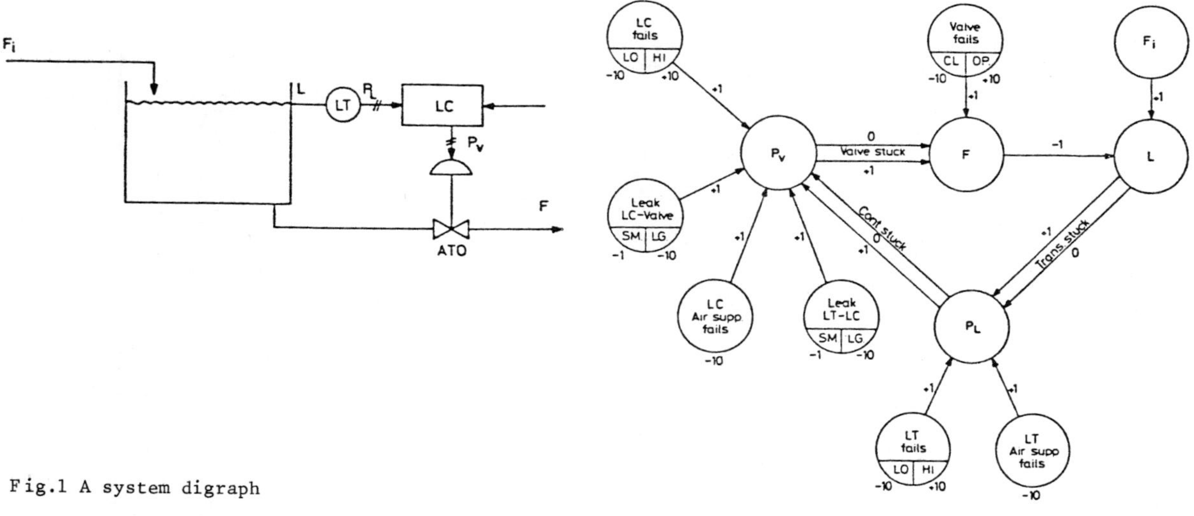

Fig.1 A system digraph

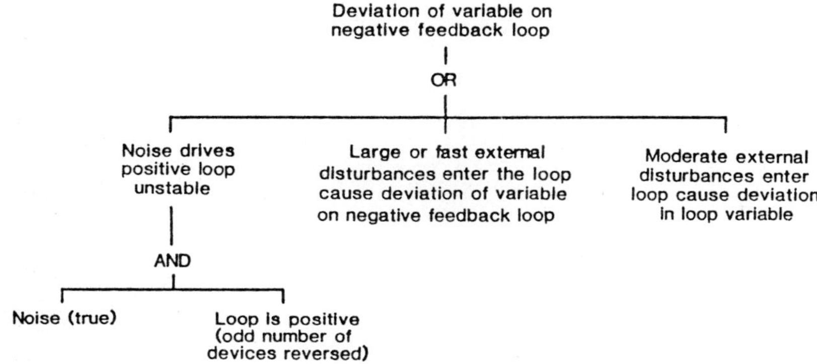

Fig. 2 Negative Feedback Loop Operator

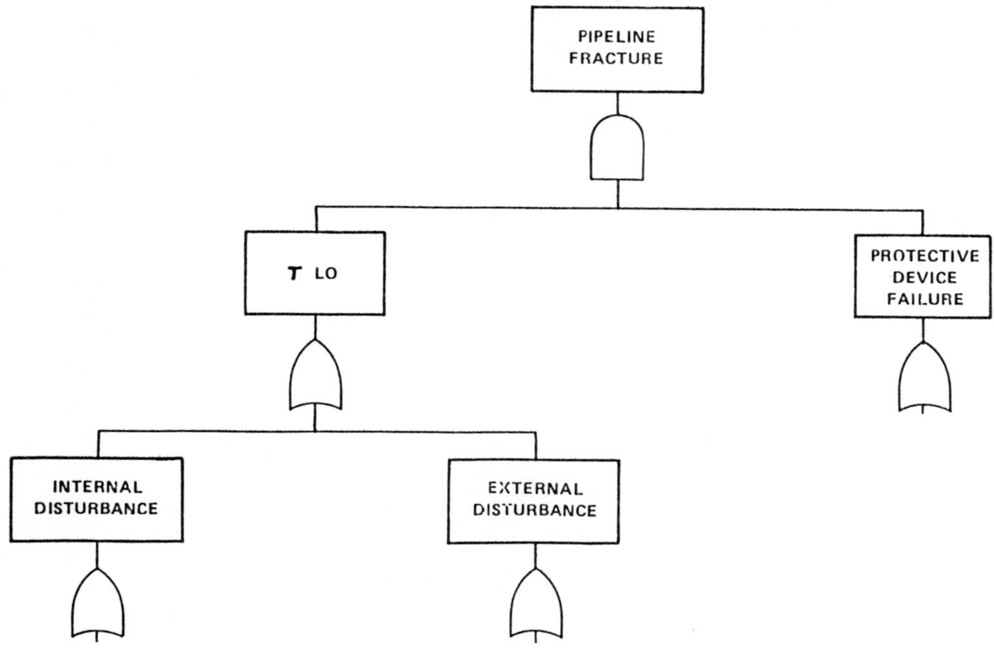

Fig.3 Top of fault tree due to Shafaghi.

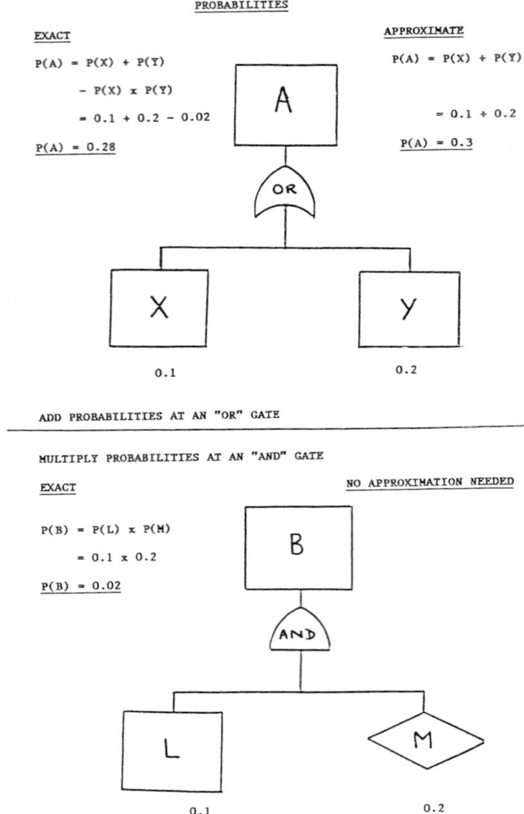

Fig. 4 Combining Probabilities.

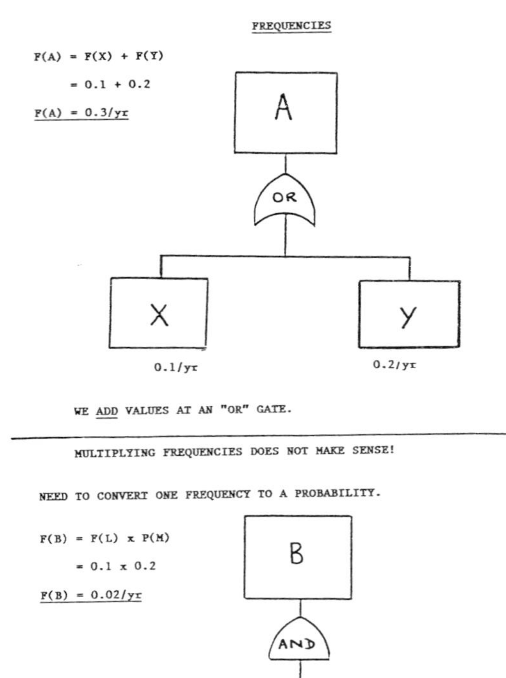

Fig. 5 Combining Frequencies.

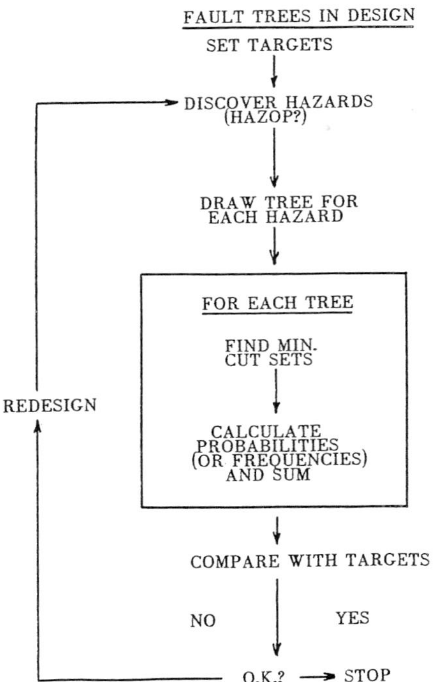

Fig. 6 Safety in Design.

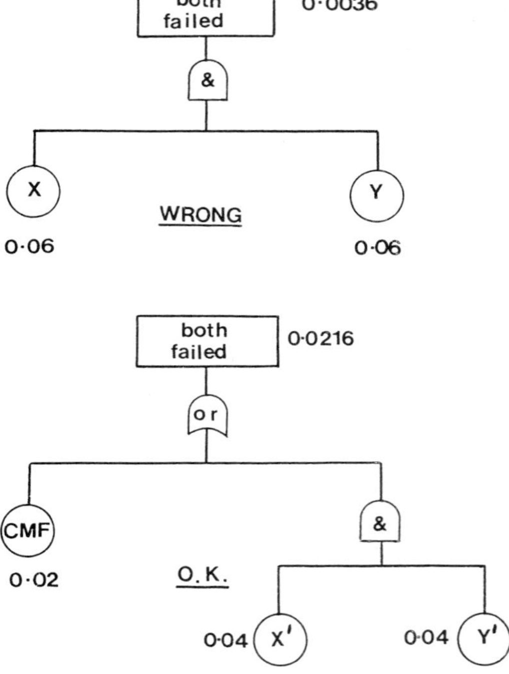

Fig. 7 Common-mode failure

DISCUSSION

Session: Reliability Engineering Tutorials

Paper: Reliability of Process Control Software

Questions: Mr. Shooman applied his software
reliability model to the integration test phase
of the software. Once a product is in the field
and has been changed many times for corrections
and to add new features, does the system then
not exhibit an increased failure rate? Is this
phenomena not analogous to the hardware "wear-out"
effects? If so, does this imply that one should
not purchase software systems, which have been
on the market for many times?

Author's reply: Yes, Lehman and Belady have found
that such effects do occur in some cases. They
base their conclusions on data from many successive
releases of an operating system. An approximation
of their results is illustrated by the following
figure for an hypothetical case.

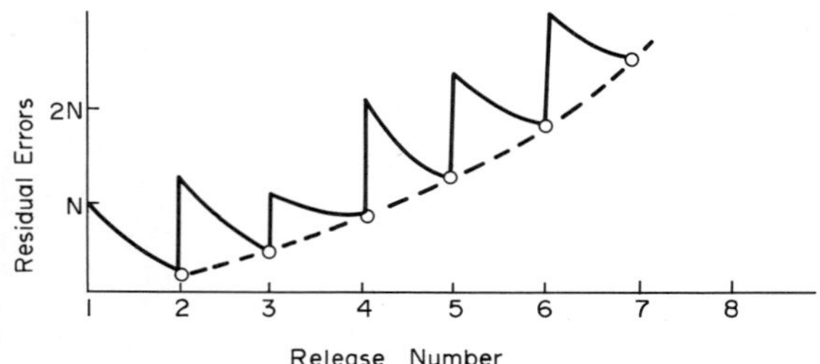

Note that the dotted line represents an increase
in the number of residual errors at the end of
field debugging.
One should however not call this a wear-out effect
because the name causes confusion. Call it
increasing failure rate with age (release number)
of the system.
Note that releases 1, 2 and 3 will have about the
same residual errors and failure rates; however,
later releases will be less reliable. Mr. Noon
cites experienced examples where an opposite effect
was observed, namely the second release was much
more reliable than the first. We may therefore
assume that both effects are possible.

Paper: Reliability analysis of systems containing
complex control loops.

Questions: In your presentation you ask yourself
some simple questions (What? How often? So what?)
for your tree construction. You seemed to imply
that only the first two mattered.

Answer from the author: The point of asking the
three questions was to note that *all* are important,
but that fault-trees are mostly applied to the
first two. The "So what?" is very important, but
is usually treated by other means, e.g. Event tree.
I would prefer to use a "Fault Tree" to find the
causes for a release of e.g. toxic material and
an "Event Tree" to see the effects of that release.
In my opinion this is better because it will show
all of the effects (dependence on wind directions
etc.) rather than searching for the cause of one
effect, which could be done very well by a fault
tree.

METHOD FOR COMPARISON OF COMPUTER CONTROL SYSTEM STRUCTURES IN THE FUNCTIONAL-RELIABILITY ASPECT

P. Wasiewicz

Institute of Industrial Automatic Control, Warsaw Technical University, Warsaw, Poland

Abstract. The paper presents a method of evaluation of reliability of computer control systems which are basic systems of the lowest level in hierarchic computer automatic control systems. Adopted conception for raising reliability of such systems resolves itself into proper choice of hardware and software structure of the system and also into realization of internal protections in order to eliminate or limit concequences of failures of particular devices of the system. As a measure connecting functional and reliability aspects, function of mean losses W has been adopted for description of the failures' effect on degree of realization of the system's task in presented functional-reliability model of computer control systems. A method for calculation of the function W according to functional states has been worked out. The method takes into consideration changes being performed in the structure of the system's internal protections. A method of comparison of the system's structures has been presented on an example of the computer cascade control system.

Keywords. Computer control; Error compensation; Adaptive systems; Control system analysis; Cascade control; System reliability.

INTRODUCTION

Among modern structures of computer automatic control systems, the main group consists of decentralized systems. The system's particular functions are being realized by separated groups of devices operating as independent subsystems. The subsystems which are directly responsible for the technologic process course belong to the group of computer control systems as shown in Fig. 1. Systems are composed of interface devices (IDS) and measuring devices (MD) and can also include analogue controllers (AC) and computer control units (CCU). The computer in these systems denotes a higher level control equipment having no matter what calculating power (i.e. it can also be a minicomputer or a microcomputer) and being able to operate in the DDC or SPC modes. Particularly damage states caused mainly by failures of technologic devices and also by measurement and control equipment are difficult to handle. Therefore inclinations to raising reliability of these devices are fully justified.

Realization of protections in cases of the failures occurences is one of effective ways of improving reliability in industrial automatic control systems. The paper considers in the first place such protections, realization of which does not require (from economic reasons, among others) any redundant devices, identical to the ones already existing in the system's basic structure (i.e. "k-out-of-n" type static redundancy has been eliminated). Instead of this the measurement circuits of substitute process variables having already their particular assignments in the automatic control systems are being utilized.

Considering any control mode of operation (e.g. stabilizing control, cascade control-as shown in Fig. 1, cascade-ratio control etc.) it is easy to notice existence of many different hardware and software structures being functional equi-

valents but having different possibilites of realization of the protections. Therefore available effects of reliability raising in different structures will be different. Thus the process of designing should be performed simultaneously for different basic structures and possibility of quick quantitative evaluation of effects achieved after consecutive protections should be ensured. For this reason it is necessary to elaborate suitable functional-reliability measure.

FUNCTIONAL-RELIABILITY MODEL OF COMPUTER CONTROL SYSTEMS

Reliability states of the system. The system's reliability state is described by a vector of reliability states of the system's particular devices:

$$R = \{S_j\}, \qquad j = 1, 2, ..., |D| \qquad (1)$$

where S_j denotes the reliability state of j-numbered device d_j: $S_j = 0$ when the device is efficient, $S_j = 1$ when the device is inefficient, and $|D|$ denotes the number of all of the system's devices D. The number of all possible reliability states is therefore equal to:

$$|R| = 2^{|D|} \qquad (2)$$

If for any given system, following inequality is fulfilled:

$$|D| \cdot \max \lambda_j \leqslant \min \mu_k; \qquad j, k \in 1, 2, ..., |D| \quad (3)$$

where:
λ_j – failure rate of j-numbered device,
μ_k – renewal rate of k-numbered device.

One can assume in all probability that the system exists in any moment of time in one of the reliability states with no more than one device being out of order. The condition is always

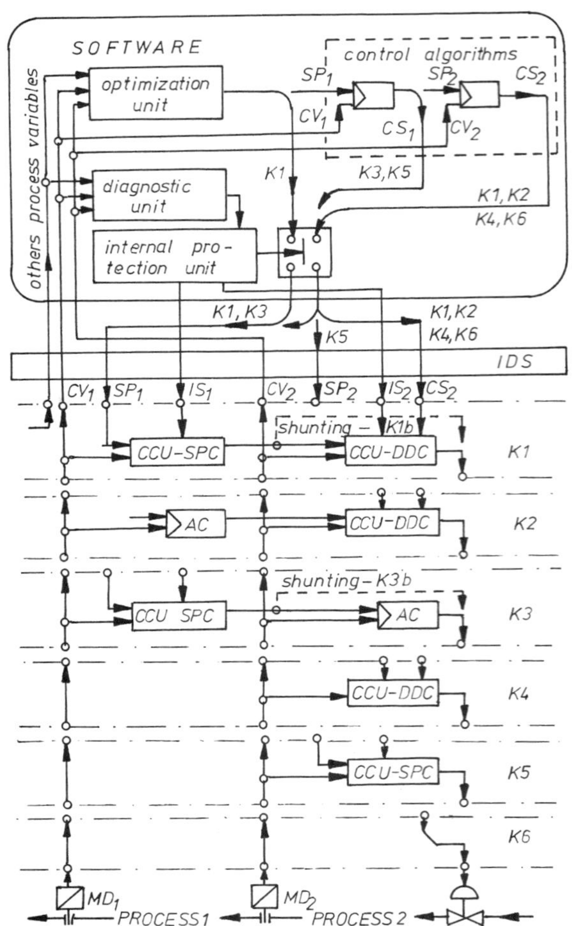

Fig. 1. Diagram of exemplary structures of the
computer cascade control system, IS -
signal of isolation of CCU from the computer, CV - controlled variable (1-main,
2-auxiliary), SP - set-point value, CS-
control signal.

fulfilled for considered systems because they
contain no more than several dozens of devices having failure rates of few orders of magnitude smaller than renewal rates. Thus it is
assumed that only single failures can occur
and the number of reliability states of incomplete efficiency is equal to the number of the
system's devices:

$$|R| = |D| \qquad (4)$$

Functional states of the system. Incorporation
of functional aspects to reliability analysis of
control systems requires that following interpretation of the reliability state of the system be
accepted. When the system remains in j-numbered reliability state of incomplete efficiency R_j,
one can distinguish (see Fig. 2):
- period of time $t_{\mu j}$ when the system's operation is incorrect due to failure of j-numbered
device but the system is being operated in
the same way as before occurence of the
failure,
- period of time $t_{\lambda j,i}$ when the system's opera-

tion is correct in spite of the failure of j-
numbered device (the failure is being tolerated). This is possible thanks to operation
of suitable internal protection even before
the system's renewal. The system's state

during the time $t_{\lambda j,i}$ will be denoted by F_i and
will be referred to as the functional state of
incomplete efficiency of the system.

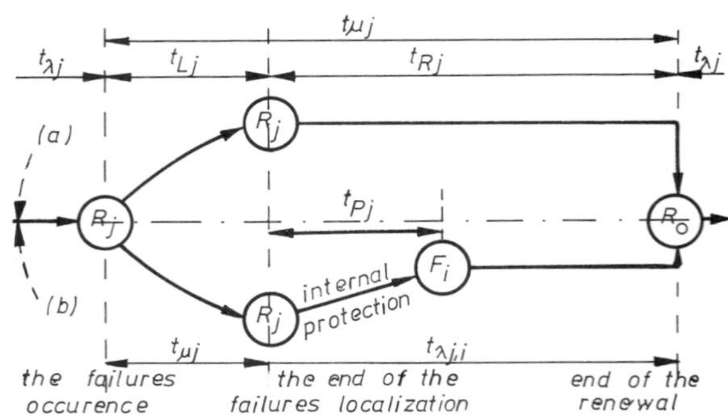

Fig. 2. Simplified time diagram of states of operation of the control system without
(a) and with (b) internal protection.
Interpretation of particular time intervals for failure of the j-numbered device:
$t_{\mu j}$ – time of incorrect operation of the system
with failure,
$t_{\lambda j,i}$– time of correct operation of the system
with failure (after protection),
t_{Lj} – time for detection and localization of the
failure,
t_{Pj} – time for realization of the internal protection,
t_{Rj} – time for renewal of the system,
$t_{\lambda j}$ – time of correct operation of the system
(after renewal).

Because of the fact that the control system
contains subsets of devices, failures of which
have identical effect on way of the system
operation, i.e which require the same protections,several different reliability states answer
usually to one functional state of the system:

$$|F| < |R|$$

where $|F|$ denotes the number of all functio-
nal states F of the system.

Projection of reliability states into functional
states of the system can be performed by
means of the graph GO which has been shown
in Fig. 3. Particular functional states of the
system answer to vertexes of the graph, when
F_o denotes the state of efficiency, $F_1...F_{|F|-1}$
denote states of incomplete functional efficien-
cy and $F_{|F|}$ is the state of shut-down of the
system. Branches of the graph are described
by the subsets of devices (they can contain
only one element in specific cases), failures
of which cause changes of functional states
determined by these branches. The set Do
contains devices, failures of which allow for
the system to be operated still by the algorithm
prowided   for the state of complete reliability efficiency. As examples of such devices
one can mention devices of measuring circuit
of substitute process variable, analogue controller of the computer DDC unit etc. The functional state of the system is defined therefore
most of all by the set of reliability states (it
can contain only one element in specific cases), by the type of control (performed by

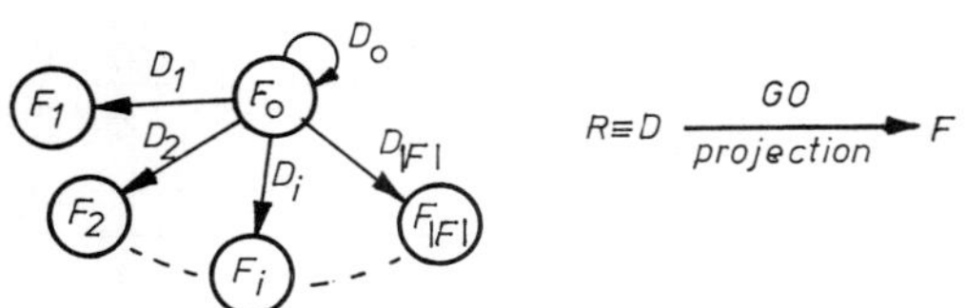

Fig. 3. Graph GO for projection of reliability states into functional states. The following equations are fulfilled:

$$\bigcup_{i=0,1,...,|F|} D_i = D \quad , \quad \sum_{i=0,1,...,|F|} |D_i| = |D|$$

the computer´s program, by the analogue controller or manually), by the structure and algorythm of the computer´s program and by the process variable being applied in this algorythm.

Function of mean losses W as the functional-reliability measure of the system.

For the functional-reliability evaluation of the computer automatic control systems, well-known (e.g. Krüger, 1978) function of mean losses W has been adopted. Because of the assumed interpretation of the reliability state of the system as shown in Fig. 1, this function will be calculated accordingly to following equation:

$$W = \sum_{j\in\{j:d_j\in(D-D_o)\}} w_j PR_j + \sum_{i=1,...,|F|} w_i P_i \quad (6)$$

where:

P_i – denotes probability of the system existence in the functional state F_i,

PR_j – probability of the system existence in the state of incorrect operation,

w_i – denotes the factor of mean losses, defined as a ratio of losses suffered during the system operation in the functional state F_i to output results obtained during the system operation in the state of complete functional efficiency,

w_j – the equivalent of w_i defined for the state R_j.

In order to make this equation usable one should precize the method for calculation values of probabilities P_i and PR_j. For this purpose it necessary to define the graph of functional states GF of the system.

Graph of functional states GF of the system. The general shape of the graph GF has been presented in Fig. 4. The graph is being constructed with help of the graph GO for projection of reliability states into functional states.

Following states of the system are attributed to vertexes of the graph GF:

R_o – the state of complete reliability efficiency;

R_j, $j\in\{j:\ d_j\in D_o\}$ – reliability states which do not have any effect on the system operation;

R_j, $j\in\{j:\ d_j\in(D-D_o)\}$ – reliability states causing incorrect operation of the system;

F_i, $i=1,...,|F|$ – functional states of the system.

The state of complete functional efficiency is composed of states R_o and R_j, $j\in\{j:d_j\in D_o\}$.

Branches of the graph GF are attributed to appropriate reliability parameters:

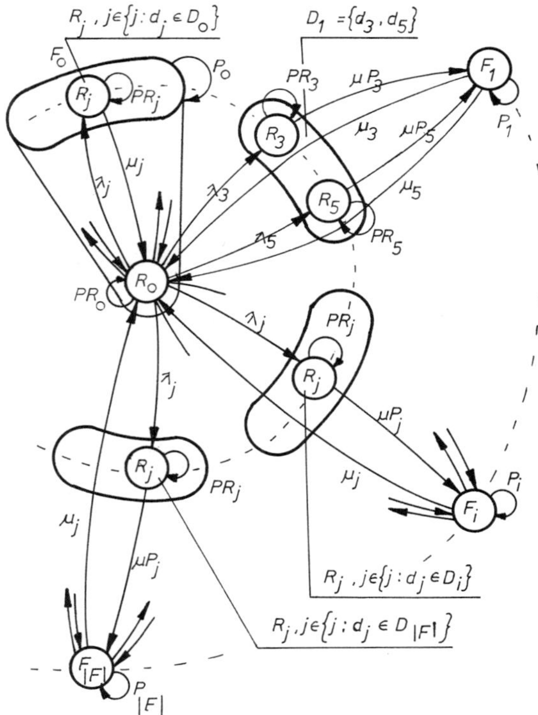

Fig. 4. Graph of functional states GF of the system

$\lambda_j = \dfrac{1}{t_{\lambda j}}$ – failure rate of j-numbered device; for branches connecting the vertex R_o to R_j, $j\in\{j:d_j\in D_o\}$ and also to vertexes R_j, $j\in\{j:\ d_j\in(D-D_o)\}$,

$\mu P_j = \dfrac{1}{t_{Lj}+t_{Pi}} = \dfrac{1}{t_{\mu j}}$ – protection rate after occurence of the failure of j-numbered device; for branches connecting vertexes R_j, $j\in\{j, d_j\in D_i\}$ to vertex F_i, $i=1,...,|F|$;

$\mu_j = \dfrac{1}{t_{Rj}-t_{Lj}-t_{Pi}} = \dfrac{1}{t_{\lambda j,i}}$ – renewal rate of the system after the protection being performed on accound of the failure of j-numbered device d_j, $j\in\{j:\ d_j\in(D-D_o)\}$; for branches connecting vertexes F_i, $i=1,...,|F|$ to the vertex R_o.

Loops of the graph are attributed to values of probabilities of the system existences in the states mentioned above, PR_o, PR_j $(j=1,...,|R|)$ and P_i $(i=1,...,|F|)$ respectively. The probability of existence of the state of complete functional efficiency F_o is equal to:

$$P_o = PR_o + \sum_{j\in\{j:\ d_j\in D_o\}} PR_j .$$

Values of the probabilities can be calculated from the set of equations which results directly from the set of the Kolmogorow differential equations describing the graph GF after their derivatives have been equaled to zero:

$$\lambda_j PR_o - \mu_j PR_j = 0, \quad j\in\{j:\ d_j\in D_o\}$$

$$P_o = PR_o + \sum_{j\in\{j:\ d_j\in D_o\}} PR_j \tag{7}$$

$$\lambda_j PR_o - \mu P_j PR_j = 0, \quad j\in\{j:\ d_j\in(D-D_o)\}$$

$$\mu P_j \cdot \sum_{j\in\{j:\ \overline{d_j}\in D_i\}} PR_j - P_i \cdot \sum_{j\in\{j:\ d_j\in D_i\}} \mu_j = 0, \tag{7}$$

$$i = 1,...,|F|$$

$$P_o + \sum_{i=1,...,|R|} PR_j + \sum_{i=1,...,|F|} P_i = 1$$

<u>Simplified functional-reliability model of computer control systems.</u> For majority of failures the following equation is true:

$$t_{\mu j} = t_{Lj} + t_{Pi} \ll t_{Rj} \tag{8}$$

Therefore one can assume that the system exists in proper reliability state immediately after occurence of the failure: $t_{\mu i}=0$. The graph GF takes with this assumption the shape of well-known graph of the reliability states – as shown in Fig. 5. The function of mean losses W can be expressed in this case as shown in Fig. 5.

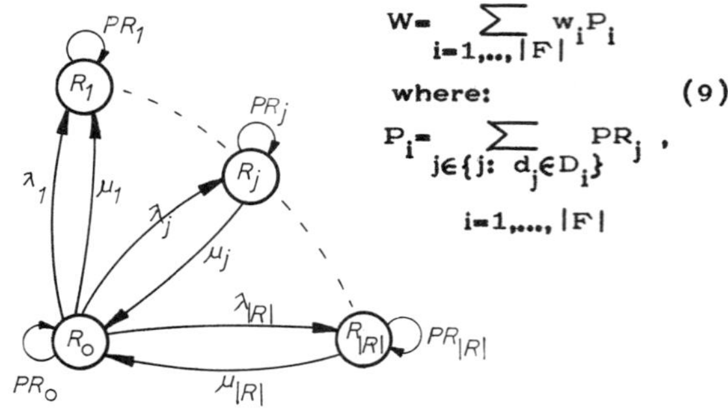

$$W = \sum_{i=1,...,|F|} w_i P_i$$

where:

$$P_i = \sum_{j\in\{j:\ d_j\in D_i\}} PR_j, \tag{9}$$

$$i = 1,...,|F|$$

Fig. 5. The graph of reliability states of the system

COMPARISON OF THE SYSTEM'S STRUCTURES ON AN EXAMPLE OF THE COMPUTER CASCADE CONTROL SYSTEM

<u>Definition of set of structures for the cascade control system.</u> The paper considers 6 structures, shown in Fig. 1, out of many possible hardware structures for cascade control system:

K1 - system with two computer control units (CCU),

K2,K3 - systems with one computer unit and one conventional control unit (K2 is a typical DDC control unit with full analogue back-up system),

K4 - basic DDC system,

K5 - basic SPC system,

K6 - basic microprocessor structure.

K4,K5 and K6 are standard control system structures widely applied in industrial control systems and the rest of them can be found in cathalogues of such manufacturers as Honeywell or Siemens. The main disadvantage of the structures K1 and K3 relies on that in the case of failure of IDS units or process variables measuring circuits, failure of auxiliary control unit prevents the computer from operating by means of main CCU unit. This disadvantage can be eliminated by means of auxiliary controller shunted in such a way that the main unit output signal could be supplied directly to the actuator (K1b and K3b structures).

<u>Description of reliability parameters of the control system devices.</u> Availability of credible values of reliability parameters (i.e. failure rate - λ and renewal rate - μ) of all devices is one

<u>TABLE 1 Reliability parameters of system's devices</u>

Description of devices	Reliability parameters λ_j / μ_j			
	I	II	III	IV
CPU,microprocessor	$1\cdot10^{-3}$			
processor / 0 unit	$1\cdot10^{-4}$			
IDS cassette controller	$1\cdot10^{-3}$			
A/D converter	$1\cdot10^{-4}$			
analogue signals selector	$12\cdot10^{-4}$	$1\cdot10^{-3}$	$1\cdot10^{-3}$	$1\cdot10^{-5}$
binary inputs unit	$2\cdot10^{-5}$			
binary output unit	$1\cdot10^{-4}$			
D/A converter	$1\cdot10^{-4}$			
CCU { analogue controller	$1\cdot10^{-5}$			
CCU { comput.oper.station	$1\cdot10^{-5}$			
process variable trans.	$1\cdot10^{-5}$		$1\cdot10^{-5}$	$1\cdot10^{-3}$
valve { controller	$1\cdot10^{-5}$			
valve { serwo-motor	$1\cdot10^{-4}$			
shunting element	$1\cdot10^{-8}$			

of main conditions for proper choice of hardware structure of a control system. But these values concerning measurement and control equipment usually are not accessible. It is because of the fact that only few manufactures run reliability experiments or publish reliability data about their measurement and control equipment. Due to great differences between published values, calculations have been performed for four sets of reliability data as shown in TABLE 1. The first set contains data accepted as real, defined after Cschornack (1975) and Unger (1980). The second set contains data values equal for all devices. Remaining sets assume high reliability of measurement and control equipment with low reliability pf processor and IDS units – set III and vice versa – set IV. Considered sets of values contain therefore all particular practical cases with the fourth set only departing most of all from the reality.

<u>Description of set of functional states for particular control system structures.</u> Depending on the hardware structure for the computer cascade control system one can distinguish following states of full efficiency of the system:

F_0 - computer control with analogue cascade control system (the computer's aim is in this case evaluation of the main controlled variable's set point),

F_3 - computer control with analogue auxiliary controller (the computer control algorythm in this case performs the function of the main controller),

F_6 - computer control with analogue main controller (no possibilities of compensation for disturbancies of auxiliary controlled variable neither in hardware nor in software way),

F_9 - computer control without any analogue controllers (computer cascade control).

There exist also states of partial functional efficiency. In such states the computer:

- does not have any information about the value of the main controlled variable. Accordingly to the states mentioned above, these states have been labeled as F_1, F_4, F_7 and F_{10}

- does not have any information about the value of the auxiliary controlled variable. Accordingly to the states mentioned above, these states have been labeled as F_2, F_5 and F_{11}

- is deprived of supervision of the control system:

F_{12} denotes the analogue control in the cascade system,

F_{13} denotes the stabilizing control with analogue auxiliary controller,
F_{14} denotes the stabilizing control with analogue main controller,
F_{15} denotes the manual control with operational station from the auxiliary control unit,
F_{16} denotes the manual control with operational station from the main control unit or the control by the microprocessor controller operator's unit – in the case of the structure K6,
F_{17} denotes the shut-down of the system.

Sets of functional states possible in particular structures of considered control system have been presented in TABLE 2 .

TABLE 2 **List of sets of functional states possible in particular structures of the computer cascade control system**

F_i, i=	0÷2	3÷5	6÷8	9÷11	12	13	14	15	16	17
K1	+			+	+	+		+		+
K1b	+		+	+	+	+	+	+	+	+
K2				+	+	+		+		+
K3		+			+	+		+		+
K3b		+		+	+	+	+	+	+	+
K4				+		+		+		+
K5		+				+		+		+
K6				+					+	+

Description of values of mean losses index in particular functional states of the control system. The mean losses factors w_i are in the presented functional-reliability model of the computer control systems the only quantities dependent on the controlled technological process properties. Therefore availability of credible values of these parameters is extriemely important for choice of the control system structure adequately to this process. The value of every factor w_i belongs to the $\langle 0,1 \rangle$ range and should reflect the degree in which any given system contributes to losses caused by the technologic process being in i-numbered functional state. Thus the values of the factors can be evaluated after long and honest investigations and observations of control systems operating within different plants. As rare an example of papers describing results of such investigations one can mention the work of Jublance (1975). Values of the mean losses factors taken for calculations have been presented in TABLE 3.

TABLE 3 Values of the mean losses factors w_i

F_0, F_3 F_6, F_9	F_1, F_4 F_7, F_{10}	F_2, F_5 F_7, F_{11}	F_{12}	F_{13} F_{14}	F_{15} F_{16}	F_{17}
0.	0.005	0.01	0.02	0.1	0.7	1.

Selection of functional states of the control system. In order to simplify the functional-reliability analysis and in effect the diagnostic-protection software it is advisable to take into consideration the possibility of elimination of certain functional states out of further considerations. For example, when appropriate functional states have identical or similar values of the factors of mean losses w_i. Thus, assumption that $w_4 \geqslant w_{12}$ (TABLE 3) allows to eliminate the functional state F_4 out of further considerations. It means that in the moment of occurence of such a failure which could be tolerated after the system (for example K3 in Fig. 1) has been taken to the functional state

F_4, a "deeper" protection will be executed which will take the system to the state F_{12}.

Such proceeding can be also necessary to perform, for example when on-line software diagnosis is being performed with insufficient accuracy. Thus, when the diagnosis of any failure being localized in the system K1 contains at least one device, failure of which can be tolerated by the system in the functional state F_{10} and at least one device, failure of which can be tolerated in the functional state F_{15}, then the functional state F_{10} should be eliminated out of further considerations.

Calculation of the function of mean losses W values for particular structures of the control system. Choice of the optimal structure.
TABLE 4 presents values of the function of mean losses W calculated for particular structures of the computer cascade control system. The choice of the optimal reliability structure of the control system is being performed accordingly to the criterion W=min. The optimal structure of the system for each set of reliability data has been indicated by thick lines. For the first set of reliability data, the structure K2 is optimal because for this structure value of W is over four times lower than for the structures K4 and K5 which are nowadays the most commonly applied in the industry. Inferiority of the structure K6 of the control system is caused by the fact that failure of the microprocessor requires the system to be taken into manual mode of the operation. All other structures however in the case of the processor failure enable control in the local automatic mode. Therefore it is advisable to apply proper hardware redundancy in the microprocessor-based cascade control system.

TABLE 4 Values of the function of mean losses W for structures of computer cascade control system. W – with realization of internal protection, W´ – without realization of internal protection.

	$W(\times 10^{-4})$ for data listed in the TABLE 1 and TABLE 2							
	I		II		III		IV	
	W	W´	W	W´	W	W´	W	W´
K1	3.7	4.2	30.6	48.8	1.7	3.8	30.	46.6
K1b	3.6	4.0	30.4	37.3	1.7	3.6	30.	36.
K2	3.0	4.2	32.0	40.4	2.9	3.7	30.	38.
K3	4.1	4.2	46.5	48.2	3.0	3.8	32.	45.
K3b	3.9	4.0	33.2	37.4	2.9	3.7	31.	36.
K4	14.5	114.	40.7	152.	11.5	115.	30.	137.
K5	14.6	15.8	54.2	60.6	11.7	17.2	31.	40.
K6	88.2	96.6	85.3	137.	72.5	112.	14.	27.

For example, the microprocessor's dynamic redundancy (as in the Honeywell's TDC 2000 system) allows to lower the value of W to 9.2×10^{-4}. The effect of redundancies of other devicies is also easy to calculate in the presented method.

In order to show the effect of internal protections realization on reliability of the system, calculations have been performed for two cases:
- when all functional states possible in every structure of the system have been taken into consideration (values W). It is the case in which all functional, diagnostic and protection possibilities of the system have been completely utilized,
- when only the protections most commonly applied in the industry, i.e. switching for manual control or automatic local control, have been taken into consideration (values W´).
For the latter case one can notice unfavour-

able increase of .value the mean losses function: $W') W$. In both cases the minimal value
of the function is being achieved for different
structures of the system. It means that the result of similar protections applying depends on
the hardware structure of the system. Therefore
different structures of control systems, similar
as for their functional properties, are not iden-
tically suitable for particular applications.

SUMMARY

Presented in the paper functional-reliability model of the computer control system enables to
describe the effect of its structure, reliability
parameters of particular devices and properties of the process being under its control on
the value of the function of mean losses W.
(It is also possible to define effects of using
various kinds of hardware redundancies in the
structures of these systems). It allows for
choice of the structure being most advantage-
ous from effectivity of operation point of view.

Presented algorithm for choice of the system
structure can be applied also to design of diagnostic and protection tasks in automatic control systems. It takes into account the system
ability for structural changes and therefore the
possibility of the internal protections software
realization. Requirements for the accuracy of
the failure localization are being simultaneously specified. If the accuracy of the failure discrimination is too low it is necessary to modify
the functional graph GF in order to eliminate
states which can not be discriminated. As a
result one obtains an increase of the value of
the function of mean losses W. When this value
change is too high one can decide whether the
set of diagnostic checkings requires to be greater in order to obtain higher discrimination of
failures occuring within the system.

REFERENCES

Cschornack,P. (1975). Ein Beitrag zur Quan-
    tifizierung der Zuverlässigkeit von Auto-
    matisierungeinrichtungen der chemischen
    Industrie. Dissertation TH Carl Schorlem-
    mer, Leuna Merseburg.
Jublanc,P. (1975). Les passages a charge
    nulle non programmes des tranches ther-
    miques a Electricite de France. Revue
    Generale Thermique, 158.
Krüger,J.B. (1978). Bewertungsmassstäbe für
    den Zuverlässigkeitsvergleich von Zentral
    und dezentral arbeitenden Prozessautoma-
    tisierungssystemen. Rtp, 8 .
Unger,E. and T.Stumpf (1980). Die Kosten der
    Zuverlässigkeit in der Messtechnik.
    INTERKAMA 80 Springer-Verlag, Berlin,
    Heidelberg, New York.

DESIGN CONSIDERATIONS FOR A FAULT-TOLERANT DISTRIBUTED CONTROL SYSTEM

Y. Wakasa

*Yokogawa Hokushin Electric Corporation, 2-9-32 Nakacho,
Musashino-shi, Tokyo 180, Japan*

1. ABSTRACT

This paper presents practical design considerations for a fault-tolerant distributed
control system. Dual redundancy is a typical method of implementing fault tolerance.
The following "3C" functions are fundamental to realizing dual redundancy of an
intelligent component such as a processor, memory, etc..
(i)   CHECK: diagnostics
(ii) CHANGEOVER: Switchover to standby unit operation
(iii) COPY: Memory equalization

There remains the so-called "common section" in a dual-redundant system. This section
is the most significant factor in total reliability as most redundant system failures
are caused by part failures in this section. Thus, the key to high reliability is
minimizing the parts count in this common section. This papar also introduces an
in-house RAPS (Reliability Analysis and Prediction System) which has been developed
based on MIL-HDBK-217D and modified from a practical standpoint.
A qualitative estimate of dual redundant system availability calculated using RAPS
is presented.

Keywords: fault tolerance, dual redundancy, reliability, distributed control system.

2. DUAL REDUNDANCY

Fig.1 shows an example of a fault-tolerant
distributed control system. Dual redundancy is
introduced in each layer of the system such as
the Operator Station, Communication Highway
(HF-Bus), and Field Control Station (FCS). The
FCS consists of Station Control Unit (SCU),
Station Internal Bus (SI-Bus) and up to 5 I/O
Nests. SCU is the intelligent portion of the
FCS, with a processor, memory, Highway
Communication Adapter Card (HCA), Nest Control
Unit (NCU) and Power Supply (PS).

2.1 CHECK: Diagnostic Routines

It has become common practice to design hardware
using such intelligent components as microprocessors, memory and other state-of-the-art
devices. Such intelligent components make it
easier to provide self-diagnostic functions. It
is well known that self-diagnostic routines are
not necessarily all-inclusive even where all
instructions in the computer can be used to
compute a known result. Thus, a watch dog timer
is used to monitor this intelligence. Fig.2 shows
the dual redundant configuration of the SCU. DXS
constantly monitors the ready status of the two
processors and decides which processor has the
control initiative to the I/O Nests. The ready
status represents the result of self-diagnostic
functions such as processor run/halt, memory
response, memory parity error, NCU and HCA
responses, etc. CPU has a non-maskable high-
priority exception interrupt which can be
generated by a CPU HALT, No Memory Response
(Invalid Address or Memory Failure), Parity
Error, No P I/O Response (Invalid Address or
P I/O Failure), etc. The exception interrupt

can be generated by hardware failures and also
by some erroneous software operations. When
this interrupt occurs, the program immediately
forces the processor to halt, and the processor
RUN/HALT status is sent to DXS on a dedicated
signal line. It is expected that most of the
possible failures will be detected by those
self-diagnostic functions. In order to improve
the error detection probability, a watch dog
timer is provided to monitor the processor
operation. As shown in Fig.3, the watch dog
timer is reset only when process control programs
such as DDC and sequence control are completed.
Those programs are executed periodically according
to the control period (normally 1 sec). When the
periodic execution of a process control program
is delayed more than 4 seconds due to a hardware
failure, transient error due to electromagnetic
noise, software bug, or system overload, the watch
dog timer times out and generates an interrupt.
  As shown in Fig.1, the Communication Highway
  and SI-Bus are dual redundant buses. Those
  buses are used alternately and tested for
  communication frame error, data coincidence
  errors, no timer responses, etc. Alternate use
  of dual redundant buses enables fast detection
  of failures on the standby bus.

2.2 CHANGEOVER: Switch over to the standby unit

DXS allows only one of two processors to communi-
cate with I/O Nests by enabling one of the I/O
Control Enable signals (IOCEI or IOEC2). When the
active processor fails, this information appears
in the processor status and is reflected in the
IOCE signal from DXS resulting in automatic
changeover to the backup processor.

2.3 COPY: memory equalization

A copy of the memory of the active processor should be available in the standby processor in the event that switchover to the backup processor is necessary. Fig.4 shows the memory copy function of DXS. The active processor's memory is copied at every memory write cycle on a word-by-word basis. This method provides a nearly instantaneous copy of the memory contents and enables a smooth switchover to the backup processor for subsequent control. The memory space of SCU is divided into program and data areas. ROM is used for the program area to provide protection against noise and program failures.

The data area in RAM is used for control data bases such as instrument specifications, loop configurations, tuning parameters, process variables, etc. The control processor is constantly updating its data base using the feedback or sequence control functions at every control period. This is done by a specific updating program. DXS constantly monitors the memory bus, and checks the address. Memory copy can be done only when the address is in the specific control data area and the write cycle is performed by the specific updating program. DXS monitors this memory equalizing operation. A check pattern is set in some pre-determined area from the control processor into the standby processor via the equalization program of the control processor at 1 second intervals during this equalizing operation.

The standby processor checks wheter the check pattern for the specific address is updated at 1 second intervals.

3. MINIMIZING THE COMMON SECTION IN DUAL REDUNDANT SYSTEMS

Even in an ideal dual redundant system, there remains so called "common section" which is in series with the two redundant sections in the reliability block diagram as shown in Fig.6 (a). Total reliability (availability) ADUAL is given as follows.

$$A_{DUAL}= A_{CMN} \cdot \{1-(1-A_{OL}) \cdot (1-A_{STB})\} = A_{CMN} \cdot (A_{OL}+A_{STB})$$

where A_{OL} and A_{STB} represent the availability of the on line and standby redundant sections, and A_{CMN} represents the availability of the "common section".

3.1 Communication Highway

From the viewpoint of operation from a CRT screen, the Communication Highway is the common section of the total system.

In a centralized communication system, which relies on a single master station to direct the Highway traffic, failure of the highway traffic director causes failure of all communications in the system. In this situation, all the operator stations will be blind. In a decentralized multi-master communication system, any station can be the master by taking control with the baton (token). The current master passes the "baton" to the next station after completing an operation. This eliminates the necessity for a "common section" in the communication system to determine mastership and direct the data highway traffic. If a station in the system fails, the watch dog timer in one of the other stations monitoring the Highway communications detect this failure as a communication timeout, and the "baton" passes to the next station in the sequence. Another station will continue trying to access the failed station and, when it is restored to normal operation, the system will automatically place it back in service. Fig.5 shows baton (token) passing and automatic recovery in a decentralized communication system.

3.2 Duplex Control Supervisor

DXS generates the control initiative signals IOCE1 and IOCE2. Those signals are sent to each SCU through a separate maintenance mode switch which allows us to replace a DXS while the system is on line by locking the mode switch to one side or the other. Due to this on-line maintainability feature of DXS, the reliability block diagram for a dual redundant SCN can be presented as shown in Fig.6 (b). SCN common section (A_{SCNCMN}) includes only an output circuit with a limited numbers of components, which can cause failure of the IOCE signals. Thus, major portion of the DXS card can be counted in A_{DXS}. Fig.6(c) shows the distribution of A_{SCNCMN} and A_{DXS}. Table I gives the failure rate and availability values. Thus the common parts count (A_{SCNCMN}) has been so minimized that total availability of the dual redundant SCN is nearly equal to that for A_{SCNCMN}, which is extremely high.

4. RELIABILITY MODELS

As discussed in Section 3.1, the data highway communication system is designed to be a decentralized multi-master system, so that the reliability model for this dual redundant data highway can be presented without any common section as shown in Fig.7 (a). Figures 7(b) and (c) show the reliability model for FCS and the I/O Nest.

Table I shows the failure rates for the hardware units. Table II shows availability figures for the duplexed FCS and the single FCS.

These figures were calculated by our in-house RAPS system.

5. RELIABILITY ANALYSIS AND PREDICTION SYSTEM (RAPS)

RAPS is an in-house system for reliability analysis and qualitative prediction which was developed based on MIL-HDBK-217D and modified from a practical standpoint.

5.1 The MIL System

MIL-HDBK-217D describes two methods of failure rate calculation.

o Component stress analysis model: This model uses environmental stress factors (environmental conditions, power dissipation, complexity etc.) to predict the failure rate. This method of computation is precise but quite complex, and varies depending on the types of components used. For discrete transistors, the failure rate is given by the following equation.

$$\lambda_p = \lambda_b (\pi_E \times \pi_Q \times \pi_A \times \pi_R \times \pi_{S2} \times \pi_C)$$

λ_p: Component failure rate
λ_b: Basic failure rate (obtained from a table based on the types of component, ambient temperature, applied voltage and other factors)
π_E: Environmental factor (stationary or land, maritime, or aircraft mobile use)
π_Q: Quality factor (from MIL standard)
π_A: Application factor (amplifier circuit, switching circuit, etc.)
π_R: Power rating (corresponds to the power rating for the component)
π_{S2}: Voltage stress factor (determined by working voltage/rated voltage ratio)
π_C: Complexity factor (related to complexity of the circuit containing the component)

o Parts count model: This is a simplified model, based on the following expression.

$$\lambda_{EQUIP} = \sum_{i=1}^{n} Ni(\lambda_G \; X \; \pi_Q \;)_i$$

λ_{EQUIP}: Total failure rate for the equipment
λ_G: Failure rate of components in the i-th category
π_Q: Quality factor of components in the i-th category
Ni: Number of components in the i-th category
n: Number of categories of components

5.2 The RAPS

MIL component stress analysis model is convenient for the reliability analysis at the component level, but too complex for system level analysis due to the many types of stress factors (π_X) in the mathematical expression. On the other hand, a parts count model is very simple to calculate, but it is difficult to predict the effect of temperature and electrical stress on reliability. RAPS combines these two methods. This simplifies computations for system level analysis and practical predictions of factors including stress factors.

5.2.1 Mathematical model for RAPS

A mathematical model for RAPS is provided by the following equation based on the MIL method. This model includes practical, significant stress factors, and gives the same expression as the MIL parts count model when $\pi_T=\pi_S=\pi_E=1$.

$$\lambda_{EQUIP} = \sum_{i=1}^{n} Ni(\lambda_B \; X \; \pi_T \; X \; \pi_S \; X \; \pi_E \;)i$$

λ_{EQUIP}: Total failure rate for the equipment
$\lambda_B = \lambda_G \cdot \pi_Q$: Basic failure rate (the most practical MIL grade is used for π_Q based on our actual field data)
π_T: Ambient temperature stress factor
π_S: Primary electrical stress factor
π_E: (Installation) Environmental stress factor
Ni: Number of components in i-th category
n: Number of categories of components

5.2.2 RAPS Database Versus MIL

While the MIL standard is the ultimate system for predicting reliability, it is difficult to apply it without modification to a product for the following reasons.
There are many reliable components which have not been submitted for MIL approval. For components that are not MIL-approved, MIL standards require that a weighting factor π_Q = max be used. This gives an unrealistically pessimistic failure rate. For these components, RAPS applies the most practical π_Q from MIL specifications based on our actual field experience. Table III shows an example of the comparison between RAPS failure rate data and equivalent MIL parts count model π_Q values.
Fig.8 shows statistically summerized failure rate data collected from answers to a set of questionnaires. In order to collect failure rate data on various components, JEIDA (Japan Electric Industry Development Association) forwarded a set of questionnaire to 114 electronic equipment manufacturing companies of Japan in 1984. 46 companies out of 114 responded to the questionnaire. Fi.8 is an excerpt from the questionnaire.
Average values of those components are listed in Table III as λ (Japan). It can be seen that the RAPS failure rates are almost the same as these values.

6. Conclusion

This paper has presented practical design considerations for a fault tolerant distributed control system, and a practical method of reliability prediction.
Qualitative prediction of reliability are apt to be primarily a matter of theory due to the lack of up-to-date practical and universal figures for failure rate that take into consideration quality improvements due to recent advances in manufacturing technology.
It is necessary to find some more practical and universal method to predict system reliability. System failures can be due not only to parmanent failures of hardware components, but also to transient errors, software bugs, and even human error. It is becoming necessary to have some new measure of reliability which takes these factors into consideration.

References

1. Y. Wakasa et al.: User-configurable Fault-tolerant Architecture for a Distributed Control System — ISA '84 C.I. 84-R770 pp.1193
2. Y. Wakasa et al.: Enhancing Reliability of CENTUM Distributed Process Control System — Yokogawa Technical Report, No. 1 (1984)
3. H. Tamura, et al.: A Reliable Duplex Station for Distributed Control Systems — Proceedings of 8th Triennial World Congress of the IFAC, p.89 to 94 (Kyoto, Japan 24 to 28 August 1981)
4. T. Ogawa, H. Akai: A Reliable Duplex Station for Distributed Control Systems — Yokogawa Technical Report, Vol, 24, No.4, PP.158 to 164 (1980) (in Japanese)
5. JEIDA: Technical Report on Industrial Computer Systems - Reliability Prediction of Electronic Equipment (in Japanese), Aug., 1984.

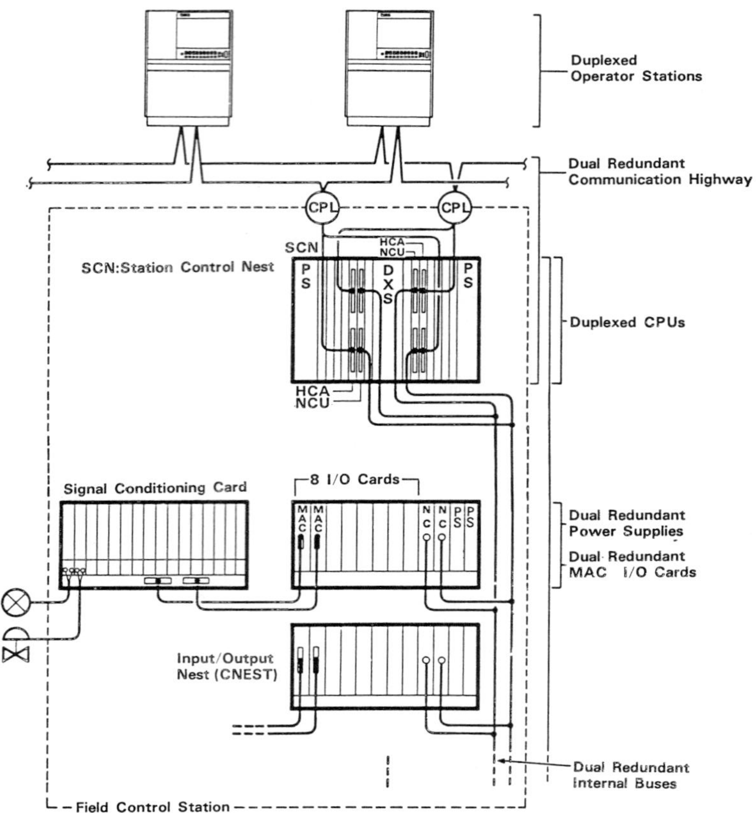

Figure 1   Redundancy in a Distributed Control System

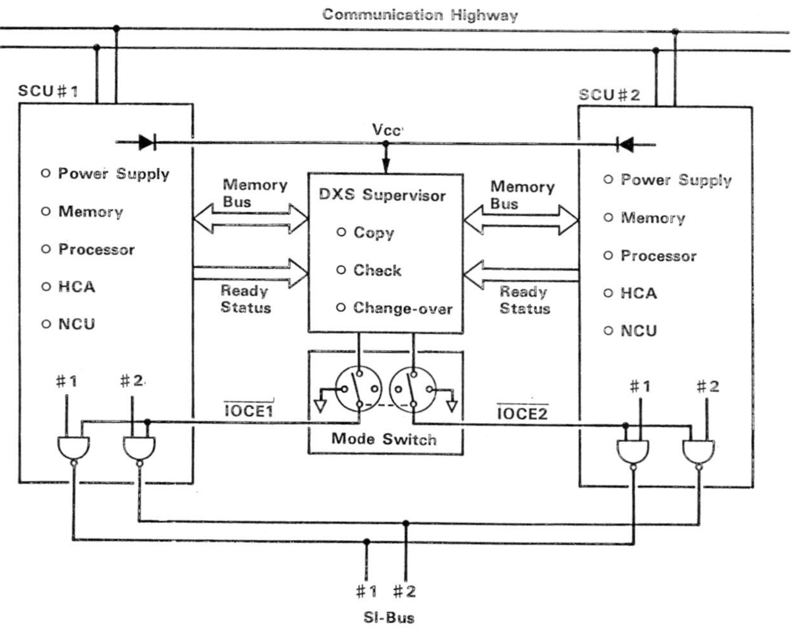

Figure 2   Duplex Configuration of Station Control Nest

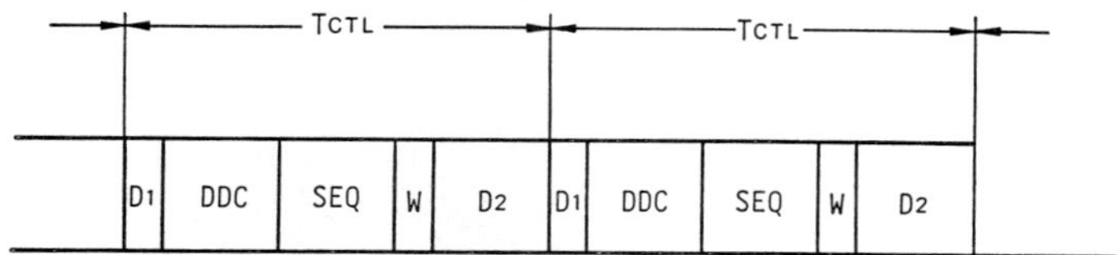

TcTL   :   Control Period   (Normally 1 sec.)

D1 , D2 :   Diagnosis

DDC   :   DDC Program

SEQ   :   Sequence Control Program

W   :   Watch Dog Timer Reset

Figure 3

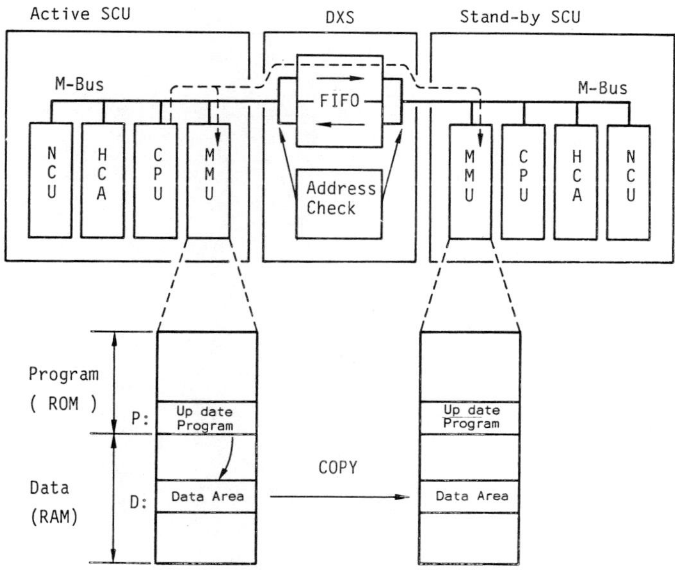

Figure 4   Memory COPY function by DXS

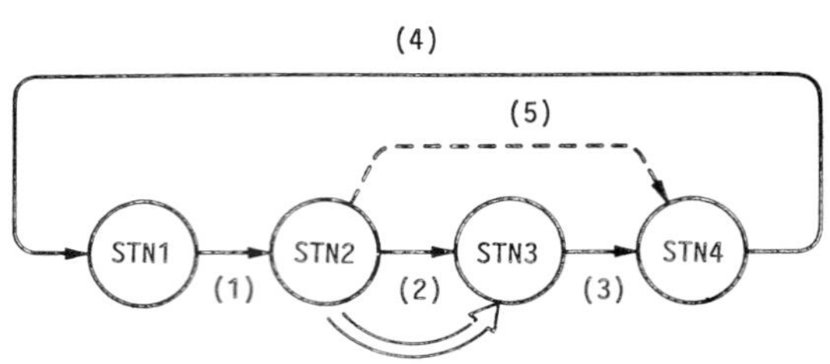

(1)(2)(3)(4)   Baton passing sequence when all stations are normal

(1)(5)(4)   Baton passing sequence when station 3 is abnormal

⟹   Communications to automatically revert system to normal

Figure 5   Baton Passing and Automatic Recovery

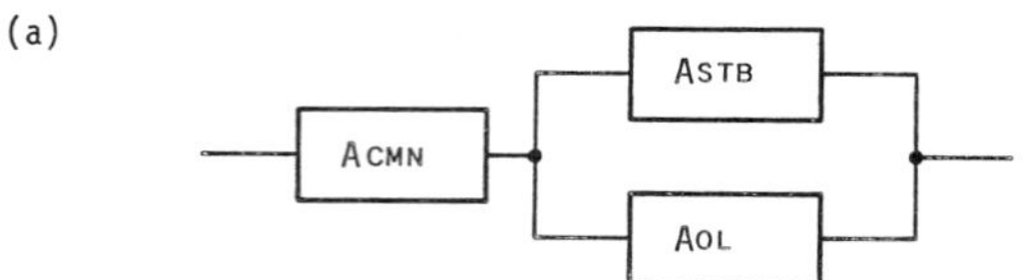

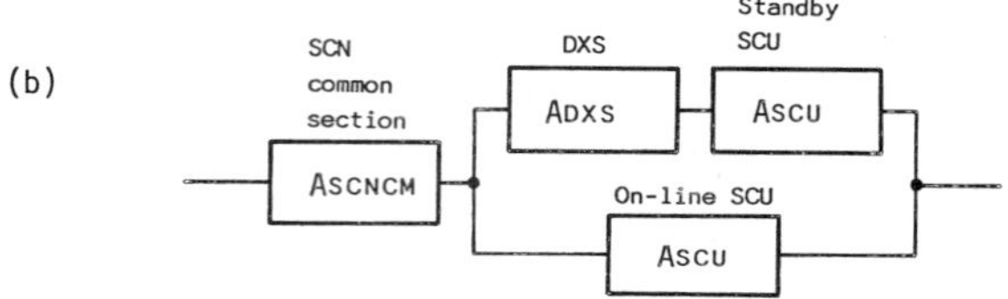

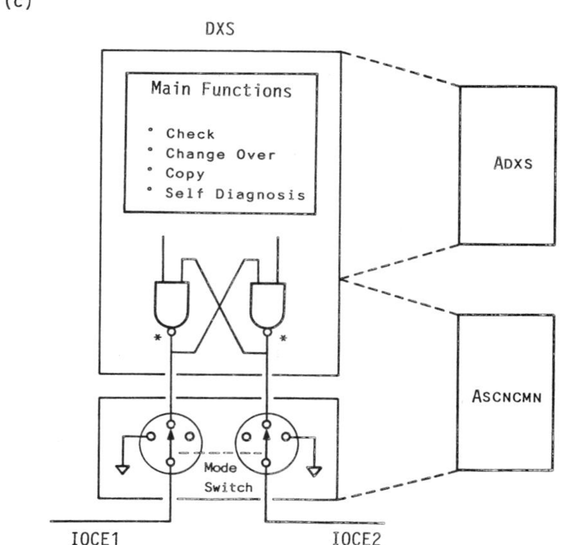

Figure 6   Minimizing the Common Part of Dual Redundancy

(a)

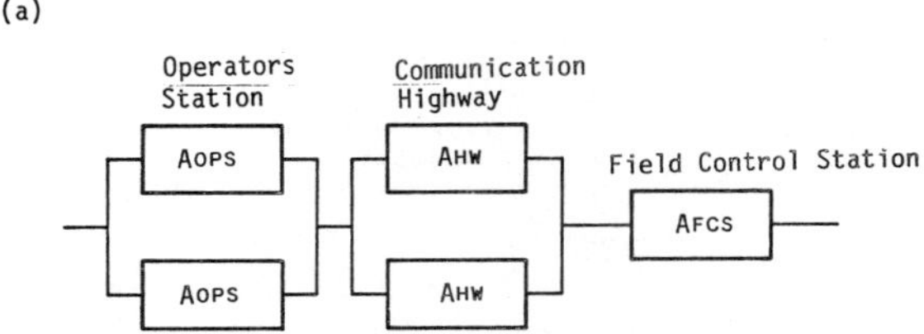

(b) Reliability Model for FCS

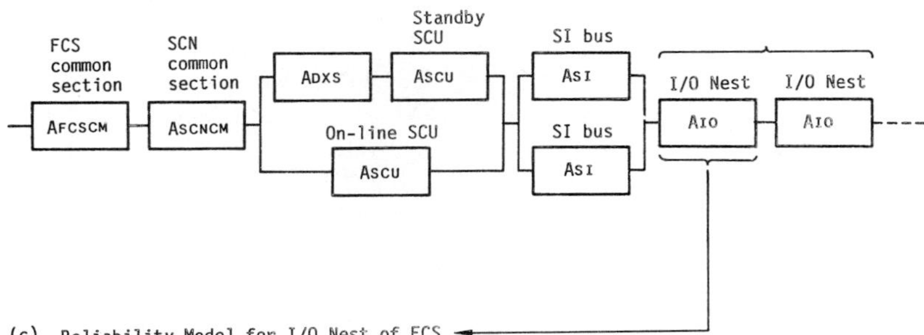

(c) Reliability Model for I/O Nest of FCS

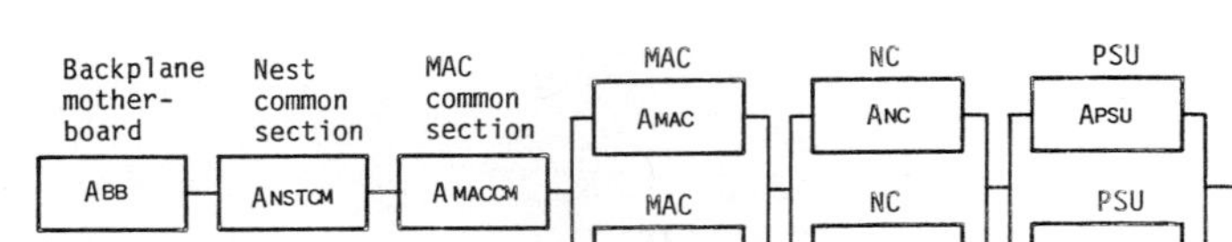

Figure 7   Reliability Models

fit = 10^{-9} (1/hour)

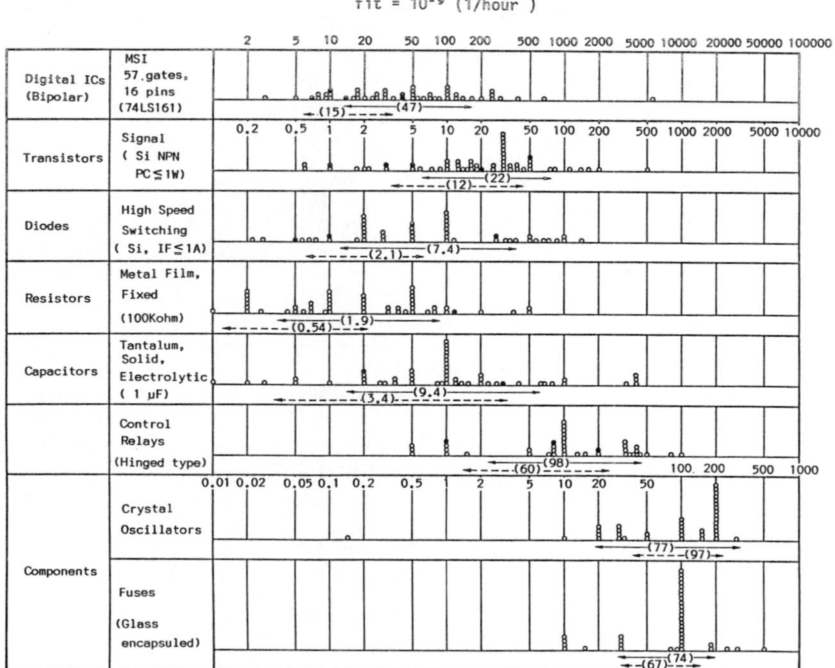

Figure 8   **Distribution of Predicted Failure Rates Data**

	Failure rate (fit)	Availability	
FCS common section	300	A_{FCSCM}	0.999 998
SCN common section	150	A_{SCNCM}	0.999 999
DXS	5 700	A_{DXS}	0.999 954
SCN	39 100	A_{SCN}	0.999 7
SI bus	340	A_{SI}	0.999 997
I/O NEST	1 000	A_{IO}	0.999 992

Table I   Failure Figures for Hardware Units

No. of I/O Nests	Duplex FCS	Single FCS
1	0.999 989	0.999 67
2	0.999 981	0.999 64
3	0.999 973	0.999 61
4	0.999 965	0.999 58
5	0.999 957	0.999 55

Table II   Availability Figures for Duplex FCS and Single FCS

| Component | | | (X 10^{-6}/H) | | πQ value / grade (dγ) : equivalent of RAPS | | | | | | | | | λ (Japan) |
	Name	Type	MIL-HDBK-217D	RAPS Data base										($\times 10^{-6}$/H)
ICs	74LS type TTL	LS193	0.021 x πQ	0.02	S 0.5	B 1	B-0 2	B-1 3	B-2 6.5	C 8	C-1 13	D 17.5	D-1 35	0.015
Discrete	Signal transistor	T018	0.028 x πQ	0.02	JANTXV 0.12	JANTX 0.24	JAN 1.2		LOWER 6	PLASTIC 12				0.012
Semiconductors	Signal diodes	200mA 35V	0.0006 to 0.0025 x πQ	0.002	'' 0.15	'' 0.3	'' 1.5		'' 7.5	'' 15				0.0021
Resistors	Metal film	1/4W, 100ppm,F	0.00216 x πQ	0.001	S 0.03	R 0.1	P 0.3	M 1	10509 5	22684 5	LOWER 15			0.00054
Capacitors	Tantalum	25μF 10V	0.0437 x πQ	0.01	S 0.03	R 0.1	P 0.3	M 1	L 1.5	LOWER 10				0.0034
Magnetic Parts	Relays		0.126 to 1.515 x πQ	0.05	R 0.1		P 0.3		M 1	L 1.5				0.060
Miscellaneous	Oscillator crystal	HC18-U	0.2	0.05	None									0.097
	Fuse		0.1	0.1	None									0.067

Table III   Comparison of RAPS Failure Rate Data and Equivalent MIL Parts Count Model πQ Values.

RELIABLE AND INTEGER NETWORKS IN CONTROL SYSTEM

P. van Damme and J. Verploegen

*Rosemount Benelux, 's-Gravelandseweg 256, 3125 BK Schiedam,
The Netherlands*

Abstract. In this paper the author describes the selection criteria which could be
used when evaluating a network of a distributed process control system. Also guidelines
are proposed to guarantee a safe and comfortable use of the control system local
network.

Keywords. Integrated plant control; computer control; system integrity; hazard and
race conditions; plant communications; error correction codes.

INTRODUCTION

The neccesity of local networks in process control
systems appears evident:
- distribution of control units is still
  recognized as the only safe approach to process
  control.
- plant (production) integration is very desired.
- plant (production) information is becoming
  vital.

DEFINITION OF A PROCESS CONTROL NETWORK
AND ITS LEVELS OF COMMUNICATION

In this workshop the distributed control network
is defined as:
A communicating device interconnecting
geographically separate, partially autonomous
controlling devices with operator accessibility
from a central control room (1). This definition
implies, that we are not simply discussing a
peace of wire (i.e. coax) but a device including
its communicating interfaces and its protocols.

Two communications phenomena are subject of this
workshop.
- protocol strategy in a particular bus system.
- communication control strategy in a particular
  (production) company.
a) Protocol strategy in a particular bus system.
   First steps have been taken for standardizing
   protocols. ISO (International Standard
   Organization) is taking a leading role.
   Modern bus systems are structured according
   this standard in layers of functionality,
   see Fig. II.1 (2). The objective of this
   structure is two fold.
b) Independently from system bus structures the
   user must develop a communication strategy for
   the control of the plant.
In fact this company communication structure is a
step beyond the system bus structure. The system
bus structure must be considered as one of many
devices to achieve the company's communication
structure.

This "device" however can not have simple
specifications and application description like a
chemical reactor, a product-transport-unit or an
analyser. The system bus(ses) is, or will be used
throughout the complete company in a horizontal
way (integrating production units) and vertical
way (from plant control to process control).

Plantwide use means plantwide environment from any
standpoint, see Fig. II.2 (3). Usage of the
system bus(ses) therefore needs users
considerations depending on environmental
conditions.

SELECTION CRITERIA FOR CABLE
MECHANICAL AND ELECTRICAL
PROPERTIES OF THE SYSTEM BUS

Selection criteria:
- suitable for outside use
- RFI insentive
- suitable for proper grounding
- surge energie protected
- easy (dis)connectable
- easy mountable
- easy repairable
- non sparking

Three types of busses are discussed in this work
shop; multicore busses, coax (twinax) busses and
light busses.

Multicore busses are normally only used for very
short distances. Main reason is, that multicore
busses can not reach the electric proportions of
coax without lifting the signal level to an
unacceptable level.

At this moment coax and twinax are mostly used.
In the future optical busses may be used more
frequently.

Main reasons are:
- proven consept
- reasonable insensitive to RFI
- acceptable damping
- easy mountable
- field repairable

Main disadvantages are:
- limited length of the system bus
- high energy RFI must be avoided
- limited speed (bits/sec)
- grounding needs special attention

Optical busses have some very important advantages
i.e.
- inherently RFI and surge insensitive
- very low damping (long distance without booster)
- very wide frequency band (over 100Mbits/sec)
- inherent optical isolated

Disadvantages of optical busses are:
- not yet proven reliability
- unknown technology for existing maintenance
  operations
- difficult to repair

Conclusion is that the following selection guid
lines should be used:
- At process control level normal coax (twinax)
  should be used for the bus system.
- The bus system must inherently allow a robust
  grounding strategy.
- The bus system must inherently have a lighting
  protection (surge energy).
- Light busses are only to be used for plant data
  transfer.

MOUNTING AND LOCATION CRITERIA
OF THE CONTROL SYSTEM BUS

Many considerations are involved when developing
the strategy of a control systems location. Some
considerations cooperate with eachother.

When planning a process control system one should
decide which geographical strategy will be used.
Possible strategies are:
- All process control units in a central control
  room.
- The operator unit in a central control room and
  all process control units in an adjacent control
  room.
- The operator unit in a central control room and
  controllers in local control rooms near the
  process.
- The operator unit in a central control room and
  controllers at the process.

The selected strategy heavely depends on:
- type of process
- size of location
- environmental conditions
- company policy on locating control systems
- financial situation (costs)
- existing situation in case of revamp.

Only in rare situations the process control units
are located at the process. Considering the
environmental needs of micro processors this is
unlikely to change. Therefore this strategy is
not evaluated in this workshop. For strategy "one"
"two" or "three" the user must consider above
points.

From a cost standpoint various subjects must be
taken in consideration. For stretched locations
wiring efforts may be reduced dramatically when
using geographical distribution. Experiences
however, show that geographical distribution does
not bring dramatic savings one would expect and
can sometimes even be more expensive than the
centralized distribution (i.e. due to local
control rooms etc.) (4). Although commercial
aspects are not part of this workshop, the authors
would like to underline, that costs mostly will
only be a minor criteriom for selecting a
geographical distributed control concept.

When selecting geographical distribution the
following provisions must be made for the system
bus:
- Allways use a redundant system bus.
- The bus should be layed in a cable duct with low
  voltage, low mA, low frequency, signals only.
- If high reliability is required than separate
  cable ducts with sufficient distance (see IEEE
  384-1981) for each of the redundant busses must
  be installed.
- If any fire danger in the plant is feasable,
  fire protected ducts or cables should be used
  (6).

USERS GUIDLINES FOR A
PROCESS CONTROL NETWORK

For the user it should not be a question how the
data is transferred in a network. The user is
merely interrested which bounderies he has to
respect concerning:
- quantity limitations of transferred data
- reliability of transferred data
- users transfer procedures
- status of data transmitter and data receiver
  during communication disturbance

In human conversations "significant data" is very
difficult to precise because many social elements
influence the data source and data receiver in
respect of actual accepted and/or understood
information.

When considering data system busses definition of
"significant data" appears more realistic. Still,
from an operator standpoint, differentation is
feasible which could be illustrated with the
following example:
In distributed control systems alarm settings
should be (and mostly are) stored in the distributed controllers. Two options to inform the
operator station(s) are:
- Data transfer of actual value and alarm setting
  to the operator station letting the operator
  station decide whether there is an alarm.
- The controller decides there is an alarm and
  only transfers data when this is the case.

From a process information standpoint all above
data is "significant data". The second transfer
option however is much more efficient with exact
the same result. Allthough data handling is not
part of this workshop, it must be clear that we
are dealing here with the following efficiency
improvements. If "significant data" is built-up
efficiently:
- the system bus has more idle time
- users data-quantity transfer bounderies have
  extended
- a extensive thus robust error coding system can
  be used since idle time is available
- priority structures can be simplified because
  all data can be transferred within a time limit.

Protection of data against errors.
Basically two error types influence the data
transfer, see Fig. V.1 (5):
- loss of significant data
- generation of false data
Infact, if only "significant data" was transferred
on a system bus losing one single bit of the data
or adding one single bit to the incorrect data
could make a complete message meaningless.
Therefore data transfers must be protected against
errors. For process control system busses an extensive error protection strategy is absolutely
necessary. Workgroup SC654 WG6 of the IEC
describes the requirements for system busses in
process control. Various redundancy codes are
available. Mostly used is the combination of a
Cyclical Redundancy Check (CRC) and a Longitudinal
Redundancy Check (LRC). Each of these checks are
quite capable of protecting the system bus data
sufficiently. The combination is merely, implemented from a diversity standpoint. If an error would
incidently fool one redundancy check it is very
unlikely that the other check is also fooled since
it has a totally different checking structure.

Applications for process control busses.
When doing more than convensional control the
system bus is more likely used for other functions
than operator interactions i.e.:
- advanced control
- batch tracking

- integration of raw material control
- integration of (process related) production
  units
- laboratory data generation
- management data generation

Using the process control bus for these features
is in fact quite logical; modern control systems
can do these jobs avoiding complex interfacing to
other devices. As already described above, also
the application of the control system bus needs
considerations. Looking at the high reliability of
control system busses one would wonder the author
concern. The main concern is the nature of process
control. Process control takes direct actions on
the extent, intensity and quality of the process,
see Fig. VI. 1. If the extent and intensity is
wrongly controlled, then people can be hurt,
production units destroyed and environment
damaged.

Control system bus features utilised are:
- peer to peer communications
- direct data transfer to a host computer

The following guidelines can be set:
- For peer to peer communications.
  A process control engineer configuring a distributed control system should have the awareness
  that the configuration strategy should have
  maximum compatibility with the distributed hardware strategy. (7). If the engineer would
  neglect doing this, he would actually destroy
  one of the main features of a distributed

control system namely: integrity of the distributed control units. Especially for the basic
control (handling the extent and intensity of
the process) no compromises are acceptable.
- For direct data transfer to host computer.
  If others than process control operators have
  acces to the control system bus than provisions
  must be made so that they can not influence the
  extent nor intensity of the process.

Efficiency of data transfer. Utilizing a highway
acces rotation schedule. This gives the possibility to transmit with variable message length. The
advantage with this strategy is: the more the
system bus is utilised, the higher the system bus
efficiency.

REFERENCES

(1) J.Leigh, J.R., "Applied Digital Control".
(2) An overview of proposed ANS for local
     distributed data interfaces William E. Burr,
     National Bureau of Standards.
(3) Multi-level control levels to efficiency
     Robin Brooks C&I.
(4) Centralized v.s. geographically distributed
     control and data acquisition: a cost compa-
     rison Nathany Michael Intech, Sept. 84.
(5) Introduction of information theorie by Dr.
     A.P. Bolle.
(6) Improve control system fire protection, G.K.
     Castle H.P., March 84.
(7) Avoid vulnerable distributed control system
     architectures, Steve Chocheo H.P. June 1983.

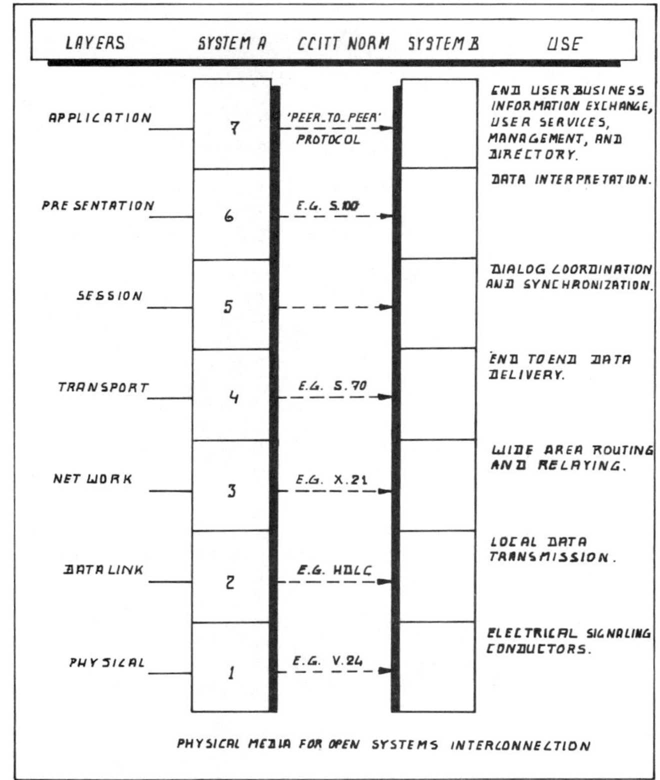

FIG. II.1. LAYERED OSI ARCHITECTURE

P. van Damme and J. Verploegen

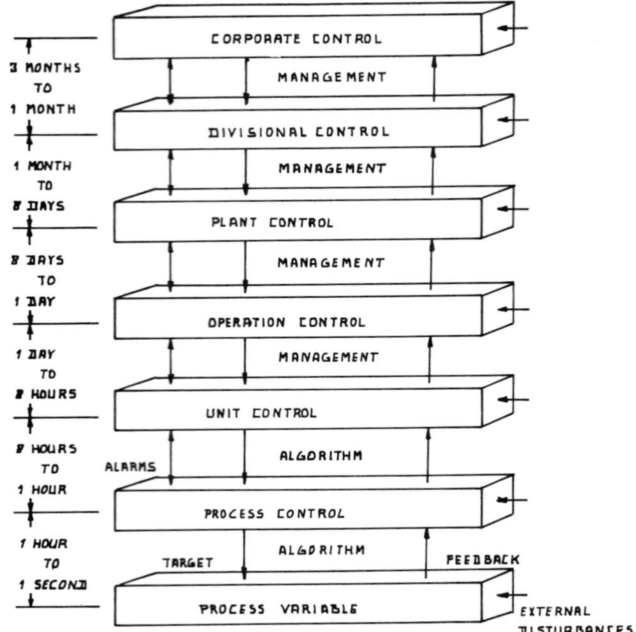

FIG. II.2.

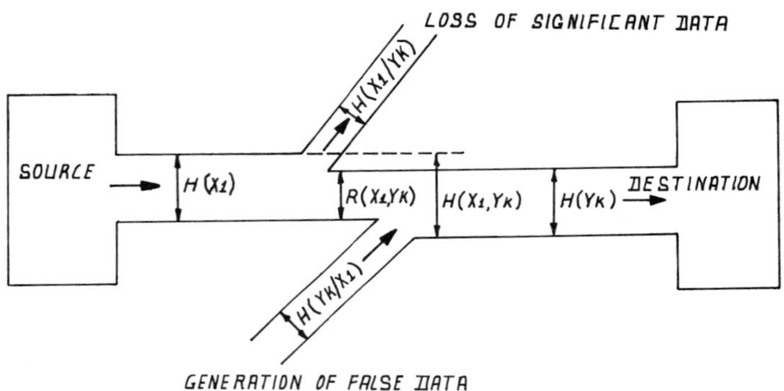

FIG V.1.

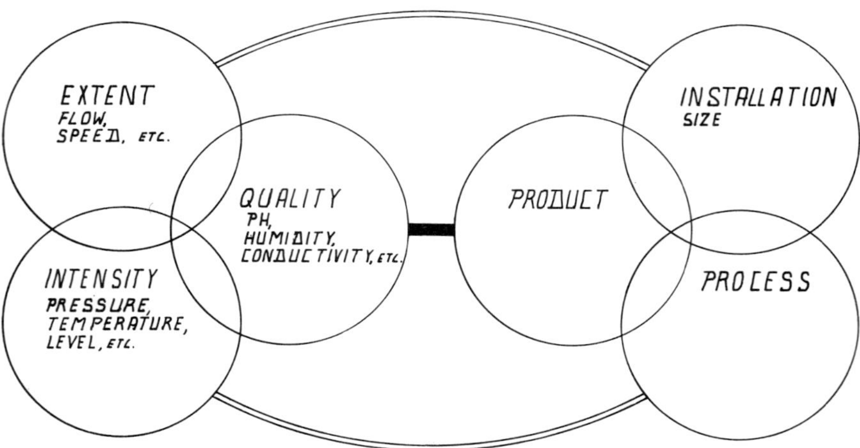

FIG. VI.1.

FUNCTIONAL STRUCTURE OF A MICROPROCESSOR-BASED CONTROLLER TOLERATING FAILURES IN MEASURING CIRCUITS

J. M. Kościelny and P. Wasiewicz

*Institute of Industrial Automatic Control, Warsaw Technical
University, Warsaw, Poland*

Abstract. The paper describes a functional structure of the microprocessor-based controller which has the ability for toleration of failures of some control system elements. A method for application of the diagnostic checkings results for the controller software structure changes in cases of occurences of the failures, mainly of measurement devices, has been elaborated. The method has been illustrated by an example of temperature control system in the heat exchanger-autoclave technologic installation. Also the effect of these protection application on this control system reliability parameters has been calculated and the results have been compared with results obtained during application of dynamic redundancy of the controller itself.

Keywords. Process control, fault-tolerant systems, Automatic testing, Error compensation, Microprocessor.

INTRODUCTION

Reliability of the automatic control systems is one of the most important features of their operation. Simple methods of reliability improving based on their elements reliability improvements are often insufficient. Therefore it becomes necessary to apply structural methods of reliability improvement, e.g. by designing fault-tolerant systems. Designing of such systems requires proper choice of the system structure (often with redundant devices), on-line diagnostics, the system hardware and/or software structure changes in failure states. This problem has been solved in a practical way only for the microprocessor-based controller (Schmidt, 1980, among others) which is the main element of the control system. But the main part of losses resulting from inefficiencies of hardware elements is due to failures of measurement equipment and actuators. It is therefore advisable to apply protections in cases when failures of these elements occur. Following kinds of protections can be applied:

a) change of structure of the process variables conversion, e.g. change of control algorithm, application of supplementary process variables or redundand variables,

b) action on the automatic control system structure by change of operation state of such devices as relays, valves, pumps etc. For this kind of protection it is necessary to know the operation state of external devices,

c) introduction of safe values of the control variables or set-point values in damage states. These protections can be realized only when suitable diagnostic information will be elaborated in an automatic way. Following text presents a conception of the microprocessor-based controller software structure which has the ability to realize the protections mentioned above.

SOFTWARE FUNCTIONAL STRUCTURE OF THE MICROPROCESSOR-BASED CONTROLER

The general diagram of the software functional structure of the microprocessor-based controller has been shown in Fig. 1. The software structure is defined by the set of data units and connections occuring between them. There are three groups of data units appearing in the controller software:

D – data unit for diagnostic procedures (described later in greater detail) and procedures of preliminary conversion of the process variable values, e.g. filtering, linearization, square-root extraction etc.,

Q – data unit for realization of procedures of arithmetic operations, calculations of statistic indexes (e.g. mean values, variances, integrals) etc.,

C – data unit for control procedures, for example proportional-integral-derivative-action control algorithm, adaptation control algorithm etc.

Single data unit contain actual parameters being applied by proper procedures appearing in the sequence of a variable value conversion and defines the sequence of these procedures applications. Particular sequences of conversion being defined by the units D, Q and C can contain any procedures attributed to particular units.

For purposes of detection of failures of the process measurement variables, a set of following diagnostic procedures has been provided:

D1. Check of credibility of the process variable based on inspection of technically admissible range of its value change:

$$x_{min} < x_n < x_{max} \quad ;$$

D2. Check of credibility of the measuring variable based on inspection of admissible rate of its value change:

$$(x_n - x_{n-1}) / \tau < \Delta \dot{x}_{max} \quad ;$$

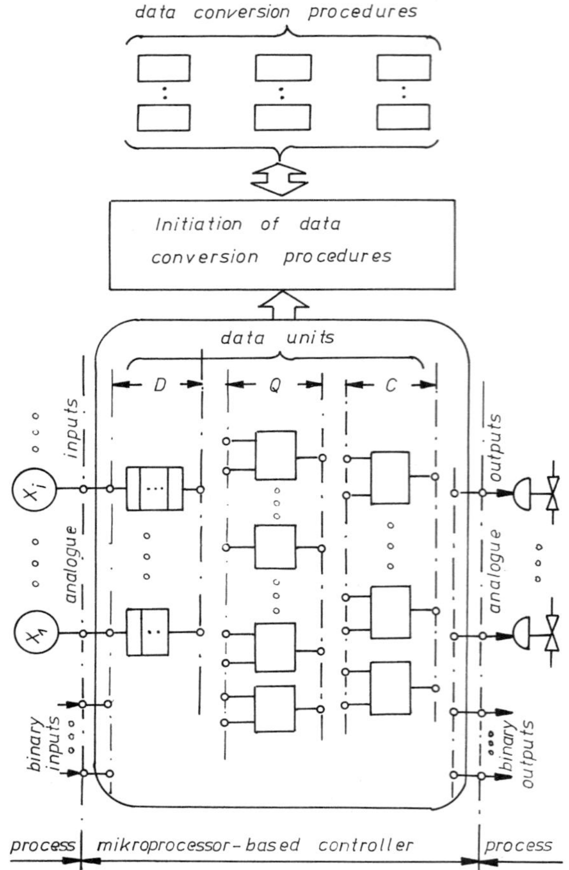

Fig. 1. General diagram of software structure of the microprocessor-based controller

D3. Check based on comparison of values of three measurement redundand variable values X_A, X_B, X_C and on choice of the intermediate value X_A for further conversion:

$$X_n = X_{An} \Longleftrightarrow X_{Bn} \leqslant X_{An} \leqslant X_{Cn} \quad ;$$

D... Check of credibility of the measuring variables based on the model binding these variables, for example:

D4. $$X_n = -a_1 X_{n-1} - a_2 X_{n-2} - a_3 X_{n-3} + b_1 U_{n-1} + b_2 U_{n-2} + b_3 U_{n-3}$$

D5. $$X = X_n - X_{n-1} = A(X_{n-1} - X_{n-2}) + B(U1_{n-1} - U_{n-2}) + C(U2_{n-1} - U2_{n-2}) + D(U3_{n-1} - U3_{n-2})$$

D6. $$X_n = A + BU_n + CU_n^2 + DU_n^3$$

The check can encompass in this case either increment values or only conformability of direction of real value changes with values calculated accordingly to the model. One out of these two variants is being chosen during elaboration of the software functional structure and parameters. Check of conformability of direction of changes is being performed in order to decrease the probability of false results of the checkings which can exist due to inaccuracy of the model. The checking result is being recognized as negative only when inconformability of the signals direction of changes in "k" successive sampling times has been discovered.

D7. Check of limiting values exceedings by the process analogue variables: $X_{LO} \langle X \langle X_{HI}$;
D8. Check of binary variables values.
D9. Check of the negative feedback loop in the control system during damage states. Efficiently operating negative feedback loop when the controlled variable excceds its admissible limiting value is strictly precized. The control signal value in this case should change in such a direction to prevent the damage state from occurence or should already achieve its proper extreme value. The check consists therefore in the check of remembered changes of the control signal values before and after the controlled variable exceeded its limiting value.

Any procedure appearing in the sequence of any process variable conversion determines so called diagnostic checking d_i, result of which can be negative (c_i) in the case when incorrect operation of the process variable conversion circuit has been detected or positive (c_i) when otherwise. The negative result of some diagnostic checkings is being accepted only after the diagnostic procedure has been n times performed in order to distinguish permanent failures from instantaneous errors.

Even before the control system containing the controller is to be realized, one should determine the functional structure of this system in the state of complete officiency and also in these states of inefficiency which enable operation of the controller with its available software actions. For each of these functional structures one should determine:
a) sets of data conversion procedures being utilized,
b) data values being stored in particular data units,
c) connections occuring between these data units.

This information help to elaborate the controller's software functional structure and parameters. This structure and parameters concern in this case not only to functions of the process variables conversion and control but also to diagnostic and protection functions. Therefore it is necessary to define the set of diagnostic checkings being realized for each of the process variable (by defining the data units D). It is also necessary to define such combinations of checkings, occurencies of which require proper structure configuration changes to be performed. These data are being declared as logic functions LF of the checking results. When the function value equals 1 it means that the structure configuration change is necessary. Thus to each of the function the set of connections of data unit is attributed which corresponds to functional structure of the control in the state of its complete or incomplete efficiency. Number of such logic functions is equal to the number of structures anticipated for realization. This method for data structure and parameters elaboration for configuration changes purposes in states of inefficiency will be investigated with help of on example a little later.

General diagram of the microprocessor-based controller functional algorithm has been shown in Fig. 2.

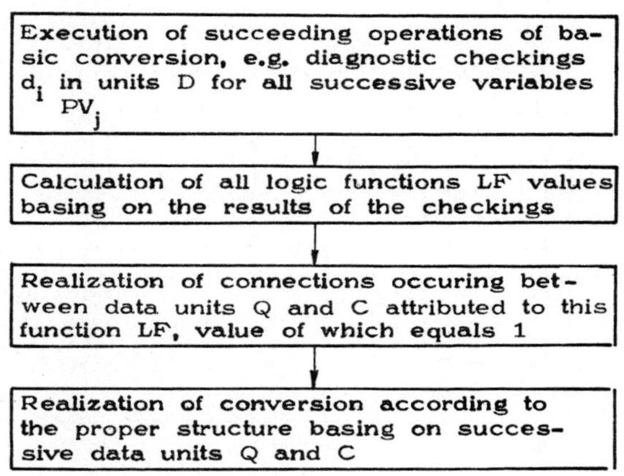

Fig. 2. Block diagram of the microprocessor-based controller functional algorithm.

EXAMPLE OF REALIZATION OF THE FAULT-TOLERANT CONTROL SYSTEM

Fig. 3a presents simplified diagram of technologic installation, which is to be equipped with temperature cascade control system containing the microprocessor-based controller.

Four functional structures of the system can be distinguished in different states of its functional efficiency as it has been shown in Fig.3b:
- basic functional structure of this system in the state of complete efficiency; variables T_1, T_2 and F are being utilized and variable P constitutes a back-up,
- closed-open cascade structure which ensures toleration of failures occuring in the variable F measuring circuit; variables T_1, T_2 and P are being utilized,
- cascade structure which ensures toleration of failures occuring in the variable T_2 measuring circuit; variables T_1 and F are being utilized and variable P constitutes a back-up,
- cascade structure which ensures toleration of failures occuring in the variable T_1 measuring circuit; variables T_2 and F are being utilized and variable P constitutes a back-up. In this case the process operator should update the set-point value of $T_{2,0}$.

Table 1 presents a set of all diagnostic checkings being realized in considered control system. Table 2 shows shapes of the logic functions LF defining conditions under which it is necessary to perform suitable changes of the system functional structure.

Software structure of considered systems can be realized basing on following set of procedures of data conversion:
DA - filtering procedure,
DB - square-root extraction procedure,
D1, D2, D5 - appropriate diagnostic procedures,
QA - addition procedure,
QB - correction procedure with proportional derivative algorithm,

CA - control procedure with proportional-integral-derivative algorithm.

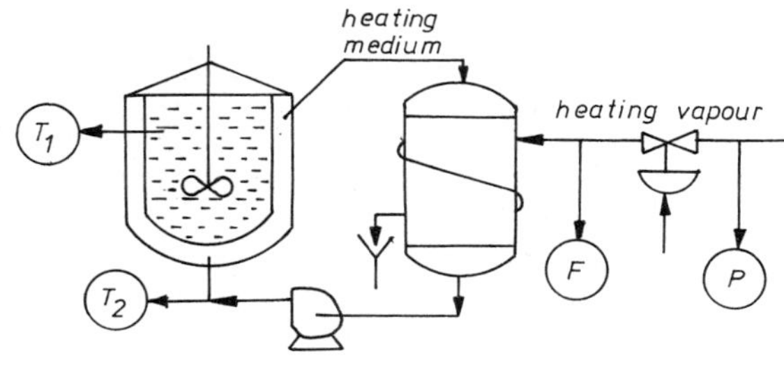

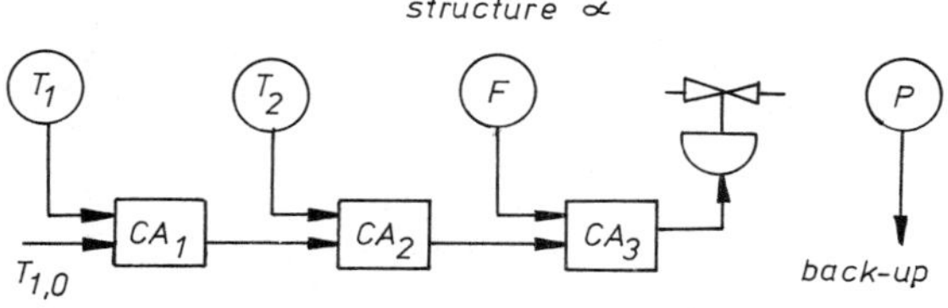

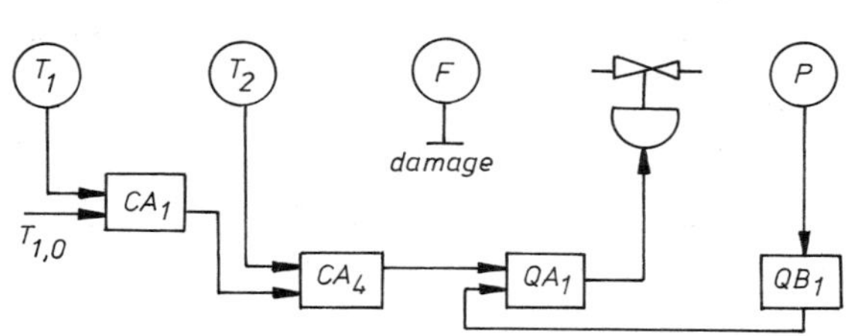

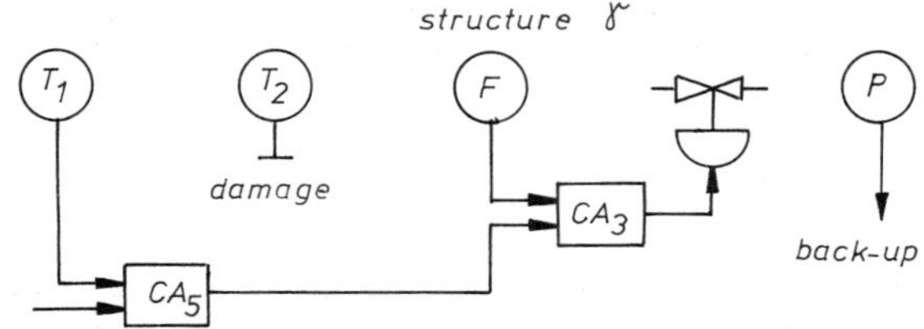

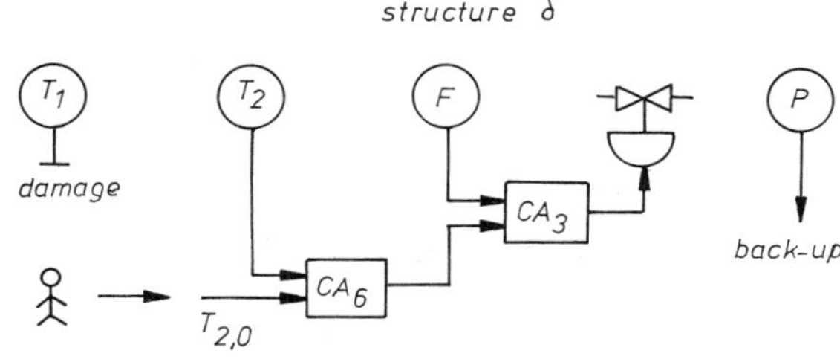

Fig. 3. a) Diagram of the technologic installation: heat exchanger-autoclave, b)functional structures α, β, γ, δ of the control system in distinguished states of efficiency of the process variables measurement circuits.

TABLE 1 <u>Set of diagnostic checkings d_i being realized in considered control system</u>

checking	diagnostic procedure	definition of checking
d_1	D1	$T_{1,min} < T_{1,n} < T_{1,max}$
d_2	D2	$(T_{1,n} - T_{1,n-1})/\tau < \dot{\Delta} T_{1,max}$
d_3	D1	$T_{2,min} < T_{2,n} < T_{2,max}$
d_4	D2	$(T_{2,n} - T_{2,n-1})/\tau < \dot{\Delta} T_{2,max}$
d_5	D1	$F_{min} < F_n < Fmax$
d_6	D6	$\Delta F = K1 \cdot \Delta P + K2 \cdot \Delta U$
d_7	D1	$P_{min} < P_n < P_{max}$

TABLE 2 <u>Set of the logic functions LF defining conditions under which it is necessary to perform suitable changes of the system functional structure</u>

functional structure	logic function LF
α	$LF = c_1 \wedge c_2 \wedge c_3 \wedge c_4 \wedge c_5 \wedge c_6 = 1$
β	$LF = c_1 \wedge c_2 \wedge c_3 \wedge c_4 \wedge (\bar{c}_5 \vee \bar{c}_6) \wedge c_7 = 1$
γ	$LF = c_1 \wedge c_2 \wedge (\bar{c}_3 \vee \bar{c}_4) \wedge c_5 \wedge c_6 = 1$
δ	$LF = (\bar{c}_1 \vee \bar{c}_2) \wedge c_3 \wedge c_4 \wedge c_5 \wedge c_6 = 1$

The set of data units being utilized by these procedures in complete cycle of the controller operation has been presented in Fig. 4.
Table 3 defines connections occuring between these units which are required in distinguished functional states of the system.

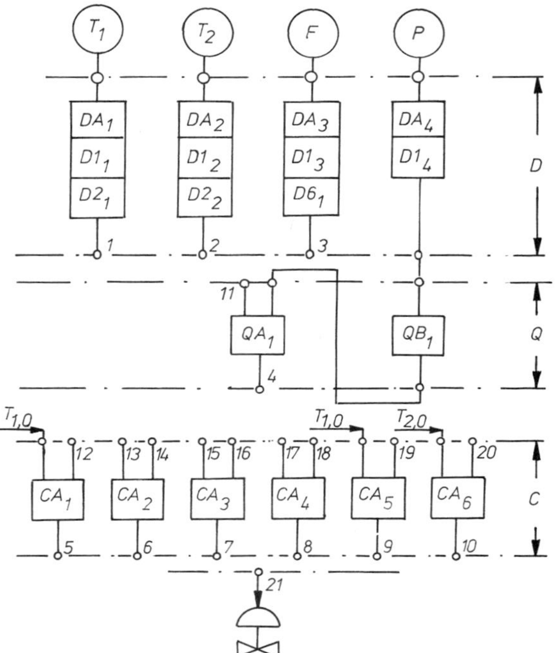

Fig. 4. Set of data units D,Q and C of the microprocessor-based controller for all functional states of considered system.

TABLE 3 <u>Diagram of connections occuring between data units in distinguished states of considered control system</u>

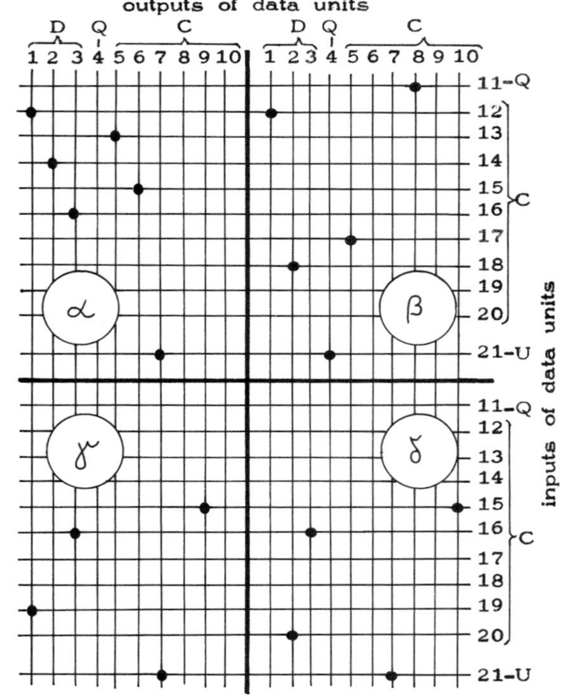

EVALUATION OF EFFECT OF THE STRUCTURE CONFIGURATION CHANGES

In order to evaluate the effects of considered structure configuration changes one should make calculations of probabilities of inefficient states existences in the control systems. Such calculations have been executed for four following cases:

I – The automatic control system without redundancy of the microprocessor-based controller and without change of structure configuration of signal conversion in states with inefficient measurement circuits;

II – The automatic control system with redundancy of the microprocessor-based controller and without change of structure configuration of signal conversion in states with inefficient measurement circuits;

III – The automatic control system without redundancy of the microprocessor-based controller and with change of structure configuration of signal conversion in states with inefficient measurement circuits;

IV – The automatic control system with redundancy of the microprocessor-based controller and with change of structure configuration of signal conversion in states with inefficient measurement circuits.

Calculations have been executed for temperature control system shown in Fig. 3. It has been assumed that the system is inefficient when the process automatic control is not possible. Calculations don't take into consideration actuator failures nor technologic devices failures because these failures are not being tolerated in any of the four cases and therefore do not influence the result of considered comparisions. It has been also assumed that the control system can exist only in the state of efficiency S_o or in states of inefficiency S_i, i=1, ... ,m with only one inefficient device of the system:

$S_1 = S_{\mu P}$ – failure of the microprocessor-based controller,

$S_2 = S_{T_1}$ – failure of the variable T_1 measuring device,

$S_3 = S_{T_2}$ – failure of the variable T_2 measuring device,

$S_4 = S_F$ – failure of the variable F measuring device,

$S_5 = S_P$ – failure of the variable P measuring device.

This assumption is justified because of small number of elements (m = 5) included in considered system and because of their high reliability parameters ($\lambda_i \gg \mu_i$, where λ_i and μ_i denote failure rate and renewal rate of i- numbered element respectively). Graph of reliability states which has been built with this assumption is shown in Fig. 5.

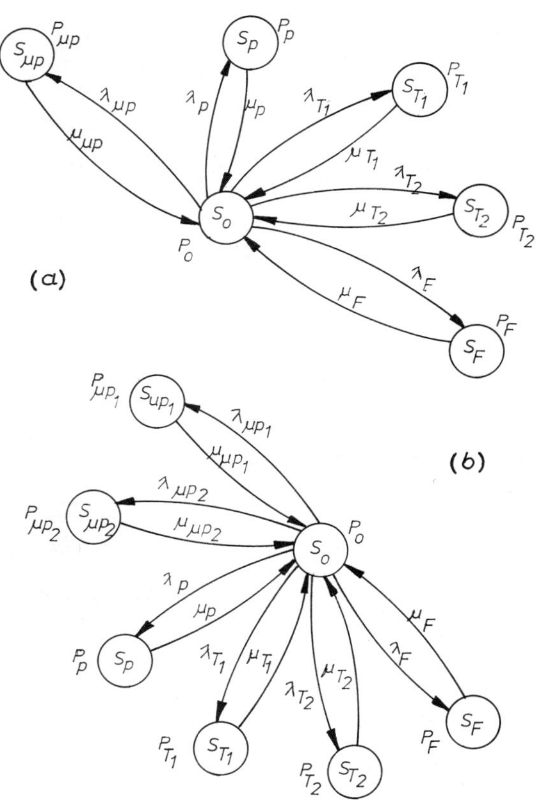

Fig. 5. Graph of reliability states of the temperature control system: (a) with redundancy of the microprocessor-based controller, (b) without redundancy of the microprocessor-based controller.

The Kolmogorow differential equations describing the graph have the shape:

$$\frac{d}{dt}P_o(t) = -\,P_o(t)\cdot\sum_{i=1}^m \lambda_i + \sum_{i=1}^m \mu_i\cdot P_i(t)$$

$$\frac{d}{dt}P_o(t) = -\mu_i\cdot P_i(t) + \lambda_i\cdot P_o(t); \quad i=1,....,m \tag{1}$$

Because the sum of probabilities equals 1, it is possible to calculate stationary values of probabilities of particular reliability states. They can be defined by a set of equations:

$$-\mu_i\cdot P_i + \lambda_i\cdot P_o ; \quad i = 1,....,m$$

$$P_o + \sum_{i=1}^m P_i = 1 . \tag{2}$$

The probability of the system existence in the state of inefficiency P_N is therefore equal to the sum of probabilities of particular states of inefficiency:

$$P_N = \sum_{i:\ S_i \in S^-} P_i = 1 - \sum_{i:\ S_i \in S^+} P_i \tag{3}$$

where S^- and S^+ denote sets of states of inefficiency and states of efficiency respectively. Sets of states of inefficiency and calculated values P_N for all four considered cases have been placed in TABLE 4. Values P_N have been calculated for three different complete sets of reliability data.

TABLE 4 <u>Calculation results of inefficiency existence probability</u>

	I	II	III	IV
Case No	I	II	III	IV
Redundancy of the microprocessor-based controller	-	+	-	+
Structure configuration change	-	-	+	+
Set of the states of inefficiency : S^-	$S_{\mu P}$, S_F, S_{T_1}, S_{T_2}	S_F, S_{T_1}, S_{T_2}	$S_{\mu P}$, S_{T_1}	S_{T_1}
Probability of the states of inefficiency P_N for different reliability data: $\frac{\lambda}{\mu}(C) = 1\cdot10^{-4}$, $\frac{\lambda}{\mu}(T) = 1\cdot10^{-4}$	0.0004	0.0003	0.0002	0.0001
$\frac{\lambda}{\mu}(C) = 1\cdot10^{-3}$ of the controller, $\frac{\lambda}{\mu}(T) = 1\cdot10^{-4}$	0.0013	0.0003	0.0011	0.0001
$\frac{\lambda}{\mu}(C) = 1\cdot10^{-5}$ of the measuring transducer, $\frac{\lambda}{\mu}(T) = 1\cdot10^{-4}$	0.00031	0.00030	0.00011	0.00010

CONCLUSIONS

The calculation results indicate that practical realization of configuration changes of the signal conversion structure is vital in states with inefficient measurement circuits. Especially when reliability parameters of the microprocessor-based controller are equal or better than reliability parameters of measuring devices it brings much lower probabilities of failure existence. Considered example shows also that the organization of the controller's software structure has great effect on reliability of automatic control systems operation.

One should stress that presented method for structure configuration changes can be applied not only in damage states but also for realization of variable-structure algorithms appearing often for example during installation start-ups.

Considered microprocessor-based controller enables also to perform the control system hardware structure changes and also changes of states of different technologic devices operation by means of the controller output binary signals and basing on values of binary and analogue input variables.

REFERENCES

Schmidt, G. and W. Sendler (1980). Redundanz Konzepte in Modernen Prozessautomatisierungssystemen. Rtp., 9, 310-313.

Sendler, W. (1980). Eine Fehlertolerierende Reglerstation auf der Basis eines bus-orientierten multi-Mikrorechner-Systems. Rtp., 3, 73-80.

HOW TO SPECIFY THE USER'S REQUIREMENTS TO OBTAIN AND VERIFY RELIABLE SOFTWARE FOR PROCESS CONTROL APPLICATIONS

P. S. Scherman

Process Control Department, Comprimo BV, Amsterdam, The Netherlands

Abstract. In the first part a brief survey of the available tools for the evaluation of software reliability will be given. First a number of quantitative techniques will be described, subsequently the qualitative methods will be highlighted, i.e. quality assurance procedures followed during the design phase.

The second part will deal with the specific problems and practical methods used for industrial process installations. The following types of software will be discussed:

1) Standard software packages for Direct Digital Control (DDC), which only need configuration for a specific application. For this type checking of parameters and functional testing will apply.

2) Software for Programmable Logic Controllers (PLCs). The best way of reliability testing is to carry out carefully specified simulation tests.

3) New developments for single use. For this type checking and follow-up of the specified quality assurance procedures and error registration during all tests is necessary.

Finally some possibilities and limitations of redundancy techniques will be discussed.

Keywords. Computer control; Computer programming; Computer software; Process control; Reliability theory.

INTRODUCTION

The behaviour of software with respect to reliability aspects is different from that of hardware. This is due to some basic properties of both of them. Although they can be considered to be a matter of common knowledge nowadays, it may be useful to list them as the starting point of the discussion.

With respect to hardware :
- Performance changes with time due to physical and chemical influences (wearing out, oxidation, etc.). This process can lead to failures which did not occur earlier under the same circumstances.
- Reliability data can be obtained by statistical methods because of the large number of nearly identical or very similar products.
- Reliability data can be extrapolated for new products by comparing them to known data of similar devices.
- Reliability figures of assemblies or complex systems can be calculated from the reliability figures of the components and their mutual logical dependency.

With respect to software :
- Performance does not change with time. Failures are caused by faults which were existent all the time.
- Different specimens of the same software are "copies", they are exactly identical. The behavior will be exactly identical under the same circumstances.

- Each software product is unique. Extrapolation of statistical data for other products is meaningless.

The extended mathematical and statistical arsenal of reliability predictions is based upon the behavior of hardware devices as described above, and therefore not applicable to software. The field of software reliability needs another approach, starting from the basic properties of software.

Programs do not wear out, they remain as good as they were when put in operation. The solution appears to be trivial: put them in operation after removal of all faults, they will work perfectly to all eternity. But, alas, experience shows that it is impossible to produce software without faults. "There is always one more bug". Failure usually occurs only when a program is exposed to an environment for which it was not designed or tested. For most programs, in practice, the large number of possible states and inputs makes perfect comprehension of the program requirements, perfect implementation, and complete testing generally impossible. Thus, software reliability is essentially a measure of the confidence we have in the design and its ability to function properly in its expected environment.

METHODS FOR THE EVALUATION
OF SOFTWARE RELIABILITY

A number of tools has been developed for the evaluation of software reliability. They can be

79

roughly divided into two groups, the quantitative and the qualitative methods.

Quantitative Methods

Although the failure mechanism of software is different from that of hardware, the result for the system is the same. This is why theories that are compatible with hardware reliability can be developed.

The most useful of the quantitative methods are based upon the failure occurrence rate during the test phase.

Whenever a failure is encountered during testing, the fault, which caused the failure, will be corrected. Once a fault is properly fixed, it is, in general, fixed for all time. Muse (1980) states, that, assuming a properly planned and executed test effort, the net number of faults detected and corrected is an exponential function of the elapsed execution time (not calendar time). After a sufficiently long test period the parameters of this exponential function can be calculated and from this the number of the still existent faults can be estimated. In addition an estimate can be made of the effort which will be necessary to reduce the number of faults to an acceptable level.

Littlewood (1980) proposes a somewhat different approach. His starting point is that one is interested in measuring the operational reliability of the program rather than the estimate of the number of faults left in it. For this purpose he introduces the joint probability function, i.e. the joint distribution of future interfailure times.

Nearly all authors of quantitative software reliability establishing methods agree upon the statement that the applicability of them is very limited, they are only the first step along what threatens to be a long road.

Qualitative Methods

The qualitative methods seem to be more suitable to process control applications. Instead of trying to measure and calculate the reliability of a nearly finished product, they emphasize the effect of the design procedures on the quality of the product. The train of thought is quite plausible: A fault, which has not been built into the program, will not have to be corrected later; neither will it, if not discovered, influence the reliability.

Daniels and Humphreys (1982) state, that some 63% of faults in software are to be attributed to the design phase. In contrast some 50% of the software budget is allocated to maintenance, of which a significant proportion is due to rectification of the faults created during the design phase. A reduction of the design faults rate will give a significant cost benefit by reducing maintenance, while increasing the availability of the equipment by reducing the failure rate. As a consequence the overall system reliability can be improved. Reliability must be built into a software product from its inception at the requirements phase and through all subsequent software life cycle phases. The achievement of a reliable product on a qualitative basis is brought about by implementing controls on the managerial and technical aspects of the various phases. Some of the techniques to be employed include :

1. Optimisation of the specification to random ambiguities and ensuring that the requirements are correctly described, by discussion with all interested parties.
2. Application of quality assurance procedures to all phases of the project.
3. Application of configuration and change controls.
4. Use of structured design and structured programming techniques.
5. Development of programs in segmented or modular format.
6. Invocation of design reviews.
7. Use of recognized and accepted standards only.
8. Use of a high-level language to enable the program operation to be understood by subsequent users and modifiers.
9. Provision of program visibility by the inclusion of comments both within program listings and in operating documentation.
10. Testing is important and should be undertaken :
    (a) by designers and developers to aid development;
    (b) by validators to prove that the software meets the requirements.

Quality assurance procedures are gaining more and more importance in nearly all fields of industrial design in general, and in the field of process industry in particular. It is logical to use these procedures for the design and production of process control software. In fact, they are used already, but it will be a long way before the ultimate goal, the obligatory use of standardised and generally accepted procedures, is reached.

SPECIFIC PROBLEMS AND METHODS WITH RESPECT TO PROCESS CONTROL EQUIPMENT

Before dealing with the daily practice in greater detail, I should like to emphasise that I am representing the engineering contractor's point of view. The approach of an engineering contractor to software reliability problems is determined by the circumstance, that he neither produces, nor uses programs for process control. As a consequence, the engineering contractor is not in a position to apply his own, in-house quality assurance procedures for the design of software, neither can he rely on extended and detailed statistical performance data about operational software. Of course, these statements do not apply to software, designed and/or used by the engineering contractor, e.g. CAD-systems, used to support engineering activities. These types of software, however, are outside the scope of this paper.

In my opinion, the position of the engineering contractor offers more advantages, rather than disadvantages for obtaining reliable software. He plays the role of a bridge, he speaks two languages. He must have an exact understanding of the client's needs and requirements on the one hand; on the other hand he must be able to make a complete, clear, unambiguous and realistic specification for the supplier. Finally, he is responsible for the product's functioning satisfactorily.

Summarising : We are no scientists, but engineers. We must use a practical, efficient approach.

Many different types of software are used in process control equipment. The practical approach of the engineer has led to different methods aiming at reliable products. From this point of view process control software can be divided into three groups, which will be discussed subsequently.

Standard software packages for direct digital control (DDC)

In most cases these form part of turn-key systems. Usually they contain an operating system and standard routines for input scanning, conversion to engineering units, plausibility check, limit checking, different control functions, output generation, operator information, operator input and possibly many other functions. These types of systems need merely be tailored for a specific application, i.e. defining inputs and outputs, control functions, operator communications, etc. This activity is called "configuration" or "filling in the blanks". Essentially, it is the definition of the data flow between data base, inputs, outputs, processing routines and operator interface, and the establishing of parameters.
Assessing reliability of the standard software is impossible without considerable effort, neither is it necessary if it is a well-known product of a well-known manufacturer. The configuration activity will deliver a good product if some basic rules of good engineering are followed. The data must be clearly and surveyably written down into lists, in many cases provided by the supplier. Thorough checking against the basic requirements will bring most of the errors to light. Systematic functional checking after implementation and subsequent correction of the errors occurred will result in a reliable product.

Software for programmable logic controllers (PLCs)

A number of aspects of this type of software are identical or similar to DDC software. They also contain an operating system and standard routines for input and output handling, control functions and operator interface. The essential difference however, is, that finalizing of the system needs real programming work, not merely "filling in the blanks". Although the programming language is quite simple, it is still programming, and this means that the same faults can be made (and will be made) as in any other type of software. Besides the usual checking of the coding a thorough functional test is necessary with the help of input and output simulation. The test procedure must be carefully set up and clearly documented. It is the quality of the testing which will determine the reliability of the product. A surveyable documentation of the program is necessary for quick fault correction during testing and operation, and will also prevent generation of new faults.

New developments for single use

This type is the most difficult one, because it cannot be assembled from building blocks of proven quality.
Fortunately they are relatively seldom used for process control applications. The best way of avoiding reliability problems with them is, indeed, not to use them. This can be done in many more cases, than one should think at first glance. Experience shows that people frequently decide to make a grass-root program without seriously looking for a usable product on the market. The necessary resources for the development of reliable software are heavily underestimated in most cases.
If there is no other way out, a reliable product can be obtained by establishing design rules and quality assurance procedures. These procedures must be agreed upon between buyer and supplier before the work begins. The qualifications of the members of the project team must also be guaranteed. Last but not least : the limits of the responsibilities of each party must be clearly defined to avoid attempts to put the blame on somebody else.

SOFTWARE REDUNDANCY

Finally a few words about redundancy.
From the basic properties of software it follows, that commonly used hardware redundancy techniques do not improve software reliability. If the same programs are executed in two or more channels in parallel, a software fault will lead to a common mode failure in all channels at the same time. Increase of reliability can only be obtained if either the channels are not working in the same input environment, or the programs are not identical. The first requirement can generally not be met in process control applications. The second approach is a possible way. It means that two (or more) programs must be developed by two (or more) independent project teams. A critical application with very high reliability requirements can justify the necessary expenses.

CONCLUSIONS

The available mathematical tools for hardware reliability assessment are not applicable to software. Some quantitative software reliability assessment methods have been developed during the past years, but their applicability is as yet very limited. In particular, programs for process control applications lend themselves better to qualitative methods. Strict design rules, application of quality assurance procedures and properly executed testing will result in reliable process control software at reasonable expenses.

REFERENCES

Bell, R., B.K. Daniels, and R.I. Wright (1983). Assessment of Industrial Programmable Electronic Systems with Particular Reference to Robotics Safety. Reliability '83, 3c/1/1-11.

Bell, R., B.K. Daniels, and R.I. Wright (1984). Safety Assessment of Industrial Programmable Electronic Systems. SRS Quarterly digest, April 1984, 3-9.

Daniels, B.K., and P. Humphreys (1982). Software Reliability Assessment. SRS Quarterly Digest, July 1982, 16-17.

Leveson, N.G., and P.R. Harwey (1983). Analyzing Software Safety. IEEE Transactions on Software Engineering, SE-9, 569-579.

Littlewood, B. (1980). Theories of Software Reliability: How Good Are They and How Can They Be Improved? IEEE Transactions on Software Engineering, SE-6, 489-500.

Musa, J.D. (1980). Software Reliability Measurement. The Journal of Systems and Software, 1, 223-241.

CAN SOFTWARE RELIABILITY BE EVALUATED?

W. H. Simmonds

Sira Ltd, South Hill, Chislehurst, Kent, UK

Abstract.   The differences between software and hardware reliability concepts are well known.   Considerable guidance is now available on methods and practices for generating reliable software;   techniques are also available for testing software reliability.

With very few exceptions these techniques have arisen within, and are designed for use within, the software production environment.   They assume the software designs and associated documentation are accessible for use in structuring the test programme.   They also assume the software sources are available for experiments such as error seeding.

Users of instrumentation systems containing embedded software have more restricted access to design documentation and software sources. Consequently end-users, and independent system evaluators, can only make limited use of these techniques for reliability assessment.

This paper briefly reviews the utility of these techniques for assessments of reliability carried out by end-users and suggests more practical alternative approaches.   The paper is presented as the viewpoint of an independent laboratory performing system evaluations on behalf of users.   Reference is made to the type of reliability failures found in evaluating software based systems.   Recommendations are presented for actions to improve the future situation both with regard to software reliability itself and our ability to evaluate software reliability.

Keywords.   Software reliability, system integrity, computer software, computer testing, process control, computer control.

INTRODUCTION

Modern instrumentation systems for safeguarding and control are designed and implemented as complex combinations of hardware and software components.

The problems associated with software reliability and the differences between the nature of software reliability and hardware reliability were recognised many years ago and were widely voiced as part of the so-called "software crisis" in the late 70's. Over the years, and particularly during the last decade, methods and practices have been established for generating reliable software and for testing software reliability.   Whilst these methods have an important role in reliability evaluation of software based instrumentation systems they do not provide a complete solution.

The major thrust in software reliability has treated the system as a software system, not as an integration of software and hardware.   The function or task performed by the system is viewed as being that performed by the software.

A view that contrasts with this is to regard software as simply a convenient and flexible method for configuring hardware (ie programmable devices) to perform a task.   Viewed in this way software exists only within the design;   consequently its contributions to reliability are identical in nature to those of hardware design.

Neither of these viewpoints are entirely appropriate for consideration of typical integrated software/hardware instrumentation systems.

In an integrated system the differences between the reliability characterstics of hardware and software are used to supplement each other.   The manner in which this is done leads to reliability structures involving concepts and mechanisms such as automatic self and cross checking and automatic correction.

Evaluation of these systems requires a broader approach to be taken and it is necessary to reconsider the fundamental concepts of reliability.

RELIABILITY CONCEPTS

The concept of reliability can perhaps be best expressed as the attribute of:

"performing the required tasks in an adequate manner for sufficiently long periods of time".

As the needs have arisen to consider the reliability of hardware components, hardware systems, software systems and hardware/software systems the interpretations of the basic reliability concept have taken a number of different directions, rather than a single evolutionary path.

Hardware Component Reliability.   The concept of reliability was first formalised in relation to hardware components.   The approach adopted was to apply probability and statistics to the problems of material failure.   Reliability was defined as:

"the ability of an item to perform a required function under stated conditions for a stated period of time."

The treatment of reliability assumed failures resulted from wear during use.   Items (ie single hardware components) performed single functions and these were assumed to have been correctly implemented in the component design.

Hardware System Reliability.   The concept of reliability of a hardware system developed from the concept of hardware component reliability.

The formal definition of reliability was retained in its original form and the term "item" in this definition was allowed to denote any part, sub-system, system or equipment that can be individual considered and separately tested.   The probabalistic and statistical treatment of reliability was extended to include redundancy but little attention was given to the problems associated with the changes in the concept of "function".

To analyse systems having very complex functions, perhaps better referred to as tasks or missions, new techniques were developed.   These were developed as independent approaches rather than being developed within the pre-existing reliability analysis framework.   A primary motivation for the new approaches was the need for safety assessment of complex systems and the models developed were concerned with representations of fault propogation. The simplest, and earliest model was the "fault tree", developed in the early 1960's at Bell Telephone Laboratories.   Other models subsequently developed were "event trees" and "cause-consequence diagrams".   These three types of fault propogation model differ mainly in their choice of starting point for analysis.   A fault tree (Barlow 1975) starts with an assumed fault and progressively analyses this, in a top-down manner, into the alternative causes.   An event tree (Buffham, Freshwater and Lees, 1971) starts with an assumed event and progressively develops this, in a bottom-up manner, into resultant consequences.   A cause-consequence diagram (Nielsen, 1974) starts with an assumed event and analyses this, in a middle-out manner, into both causes and consequences.

The existence of these three forms of fault propogation model is an indication of the problem that it is often impractical to consider all possible failures in a complex system.   To evaluate such systems it is necessary to identify the critical situations and hence reduce the analysis task to a manageable size.

Software Reliability.   Although software does not wear out the term "reliability" was adopted in the early 1970's to describe the ability of software to perform its assigned function for an adequate period of time.   The earliest conceptualisations of software reliability attempted to borrow from the prior concepts of hardware reliability.   Reliability models were developed (Jalinski and Villoralay, 1972) that were analogous to those for hardware.   The simplest of these was based upon the assumption that:

a)   failure rate is constant unless changes are made to the software.

b)   each change made to correct a discovered error reduces the number of errors present in the software by one.

c)   failure rate is proportional to the number of errors present in the software.

Models of this type can be used, together with recorded observations of failures, to predict the number of residual undisclosed errors.

Another class of techniques developed for estimating the error content of software is based upon analysis of the structure and complexity of the software (Halstead, 1977).

One further technique is "error seeding" (Mills, 1972). In this technique the software is randomly seeded with a number of known errors, referred to as "calibration errors".   A defined set of tests is then applied to the software.   The numbers of seeded calibration errors and original (non-seeded) errors detected by the tests can be used to estimate the total number of original errors.

These three classes of assessment techniques are primarily concerned with estimation of error content rather than with assessment of reliability.

The third error seeding technique is in fact more concerned with estimation of the efficiency of a test programme.

Hardware/Software System Reliability.   A system is used to perform a variety of mission tasks using various functions available within the system in a coordinated manner.   The systems' total mission may be successfuly accomplished even though some functions fail;   consequently the concept of system reliability must be one that allows for failures of individual functions.   The concept must also take into account the modes of failure.   A system that degrades gracefully is, for example, usually considered to be more reliable than one that may enter a stage where its behaviour is unpredictable.

No clear concepts for assessing the "reliability" of modern hardware/software systems have yet been established.   Work in developing adequate concepts and assessment methods is perhaps hindered by the existence of the previously established narrow concepts of hardware reliability.

One approach that can be taken to develop an assessment framework for complex systems is to supplement the concept of hardware reliability with further concepts and embrace these collectively within a concept of "dependability".   To be dependable a system should have the following attributes:

a)   It should be free from design or implementation errors (ie it should have the attribute of "correctness").

b)   In the event of any internal function failing, either by virtue of a design or implementation error or a component failure, the system should make the loss of this function known to the system user which may be some larger surrounding system and it should continue to function in an acceptable mode, (ie it should have the attribute of "integrity").

c)   It should reject any attempted misuse or interference from outside itself (ie it should have the attribute of "security").

d)   It should perform its designed functions for acceptably long periods of time without component failures (ie it should have the attribute of "reliability").

e)   In the event of failure the system should be capable of restoration to its initial as designed state within an acceptable period of time (ie it should have the attribute of "maintainability").

Within this framework software (as a system component) has the attribute of correctness but not reliability.

Based upon this conceptual framework a system's dependability could be assessed using a state transition model.

State transitions are triggered by: design or implementation errors being encountered, component failures, attempted misuse or interference, correction of a design or implementation error, or repair of a component failure.   The states entered following these triggers are dependent upon the system design and in particular upon the integrity and security mechanisms incorporated into the design.

To perform such an assessment it is necessary to analyse the functional and physical structure of the system and to obtain representative estimates of the nature and frequency of the various events that trigger state transitions. In practice it will rarely - if ever - be possible to perform a totally comprehensive system assessment;   fortunately this is rarely necessary. Attention can be focussed into the critical areas by

*   identifying within the state transition diagram those states that are of greatest concern (which will be dependent upon the specific mission or class of tasks to be performed by the system).

*   identifying within the functional structure of the system and the mapping of these functions onto hardware and software elements those functions and system elements that are most crucial to the systems mission.

*   identifying within the system design those portions that are most likely (by virtue of complexity) to contain design or implementation errors.

*   identifying within the system implementation those parts that are most likely (by virtue of technology) to be least reliable.

*   identifying within the system environment (ie all process, human and other domains external to the system's domain) those factors most likely to interfere with or misuse the system.

SOFTWARE CORRECTNESS

From the preceding discussion it is clear that software can contribute to system dependability as a means for implementing integrity and security mechanisms;   it can also detract from system dependability by conveying design and implementation errors into the system.   It is interesting to question whether there is reason to suppose that a software design will contain more errors than a hardware design of the same functional complexity. Any differences can be only in the respective degrees of care and control in design.   Unfortunately such differences can easily arise due to the fact that software can be modified more easily than hardware. Ease of modification can tempt designers to use less disciplined practices; it can also increase the frequency of design changes and hence increase the risk of failures in change control.

Two approaches can be adopted for assessment of design and implementation correctness;   the quality of the design and implementation processes can be examined; or the resultant system can be tested.   In practice neither of these approaches alone can provide total confidence of correctness.   It is recommended that a combination of the two approaches should be used.

Assessment of Design and Implementation   Techniques for assessing the procedures and practices used in design and implementation of software have been developed within the framework of quality system evaluation.

Requirements of a software quality system are laid down in various standards, the most widely used being the NATO Applied Quality Assurance Publication, AQAP 13, "NATO Software Quality Control System Requirements".   Companion standards provide guidance for evaluation of such systems, eg AQAP 14, "Guide for the Evaluation of a Contractor's Software Quality System for Compliance with AQAP 13".

Whilst these standards and guides are aimed   at definition and evaluation of the quality control system, rather than the application of the system to a specific software product, they are valuable as a starting point for evaluation of a product. However, to perform such an evaluation it is necessary to have access to the design and implementation records generated within the quality control system during the design and implementation of the product.

A further disadvantage of this approach is its emphasis upon comprehensive coverage of procedure and system.   Attention is distributed uniformly across all elements of the quality control system rather than being focussed upon those aspects most crucial to software correctness.

A more targetted and cost effective approach can be derived by firstly examining the structure and composition of the software components within the product to identify the critical areas.   Directed questions can then be asked regarding the measures taken to ensure correctness of software in these areas.

System Testing.   Depending upon the nature of the product it will be possible to access the internal elements of the system, in a manner sufficient to perform tests, to varying degrees. Beyond a certain point it will be possible only to seek information on the tests performed by the supplier;   this then being a line of investigative assessment falling within the scope of the above "assessment of design and implementation".

Where software components can be accessed it is necessary also to have access to the functional specification of that component if any meaningful tests are to be performed.

Recommended Approach.   In planning an evaluation of the "reliability" of the software within an instrumentation system there are many special considerations, as discussed above and summarised below.

1.   The extent to which the "reliability" of software elements can jeopardise the system's performance of its mission may be extremely dependent upon the design of the total software/hardware system.

2.   Software "reliability" is determined by its correctness, ie the absence of any errors in its specification, design and implementation.

3.   All techniques for assessing the error content of software prior to field use require access to either or both the source software and the design documentation.

4.   Models are available for prediction of "reliability" growth based upon observations of operational disclosure (and removal) of errors.

5.   Practices are available to minimise the risks of errors in software specification, design and implementation. Carefully directed questions and investigation regarding these practices can assist in assessment of software quality. However full assessment of the quality system may not be practical and may not be proportionally beneficial.

6.   Testing (by the user) of the software elements of a system, which requires access to the software elements and their functional specifications, is unlikely to be practical. Where this is practical there may be little benefit unless the software element happens to be crucial to total system dependability.

In view of these considerations the approach recommended is as follows:

1.   Determine the critical software elements by performing a system dependability analysis using techniques such as state transition models. As discussed earlier it will rarely be practical for this analysis to be total comprehensive. A considerable degree of careful judgement must be used in directing attention into the critical areas. To make these judgements prior studies should be made of:

*   the mission tasks
*   the structure and composition of the system
*   the operational environment of the system

2.   Assess the measures taken by the system supplier to assure the quality of the critical software components. Whilst published guidelines for assessment of software quality control systems can help in planning this activity it is recommended that attention should be concentrated upon establishing the sources of the software components, the reasons for confidence that the software is error free, the maturity of the software and the effectiveness of change control.

3.   For any software components that are critical and for which the supplier cannot provide sufficiently strong objective assurance plan and conduct systems tests. The plan for these tests is set by the dependability analysis; the tests being designed to check state transitions triggered by defined events. The tests are performed on the total system, not just the software component, and they check that state transitions are never into states triggered by software errors.

SOFTWARE RELIABILITY IN PRACTICE AND RECOMMENDATIONS FOR IMPROVEMENT

Little empirical data has been published regarding the reliability of software within instrumentation systems. A significant body of such data does however exist. Major instrumentation suppliers operate systems for qualificiation of software and for recording errors reported subsequent to product release. Similarly many instrumentation users operate systems for recording failures, including those diagnosed to be the result of software errors.

In theory the value of this distributed data could be greatly increased by gathering it together. In practice however it would be difficult to draw meaningful conclusions from the data. To appreciate these difficulties it is useful to consider the types of fault actually found in instrumentation systems. The experiences described here are drawn from results obtained by Sira in evaluating programmable controllers. A general conclusion of these evaluations is that in relation to other software products the software within programmable controllers is relatively error free.

The largest proportion of errors found are discrepancies between the documentation and the actual implementation. In most instances, but not all, the error could be corrected by changing the documentation. In one controller for example it was found that two characters used to invoke two distinct but related functions were stated in the documentation to be assigned in reverse to the actual implementation. This error could be removed by changing the documentation. In this same controller two further distinct but related functions defined in the documentation had not been implemented. It is debatable whether removal of this discrepancy by amendment of the documentation should be regarded as correction of the error.

The second most frequent types of error found are errors that cause difficulties in configuring the system or in application programming. This may or may not jeopardise the system in normal operation. In one system it was found that a tape verification routine always reported at least one error even after all errors had in fact been removed. In another system it was found that if parameters were accidentally omitted from a particular function this omission was not detected by the entry routine.

Very few errors are found which would remain undetected until the system is put into normal operation, however those that are found do have the most serious consequences. A feature of these errors is that they are difficult to locate and are often a result of interactions between different portions of the software or between software and hardware.

To make software failure records such as these and those held within individual companies more generally useful for reliability analysis and prediction some means is required for classification of the errors and for classification of the software components within which the errors are found. If software components were standardised, in the manner that many hardware components are, then reliability data could be accumulated for each component. Whilst such a policy of standardisation would reduce the flexibility benefits that software provides to designers, it is important to recognise that a trade off must be made between innovative design and product reliability. Instrumentation suppliers are conscious of this trade off and standardisation policies are observed - to varying degrees - within individual companies.

The major part of any instrumentation system can be assembled by building upon a number of basic functions, such as filtering, PID algorithms, etc. The utility of the reliability data presently distributed amongst many individual companies could be greatly increased if common software component standards (including interface definitions) were developed and reliability data for such standard software components was pooled.

The benefit of this approach might be questioned on the basis that experience shows errors are more frequently found in the software that binds the components together than in the basic software components themselves. The use of standard software components, having defined interfaces, can however contribute to reliability of the binding software; it would encourage the use of good, modular, design practices and it would provide a firmer basis for application of the dependability assessment approach recommended above. This approach becomes far more practical if systems are designed using standard base-function level software components. the software and hardware functional structure of the system can be presented in a more visible manner, and base level reliability data would become available for software components to supplement that already available for hardware components.

CONCLUSION

Returning to the question posed in the title of this paper the answer is a qualified yes.

The reliability of software components can be assessed and increased standardisation can increase the knowledge base available.

At the system level we should be concerned with a broader concept than that generally understood when the term reliability is used. This broader concept is embraced by the generally understood meaning of the term dependability. Further work is required to clarify such a concept and define its relationships with reliability. A wide range of techniques developed against the needs to assess hazards is available. Work is required to select from these the best techniques for dependability analysis and to mould them into a consolidated form that can be applied in a practical manner.

REFERENCES

Barlow, R.E. (1975). In J.B. Fussell and N.D. Singpurwalla (Eds). _Reliability and Fault Free Analysis,_. Society for Industrial and Applied Mathematics, Philadelphia, Pa.

Buffham, B.A., D.C. Freshwater and F. P. Lees (1971). Reliability engineering - a rational technique for minimising loss. _Major Loss Prevention in the Process Industries._ Institution of Chemical Engineers, London, p.87.

Halstead, M (1977). _Elements of Software Science._ North Holland Publishing Company, p.84.

Jelinski, J. and P.B. Moranda (1972). Software reliability research. In Frieberger (Ed), _Probabilistic Models for Software._ Academic Press, New York, pp.485-502.

Mills, H.D., (1972). On the statistical validation of computer programs. _Report FSC 726015._ IBM Federal Systems Division,

Nielsen, D.S. (1974). Use of cause-consequence charts in practical systems analysis. _Report Riso-M-1743,_ Atomic Energy Commission Research Establishment, Riso, Denmark.

PRACTICAL SOFTWARE RELIABILITY

D. W. Noon

The Foxboro Company, MA 02035, USA

Abstract. Virtually all software programs contain errors. Personal estimates
suggest that there can be initially one to ten errors per 1,000 lines of
program code. Large industrial process control system programs are currently
being designed with more than one million lines of code. Program failures in
the industrial process control arena can mean loss of production, bad product
produced, damage to process plant and even loss of life.

Given such a negative scene: What, if anything, can be done to quantify
software reliability?

Several quantifiable reliability measures of goodness exist[1]. Probabilistic
techniques are discussed and then applied to real life examples. Their
attributes and deficiencies are highlighted. One appendix example shows the
appropriate application of a probabilistic software model used to estimate
remaining software module defects at any given time. Other examples use
Duane[2] growth modeling to obtain system program error occurrence rate of
change and to make short term future error predictions. The Duane growth
model is complemented by a very simple Software Error Discovery concept.
Another technique illustrating Pareto[3] Error Profiling is examined. It is
helpful when trying to decide where effort should be expended in further
reducing program defects.

Keywords. Coding Errors; Computer Software; Mathematical Analysis; Modeling;
Models; Parameter Estimation; Pareto Profiles; Prediction; Probability;
Reliability Theory; Software Errors; Software Defects; Software Reliability.

INTRODUCTION

Our personal lives, as well as the modern
industrial world, have become increasingly
computer dependent. Current control and
management system products are predominately
microprocessor based, and the trend is towards
more powerful systems with more memory
capability, and consequently, larger system
programs. It is a fact that software errors do
occur [Figure 1]. They are a form of failure,
albeit a design failure that can be corrected,
and these failures must be considered.
Estimates of total software errors within a
system program must be made prior to product
release.

Today it is generally true that more money and
effort is directed toward software development
rather than hardware development. At the same
time, more money is spent on hardware
reliability than on software reliability. A
terrible imbalance has developed between effort
devoted to software development versus effort
applied toward software reliability. The
consequences of this imbalance are potentially
disastrous, particularly in an industrial
process control environment. Critical software
errors, those that can cause loss of life,
damage to plant, or bad product, must be
located, corrected, and eliminated prior to
product release.

RISC-G

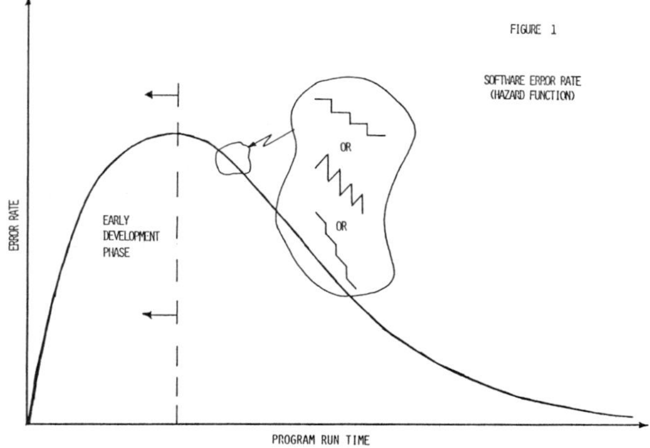

Over the past 15 years there have been millions
of words, and hundreds of probabilistic and
mathematical relationships written in textbooks
and technical papers in an attempt to describe
and quantify software reliability. Yet,
software reliability still remains a mystery.
Quantifiable reliability measures of goodness
[Table 1] are currently not contractually
demanded by the customer.

TABLE 1

SOFTWARE RELIABILITY MEASURES OF GOODNESS

* Number of Total Errors

* Number of Remaining Errors

* Program Error Density

* Time Estimate to Next Error

* Rate of Change of Errors

* Error Discovery

* Software Error Profile

Software reliability remains a mystery because much of the work done is shrouded in probabilistic statements that very few end users can understand. Some who do take the time to study the probabilistic software reliability models quickly realize that the models may be applicable at the software module level or very small program level, but they are not readily applicable during development of, and subsequent test of, the system program. A cardinal rule in structured software development, test, and evaluation is that when errors are encountered during system program testing they are recorded, but no immediate action is taken. Corrective action is only undertaken after an assessment is made as to what impact a potential code change will have on other software modules within the system program. Most of the probabilistic software reliability models have as one of their underlying assumptions, characteristics, or limitations: THE DISCOVERED ERROR IS REMOVED IMMEDIATELY, and then testing continues. Please refer to Appendix 1 for further discussion and an example of probabilistic software reliability modeling.

However, do not dismay - simple number manipulations that can easily yield software reliability measures of goodness do exist [Table 1]. The remainder of this discussion on Practical Software Reliability relates to those easily obtainable [minimal additional discipline and effort] software measures of goodness. Neither software design nor software testing techniques are going to be directly discussed other than to use their terminology.

MEASURES OF GOODNESS

Quantifiable measures of goodness can be calculated from graphically displayed data that is recorded during a given software test phase. The software test phase may be detailed design module testing, high level design integration testing, functional specification testing, or product specification operational testing. The latter two fall into the category of Acceptance Testing, and as a minimum, a prospective end user or customer should demand to see numeric measures of goodness at the acceptance level. Always remember that software testing does not prove that the software is correct. Testing merely shows the presence of errors, and hopefully a positive trend of error reduction. It is from the trend data that estimates of goodness, such as an estimate of remaining errors, are generated.

The information required to generate software reliability numerics is obtained from a problem log kept during the test phase. Most important is the recording of program run time between each or each batch of detected errors, bugs, defects, or failures. Problem log error

recording must also allow problem traceability to a specific routine, module and system program. Standard library names for all software modules and system programs must be established and used by all who do the testing. Library names must be used when describing the problem. The potential severity of the observed error[s] must be judged and recorded. Categories such as Critical, Major, Minor or High, Medium, or Low priority must be established with an appropriate definition for each class of failure. Ultimately, corrective action code changes are made. Programmer terminology describing why the code change was necessary must be recorded. This latter information can become useful in generating software error profiles.

The first measure of goodness to be examined is Software Defect Rate of Change. Its simple theory is explained and an example illustrates the mechanics of its use. A second real life example shows how it can deviate from the ideal, and still be useful. The sequence of subject discussion, simple theory, ideal example, and deviant example is then applied to Software Error Discovery. Software Error Discovery is an approach that is also useful in examining software reliability numerics. Finally there is a discussion and example of Software Defect Pareto Profiling. Error profiling is important when there is acknowledgment of an unhappy situation - too many software defects, and the question asked is: Where does one most efficiently expend corrective effort?

<u>Software Defect Rate of Change</u>

The rate of change of an entity, in the short term, is usually a precursor showing both the direction of and the speed of direction of the item being examined. The rate of change can be increasing, holding constant, decreasing, or any sequence of the preceding. If the objective of software acceptance testing is to release an error-free program, then increasing software errors per time unit is not good. Instead, it signifies the need for much more testing and debugging. A constant defect rate is also not a good measure of goodness and it too signals the need for more testing. Rapidly decreasing rate of change of software defects is a good measure. An excellent measure of goodness is a very fast reduction in rate of change of errors. It is one that, when projected into the future, yields an estimate of thousands of hours until possible discovery of the next error.

Given software problem log test data that meets the requirements previously demanded - how can we make sense out of the data so that we can see what is happening?

About 25 years ago, J.T. Duane explored an approach to examine time-based data. Codier[4] applied his technique to hardware reliability. This same thought process can be applied to software test data provided that all parts of the program are ultimately exercised. Duane postulated a simple mathematical model that is valid as long as improvement continues [in software - as long as there is a decreasing rate of change of defects]. It is very easy to apply the Duane growth concept. All available software test hours and error data is forced into a specific format and then plotted on log-log paper. Graphical integration takes place yielding information which can be used for short term forecasting, test progress evaluation, and for making program release decisions.

The mathematical version of the Duane growth
model using software terms is:

$$E/T = KT^X \qquad (1)$$

Where:　E = Summation of Errors During T
　　　　T = Summation of Program Run Time
　　　　K = Constant
　　　　X = Growth Rate (valid when X is
　　　　　　negative)

The constants K and X can be quickly evaluated
from the graphical presentation of the log-log
data. K is equal to the Y axis intercept when T
is unity, and X is the slope.

Let's look at an example. Given the data in
Table 2:

TABLE 2

Example 1

Question: Would you release this program based upon it being error free?

Time Interval, t_i (Program Run Time)	Errors in t_i
5	100
10	250
20	100
50	50
75	30
60	10
80	20
100	5
168	1
168	0
168	0
904 hours	566 Errors

$$\text{Program Error Density} = \frac{566 \text{ Errors}}{200{,}000 \text{ Lines of Code}} = 2.83 \text{ Errors/K Lines}$$

* Should the program be released based on
  the premise that it is error free?
* What is the time of expectation of the
  next error?
* How many errors might be encountered in
  the next three months of program use [if
  it is released as is to the customer]?

Take the data and force it into the format shown
in Table 3. Basically, errors are cumulated,
and program run time is cumulated. Program run
time becomes the abscissa of the Duane growth
graph, and cumulative errors are divided by the
cumulative run time and this numeric becomes the
ordinate of the graph.

TABLE 3

Example 1

FORMAT REQUIRED TO GENERATE THE DUANE GROWTH GRAPH

t_i	Errors in t_i	Σ Errors	Σ Time	$\Sigma E/\Sigma T$
5	100	100	5	20.0
10	250	350	15	23.3
20	100	450	35	12.9
50	50	500	85	5.9
75	30	530	160	3.3
60	10	540	220	2.5
80	20	560	300	1.9
100	5	565	400	1.4
168	1	566	568	1.0
168	0	566	736	0.77
168	0	566	904	0.63

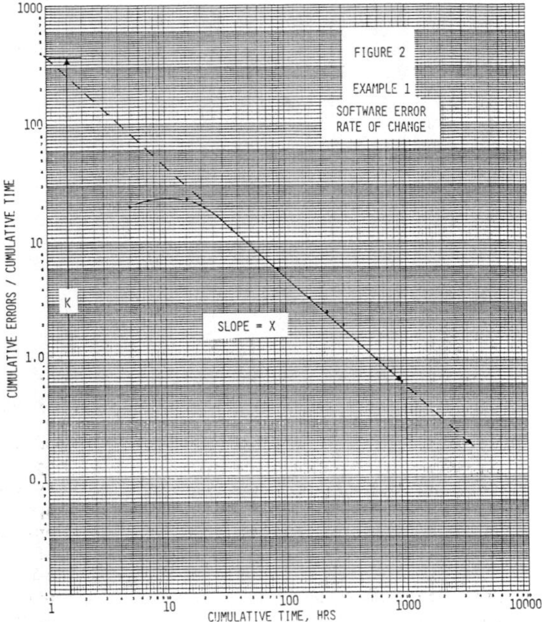

Plot the information from Columns 4 and 5 of
Table 3 on log-log paper as shown in Figure 2.

K, the Y axis intercept, is approximately 370.

The slope is:

$$\text{Slope} = X = \frac{\ln(0.57) - \ln(5)}{\ln(1000) - \ln(100)} \qquad (2)$$

$$X = -0.94$$

Therefore, the Duane growth model is:

$$E/T = 370T^{-0.94} \qquad (3)$$

When can we expect to observe the next error
[E = 566 +1 = 567] if the currently observed
trend continues?

$$567/T = 370T^{-0.94} \qquad (4)$$

$$567 = 370T^{0.06}$$

$$T = 1229 \text{ hours}$$

We can expect the next error in about:

$$T_{567} - T_{566} = 1229 - 904 = 325 \text{ hours} \qquad (5)$$

How many more errors can be expected in the next
three months [2190 hours] of program operation
if it is released as is, and the current trend
continues?

$$\text{For } T = 904 + 2190 = 3094 \text{ hours} \qquad (6)$$

What is E?

$$E/3094 = 370T^{-0.94}$$

$$E = (370)(3094)^{0.06}$$

$$E = 599 \text{ errors}$$

We can expect $E_{2094} - E_{904}$ more errors in the
next three months or:

$$E_{2094} - E_{904} = 599 - 566 = 33 \text{ more errors} \qquad (7)$$

The program in Example 1 is probably far from being error free; tens of errors should be expected in the near future. I personally would want an estimate of zero errors, particularly of a critical or high priority nature, in the next 2000 hours of operation.

Suppose the software data does not project into a singularly smooth defect rate of change, but instead has discontinuities [Figure 3]. The discontinuities might mean that there have been specification changes taking place while testing was in progress – the game ground rules are changing.

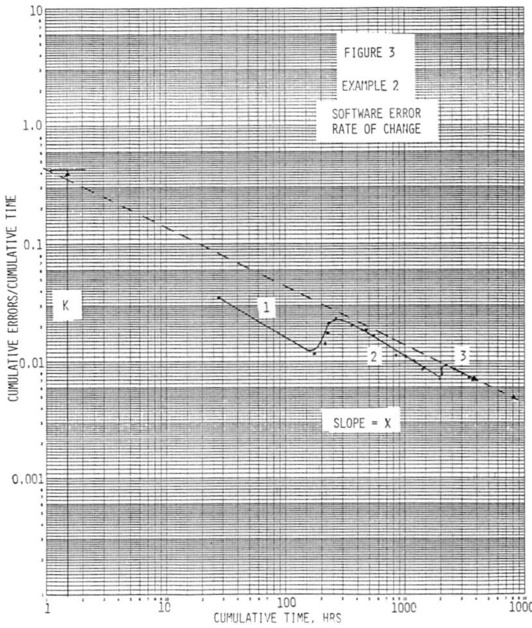

However, most likely the discontinuities are a result of not all of the program being equally exercised or tested. In Example 2 that is exactly what happened. For the first 172 hours, one portion only of the program had all of its decision parts exercised – Region 1. From about 172 to 2000 hours another portion of the program was thoroughly exercised as well as the previous portion – Region 2. From 2000 hours to 4000 hours, the final portion was extensively tested along with the previous two parts – Region 3.

How should the Duane growth graph in Example 2 be projected? Always extend or forecast from the last amount of data. The latter points have more information content and they should be given more consideration than the earlier data.

Using the preceding rules, the constant K is approximately equal to 0.43.

The slope is:
$$x = \frac{\ln(0.005) - \ln(0.0085)}{\ln(7500) - \ln(2600)} \qquad (8)$$

$$X = -0.50$$

and the Duane growth model is:

$$E/T = 0.43T^{-0.50} \qquad (9)$$

The time to the next expected error is:

$$29/T = 0.43T^{-0.50} \qquad (10)$$

$$67.4419 = T^{0.5}$$

$$T = 4548 \text{ hours}$$

So, we can expect the next error in about:

$$T_{29} - T_{28} = 4548 - 4000 = 548 \text{ hours} \qquad (11)$$

The program in Example 2 also is probably not sufficiently error free. It, too, does not meet the criterion of no expected errors in the next 2000 hours of operation.

Both Examples 1 and 2 suggest the need for more acceptance level testing prior to release even though the software error rate of change is decreasing.

The negative slope of the growth process can be forced to become even more negative by putting more energy, more effort, more resources into the testing phase. If management is satisfied with the current rate of improvement, but wishes to reduce the errors further, then they can continue the same rate of growth with the same or previous amount of effort expended. Please note that the Duane growth concept does have the limitation that no estimate of total program errors are obtainable from the technique.

The Duane growth model implies an ever continual rate of change, and therefore, a continuum of errors. In reality, every program contains a finite number of errors and, in the long term, the number of remaining errors should decrease to zero. So what can be done to estimate the total number of program errors and, subsequently, the remaining defects at any given time. Let's look at the concept of Software Error Discovery and see what it shows.

<u>Software Error Discovery</u>

Software Error Discovery is another graphical approach useful in obtaining software reliability measures of goodness. Its underlying premise is very simple. Program error content asymptotically approaches a limit. There are a finite number of defects in a software program. As time passes, the errors are detected and corrected, detected and corrected . . . . and ultimately the program is error free. Ideal software error discovery follows the curve in Figure 4. Estimates of software testing completeness, number of total program errors, and number of remaining errors can be obtained. The technique assumes that all parts of the program are equally exercised.

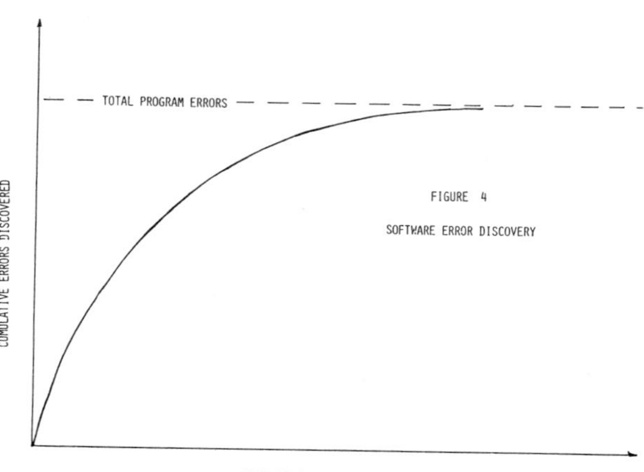

The assumption that all branches and decisions
are ultimately exercised must be forced to
occur. In reality, this can be done, but seldom
can all parts be equally exercised within the
same time frame. All decision paths can be
examined but usually the examination must be
done sequentially. When cumulative errors as a
function of cumulative time are plotted using
Example 2 data, a series of error discovery
curves appear (Figure 5). The reason for the
discontinuities was previously stated. The
program content was tested in three distinct,
non-homogeneous time frames.

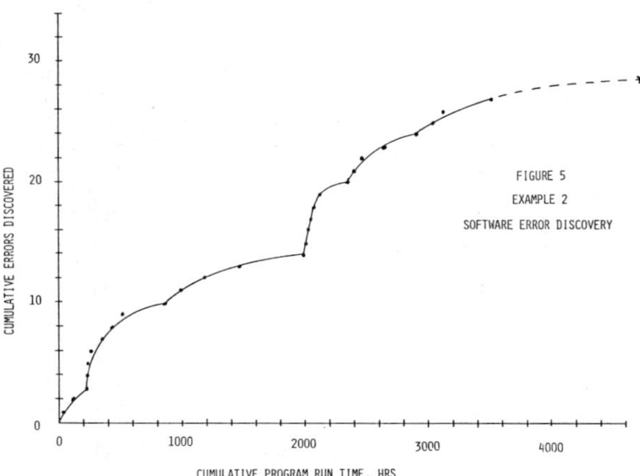

Slight projection of Figure 5 Software Error
Discovery data yields an estimate of another
error in about 500 hours. The total number of
program defects can be visually extrapolated to
be about 29 or 30. If the error discovery curve
is not yet asymptotically approaching a limit,
but is still increasing, this signifies the need
for additional testing before program release,
and before good estimates can be made.

At first glance the software error discovery
data appears to follow an exponential growth or
a series of exponential growth models. Numerous
models proposed by Musa[5], Littlewood and
Verrall[6], and Goel and Okumoto[7] are
variations of the exponential growth model that
sometimes can be used to describe Software Error
Discovery. Ohba[8] believes that an Inflection
S Shaped or a hyperexponential growth model
often has a better fit for calendar-time-based
data with non-homogeneous time distribution of
the testing effort. This paper does not explore
the validity of the various numeric error
discovery models. Their authors are referenced
for completeness and for consideration only.

Both the Software Error Rate of Change and the
Software Error Discovery concepts suggest that
another defect will be observed in the next 500
to 550 hours of program run time. Where might
testing effort most efficiently be expended to
help locate the remaining error(s)? Software
Defect Pareto Profiling might yield answers to
this question.

Software Defect Pareto Profiles

Around the turn of the century Vilfredo Pareto,
an Italian economist, noted that in most complex
entities, there are just a few items that
contribute to most of the observed influence.
He related his observations to the concept of
diminishing returns. His thinking has evolved
into what today is referred to as the "80/20"
rule or the trivial many and consequential few.
That is - most problems are often caused by a

small percentage of the possible contributing
variables. Identify the consequential few, work
on and correct them, and most of the
undersirable effects will disappear.

In software program Example 2, there are about
10,000 lines of code, 10 software modules, 28
observed errors to date, and another error
expected to occur in about 500 more hours of
program run time.

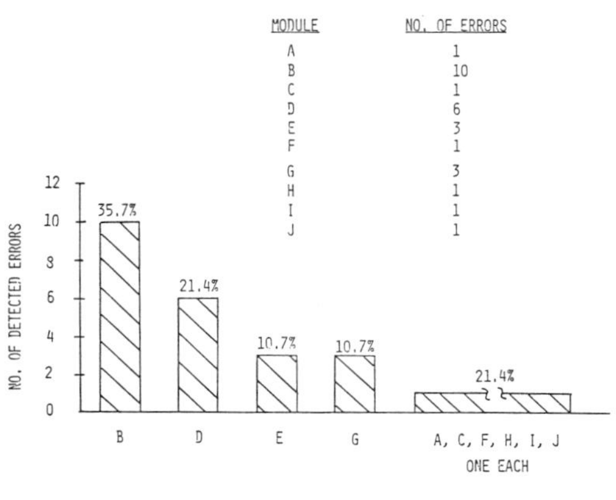

There is clearly a Pareto software defect
profile by module for program Example 2 (Figure
6). The data suggests that testing emphasis
should be more heavily directed at Modules B and
D because they have been most problemsome.
Within moments, and without ponderous reading,
all participants are brought to the same level
of awareness by presenting the information in
the form of a Pareto profile.

Suppose, however, that there were no tall poles
when the errors were profiled by module; that
all modules were approximately equal
contributors to the observed errors. What does
one do when errors are still being discovered
rapidly?

First suggestion - Is there a common mode
program error problem pattern? Maybe identical
programming errors have been made in several or
all of the software modules. If there is no
software defect Pareto profile by module, then
attempt to generate a software defect Pareto
profile by common errors. Now it becomes
evident why it is important to record the reason
for the code changes in the software problem log
generated during the test phase.

If there is neither a module nor a common mode
error pattern, then what next?

Second suggestion - Consider implementing
Ishikawa's[9,10] cause and effect diagrammatical
approach to problem solving. Almost any program
can be solved with this diagrammatical approach.

Many times problem resolution using Ishikawa's
cause and effect process requires highly
knowledgeable people and extensive time to
pursue all of the avenues revealed. Both
translate into large expenditures.

What does management do if they perceive that resources are not available to implement the Ishikawa cause and effect philosophy, and there is no apparent module or common mode error pattern? I know of only one more suggestion. Scrap the existing design, acknowledge your losses, and begin again with a brand new structured design effort.

CONCLUSION

Critical software errors that lead to loss of life, plant damage, or bad product must be eliminated prior to software release and use.

Estimates of software reliability measures of goodness must be made prior to release. Contractual demands by the end user must begin to be placed on the software supplier. These demands for measures of goodness can consist of one or more of the following program quantifiers:

* Number of Total Errors
* Number of Remaining Errors
* Program Error Density
* Time Estimate to Next Error
* Rate of Change of Errors
* Error Discovery Profile
* Software Defect Profile

Use of the Duane growth concept not only yields an estimate of the rate of change of errors, but estimates of time to next error, and number of errors to be expected in the near future. Our experience suggests that no program be released where the Duane model implies the occurrence of another error within the next 2000 hours of program run time. Please note that no estimate of total program errors can be obtained from the Duane growth concept.

Graphical representation of error discovery gives a measure of program readiness for release. This measure is obtained by judging how well the error discovery is approaching the asymptotic limit of a finite number of total program errors. The error discovery concept also yields an estimate of program run time to next error.

Should additional program error detection testing be required, then software defect profiling often suggests what parts of the program require additional testing. Pareto defect profiles also very quickly and easily bring all participants to the same level of problem awareness.

On occasion, when programs are small (or at the module level), and when their respective stringent assumptions are met, probabilistic software reliability models can be useful. The requirement that a discovered error must be removed immediately prohibits their use during system program testing. A program for using the very popular Jelinski and Moranda[11,12,13] software reliability model named RELMODEL and written in ANSI BASIC is available from the author.

A severe limitation to all of the tools discussed is the assumption that ultimately all parts of the program have been exercised, and preferably exercised uniformly. In reality this is not easy to do. However, when done, the Duane growth, Error Discovery, and Pareto Profiling techniques readily yield quantifiable software reliability measures of goodness that are useful in judging a program's readiness for release.

One final comment – never rely completely on numbers.

REFERENCES

1. Dunn, R. and R. Ullman, _Quality Assurance for Computer Software_, McGraw Hill, New York, 1982, pp. 297-333.

2. Duane, J.T, "Technical Information Series Report DF 62MD300", 1962, General Electric Company, Erie, Pa.

3. O'Conner, Patrick D.T., _Practical Reliability Engineering_, John Wiley & Sons, New York, 1983, pp. 218-220.

4. Codier, E.O., "Reliability Growth in Real Life", _Proceedings, Annual Symposium on Reliability_, 1968, pp. 458-469, IEEE Catalog No. 68C 33-R.

5. Musa, J.D., "Validity of Execution-Time Theory of Software Reliability", _IEEE Transactions, Reliability, R-28_, 1979, pp. 181-191.

6. Littlewood B. and J. Verrall, "A Bayesian Reliability Growth Model for Computer Software", _Proceedings, IEEE Symposium on Computer Software Reliability_, New York, 1973, pp. 70-77.

7. Goel, A.L. and K. Okumoto, "Time-Dependent Error Detection Error Rate Model for Software Reliability and Other Performance Measures", _IEEE Transactions, Reliability R-28_, 1979, pp. 206-211.

8. Ohba, M., "Software Reliability Analysis Models", _IBM Journal of Research and Development_, Vol. 28, No. 4, 1984, pp. 428-443.

9. Ishikawa, K., _Guide to Quality Control_, Nordica Int. Ltd., Hong Kong, 1983, pp. 152-160.

10. Kinglarski, E., "Ishikawa Diagrams for Problem Solving", _Quality Progress_, ASQC, Dec. 84, pp. 26-30.

11. Myers, G.J., _Software Reliability Principles and Practices_, John Wiley & Sons, New York, 1976, pp. 329-335.

12. Koch, H.S. and P. Kubat, "Quick and Simple Procedures to Assess Software Reliability and Facilitate Project Management", _The Journal of Systems and Software 2_, 1981, pp. 271-276.

13. Jelinski, Z. and P. Moranda, "Software Reliability Research", _Statistical Computer Performance Evaluation_, Academic, New York, 1972, pp. 465-484.

Appendix
SOFTWARE RELIABILITY MODELS

There are numerous formal, mathematical, probabilistic software reliability models. Typically they are named after their author(s), e.g., Jelinski-Moranda, Shick-Wolverton, Shooman, Weibull, Musa, Gompertz, Trivedi-Shooman, Shooman-Hatarajan, Goel-Okumoto, Littlewood-Verrall, Thompson-Chelson, Ohba, ... All of the models have a set of underlying characteristics, assumptions, and limitations which if not honored can lead to erroneous results. The probabilistic models are inexact. Most operate on calendar or program run time. Some reportedly are applicable during execution times. All assume error independence. All assume that all parts of the program are equally exercized. Each assumes that all errors are of equal significance.

One of the more popular software reliability models is the Jelinski-Moranda (J&M) model. It is a calender or program run time deutrophication model that assumes:

* Error Independence
* Discovered Errors are Removed with Certainty
* Discovered Errors are Removed Immediately
* Time of Error is Expentially Distributed
* Error Rate is Constant Between Failures
* Error Rate Decreases by a Constant Amount at Each Error Detection

The J&M model attempts to depict the software hazard function to the right of the Early Development Phase notation in Figure 1. A hazard function showing the J&M model is shown in Figure 7.

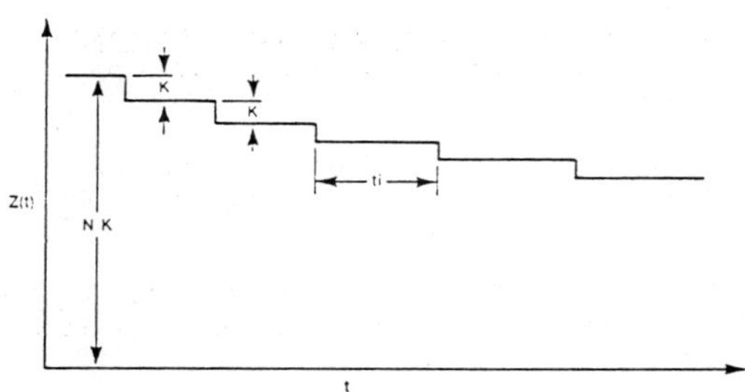

N = UNKNOWN NUMBER OF INITIAL ERRORS
K = UNKNOWN CONSTANT
t_i = ELAPSED TIME BETWEEN (i-1) st
     AND i th ERROR
t = OPERATING, CALENDAR, OR PROGRAM RUN TIME

    IN ADDITION, LET:

        n = ESTIMATE OF ERRORS AT
            TIME t_i

        t_p = TOTAL TIME PREVIOUS TO
            MOST RECENTLY DISCOVERED
            ERROR

        $\hat{M}$ = ESTIMATE OF REMAINING ERRORS

The J&M HAZARD FUNCTION, Z(ti), is:

$$Z(t_i) = K\,[N-(i-1)] \qquad (12)$$

The J&M reliability function, R(ti), is:

$$R(t_i) = e^{[-K(N-n)t_i]} \qquad (13)$$

A conservative estimate of time-to-next-error, Θ , is:

$$\Theta = \frac{1}{K(N-n)} \qquad (14)$$

N and K must be estimated through either statistical maximum likelihood estimation techniques or through iteration using:

$$\sum_{i=1}^{n} \frac{1}{N-(i-1)} = \frac{n}{N-\left\{\sum_{i=1}^{n}(i-1)X_i\right\}T} \qquad (15)$$

and

$$K = \frac{n}{NT-\sum_{i-1}^{n}(i-1)X_i} \qquad (16)$$

where

$X_1, ..., X_n$ are elapsed times between successive errors

and

$$T = \sum_{i=1}^{n} X_i \qquad (17)$$

If the two equations do not converge, or if they do not converge quickly, then the model cannot — should not be used. Non-convergence might mean that the error rate is increasing, decreasing, or both.

Finally, if there is convergence in less than or equal to five iterations, then an estimate of the remaining errors, N-n, can be made:

$$\hat{M} = \frac{n}{e^{(t_i+t_p)/K}-1} \qquad (18)$$

TABLE 4

Example 2

DATA USED TO GENERATE THE DUANE GROWTH GRAPH IN FIGURE 3

t_i	Errors in t_i	Σ Errors	Σ Time	$\Sigma E / \Sigma T$
28	1	1	28	0.0357
144	1	2	172	0.0116
43	1	3	215	0.0140
10	1	4	225	0.0178
5	1	5	230	0.0217
30	1	6	260	0.0231
90	1	7	350	0.0200
120	2	9	470	0.0191
60	0	9	530	0.0170
330	1	10	860	0.0116
140	1	11	1000	0.0110
200	1	12	1200	0.0100
200	0	12	1400	0.0086
600	2	14	2000	0.0070
40	2	16	2040	0.0078
10	2	18	2050	0.0088
150	2	20	2200	0.0091
400	2	22	2600	0.0085
400	2	24	3000	0.0080
400	1	25	3400	0.0074
600	3	28	4000	0.0070

$$\text{Program Error Density} = \frac{28 \text{ Errors}}{10000 \text{ Lines of Code}} = 2.8 \text{ Errors/K Lines}$$

Given the data in Table 5 with the implicit acknowledgement that every time an error was encountered, testing stopped, the defect was found, the necessary code change was made, and only then was testing resumed:

TABLE 5

Test Data for a Program with
7000 Lines of Code
(7 Feb 82 to 13 Sep 84)

Error	t_i	t_p
1	312	312
2	256	568
3	200	768
4	120	888
5	25	913
6	180	1093
7	225	1318
8	22464*	23782
9	12096*	35878

*As many as 30 units were simultaneously tested

Error Density:
$$\frac{9 \text{ Errors}}{14000 \text{ Lines of Code}} = 0.64 \text{ Errors/K Lines}$$

The program RELMODEL written in ANSI BASIC (by Walter White, The Foxboro Company) was used to attempt convergence, and to establish a set of reliability numerics. Results were:

* No convergence in 20 iterations for first 7 errors.
* Convergence in 4 iterations when 8 error data used.
* Convergence in 5 iterations when 9 error data used.
* Estimated time-to-next error is 7603 hours.
* Estimated number of remaining errors is 1.

Less than five minutes of elapsed time was consumed in entering the data into RELMODEL and obtaining the quoted numerics.

DISCUSSION

<u>Session:</u> System Design.

<u>Paper:</u> Design consideration for a fault-tolerant distributed control system.

<u>Questions:</u> Both systems are using identical software, therefore the system is not software fault tolerant. Even on one CPU you can have redundant software. Why is it not implemented?
Do you intend this to include in your future systems?

<u>Author's reply:</u> 1. Basically the system is a "turn-key" product. We use "hardware-embedded" software, there are no user-programs. The control software, such as DDC sequence control etc, are stored in ROM with memory protection. The system was introduced in 1975. At that time "memory" was fairly expensive and the control software was designed within a limited space. Since 1975 more than 1000 systems are in operation in the field without any trouble and hence is considered as field proven.
2. We have been and are still interested in software redundancy. However it is necessary to investigate all possible system operations for redundancy and establish a "confident technology". We are reluctant and somewhat conservative to introduce new technologies, especially for control algorithms, prior to a complete evaluation. With regard to the system software (such as Operating System, Drivers etc.) these are not software redundant, but may be stored in silicon. The "core" part of the total software is then proven and may be considered as part of the hardware. For the application software, including the user programming, the documentation, the revisions etc, the software maintenance is very important. Especially in terms of quality control. At this moment, it is felt that software redundancy may cause confusion to both manufacturers and users from the software maintenance point of view.

ANALYSING CONTROL SYSTEMS BY MEANS OF EVENT TREES

R. A. J. Badoux and R. W. van Otterloo

*NV KEMA, Dept. BBA, Risk and Reliability Analysis Group, Box 9035,
6800 ET Arnhem, The Netherlands*

Abstract. In the failure analysis of a control system one is
nearly always confronted in practice with several different
consequences as result of a disturbance in the controlled
process and/or a failure in the control system. Often enough one
performs such an analysis by means of fault trees. This may lead
to incorrect conclusions. In the author's opinion a failure
analysis of a control system should start with the construction
of event trees. A particular failure mode of the control system
can be analysed by means of a fault tree. To illustrate this
approach an example is given. A pressure regulating system is
analysed by means of event trees.

Keywords. Control system analysis; pressure control, Markov
processes; failure mode and effects analysis; hazard and
operability study; event tree; fault tree.

INTRODUCTION

Several techniques are available for
the failure analysis of systems. Some
techniques such as failure mode and
effects analysis (FMEA), and hazard
and operability study (HAZOP) only are
qualitative. Techniques with both
qualitative and quantitative aspects
are for example event trees and fault
trees. An example of a more
quantitative technique is modelling of
the failure process by means of Markov
processes. FMEA, HAZOP, event tree
analysis, fault tree analysis and
Markov analysis are the more
frequently used techniques in safety
and reliability analysis. Especially
the failure analysis of a control
system should begin with the
construction of event trees. This will
be shown later on in an example.

SOME GUIDELINES FOR CHOOSING AN ANALYSIS TECHNIQUE

Failure Mode and Effects Analysis (FMEA)

This is a qualitative technique. FMEA
is a detailed inductive analysis
carried out at the design stages of a
system. It systematically analyses all
contributory component failure modes
and identifies the resulting effect on
the system. The purpose of FMEA is to
identify areas in the design where
improvements are required to ensure
the system will be reliable and safe
for its intended use. FMEA is probably
the most widely used design review
technique for failure analysis. The
method is particularly suited to
analogue and redundant systems.

Hazard and Operability Study

A HAZOP can be regarded as an
extended FMEA, the extension being
in the direction of including
operability factors in addition to
equipment fault modes. In fact this
technique is a detailed failure
mode and effects analysis of the
Piping and Instrument (P&I) line
diagram. HAZOP is widely used in
the process industry.

Event Trees

An event tree is an inductive logic
diagram. The diagram starts with a
given initiating event and depicts
various sequences of events leading
to multiple-outcome states. To each
state is associated a particular
consequence. The event tree
approach is similar to decision
tree methodology in business
applications. It forms the
necessary framework for the overall
risk assessment by (1) providing a
basis in defining accident
scenarios for each initiating
event, (2) by depicting the
relationships of success and
failure of safety related systems
with various accident consequences,
and (3) by providing a means for
defining top events to system fault
trees.

Fault Trees

A fault tree is a deductive logic
model that graphically represents
the various combinations of
possible fault events that lead to
the top event. The top event has to
be uniquely defined, e.g. a
particular system failure or a
specific unwanted event.

The objectives of fault tree analysis are: (1) to identify systematically all possible occurrences of a given undesired event, (2) to provide a clear and graphical record of the analytical process, and (3) to provide a baseline for evaluation of design and procedural alternatives.

Markov Analysis

Markov analysis considers multiple effects and is a multiple-thread inductive analysis. In most cases, the complexity of the analysis makes hand calculations impractical. The answer to this problem is simulation. However, the performance of accurate simulations requires a lot of computer time. In a Markov process, all the mutually exclusive system states must be identified. The set of possible states in which the system is functioning properly is called the "good" set as opposed to the set of possible states in which the system is out of order, which is called the "bad" set. Of particular interest is the determination of the probability of a system making a transition from the "good" set to the "bad" set as a function of time. Of course, refinement is possible in the direction of dividing into more than just a "good" and a "bad" set. There are however, two restrictions: (1) the system as it enters each state is influenced by the occurrence in the immediately preceeding state only and does not depend on any other previous system states, and (2) the rates of transition among possible system states must be constant with time.

FAILURE ANALYSIS OF A CONTROL SYSTEM: AN EXAMPLE

Quite frequently one sees in practice safety or reliability engineers performing a failure analysis of a control system by means of a fault tree. A closer look at the produced fault trees shows that they actually are a combination of an event tree and one or more smaller fault trees. This means that the analyst has mixed inductive and deductive logic. This controdicts the proper definition of a fault tree, viz. a fault tree is a deductive logical and graphical model of the various parallel and sequential combinations of faults that will result in the occurrence of the predefined undersired event (Haasl, 1981).
Now let us try to perform such a fault tree analysis of a control system by looking at an example.

Pressure Regulating System

The pressure regulating system which is depicted in Fig. 1 at the end of this paper, controls and

reduces the pressure in a gastransmission pipeline system. The pressure at the left hand side is over 35 bar. The terminal pressure on the right hand side is under normal operating conditions equal to 20 bar. The system consists of 3 parallel trains in order to comply with a variable demand, i.e. a variable flow. Valves A, D and G are the primary regulating valves. If for any reason one of these valves fails open, a secondary regulating valve takes over. This second set consists of the valves B, E and H. For safety purposes a set of shut-down valves C, F and K are installed. These valves block the flow in case of failure of the regulating part and can only be opened by hand.
For facilitating the failure analysis we simplify the system to the one that is depicted in Fig. 2. This simplified system just consists of 3 parallel pressure regulating valves.
In Fig. 3 one finds the upper part of a fault tree with top event: 'Control system failure'.
Apart from the fact that combining logically faults such as 'system opens spontaneously' and 'system closes spontaneously' is rather doubtful, a further reason for rejecting this fault tree is that the consequences of the faults are quite different by nature. A fault tree is tailored to its top event which corresponds to some particular system failure mode, and the fault tree thus includes only those faults that contribute to this top event. This means that we have to construct several fault trees, one for each system failure mode.
The problem now is to find the different top events and to put those in a proper framework. This is done by means of event tree analysis.

Event Tree Analysis

This technique is well documented in NUREG-75/014 (1975). The first step is to find the initiating events. These evolve from the system design because the system is laid out in such a way to cope with this kind of events. A global HAZOP may also yield the relevant initiating events.
The initiating events for the simplified system of Fig. 2 are:

- R opens spontaneously
- S opens spontaneously
- S closes spontaneously
- T closes spontaneously
- Demand decreases
- Demand increases

The event tree for the initiating event 'T closes spontaneously' is depicted in Fig. 4. This tree is self explaining.
The next step is to set up a table like Table 1. This table contains

the consequences resulting from the occurence of the initiating event and the occurrence of subsequent faults. The consequences are ranked with respect to the degree of seriousness. The entries of the table are the minimal combinations of events that cause a particular consequence.
The non-minimal combinations of events in Table 1 are placed between parantheses. One should apply Boolean reduction in order to produce the definite table.
The first row of Table 1 shows the combinations of events with initiating event 'T closes spontaneously'. The second row may be the combinations of events with initiating event 'R opens spontaneously'. Only a part of the complete table is printed.
Combinations of events that lead to an explosion are interesting from a safety point of view. Concerning maintenance one should focus on the combinations of events that lead to insufficient flow.
As the structure of the failure analysis of the simplified system is clear now, we can proceed by tackling the more complex system. The approach is the same. The complete event trees and the corresponding table will take a vast amount of space. For that reason this information is not included in this paper.
In order to assess the probability of the occurrence of a combination of events it may be necessary to construct a fault tree for a single event in such a combination. For instance, a particular event tree for the system of Fig. 1 contains the event 'valve D will not close'. In order to assess the probability of this event one may have to construct a fault tree.

PRACTICAL GUIDELINES

As an aid for constructing event trees in general we give some practical guidelines. These are:
1   Write down all the possible fault events. Describe those events exactly.
2   Rank the events.
     a Initiating events at the beginning on the left. Draw an event tree for each initiating event separately.
     b Look for the possible chronological order of the events in the tree.
     c Look for other possible influences.
     d Restrict yourself to undesired events only.
3   Draw the tree and state the combinations of events.
4   Assess the consequences of each combination of events.
5   Set up a table with column headings: the possible consequences. The entries in a row for a particular initiating event are the combinations of events

corresponding to the possible consequences.

The probability assessment is the next step in the failure analysis of a system. We will not deal with that subject here.

CONCLUSIONS

As we have seen one should avoid constructing fault trees like the one in Fig. 3 when performing a failure analysis of a control system. The event tree analysis technique has shown to be a powerful tool in the analysis of control systems. This technique gives a clear picture of the things that may go wrong. It also provides a proper framework for an overall risk analysis.

REFERENCES

Haasl, D.F. and co-works (1981). _Fault Tree Handbook,_ NUREG-0492,Washington, D.C. 20555.
Nuclear Regulatory Commission (1975). _Reactor safety study. An assessment of accidents risks in U.S. commercial nuclear powerplants, Appendix I Accident definition and use of event trees,_ NUREG-75/014, Washington, D.C. 20555.

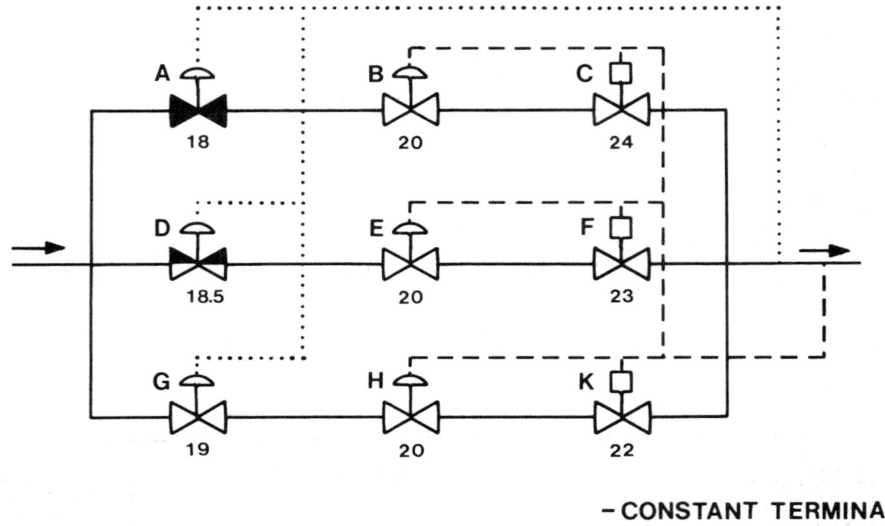

Fig. 1.   Pressure regulating system

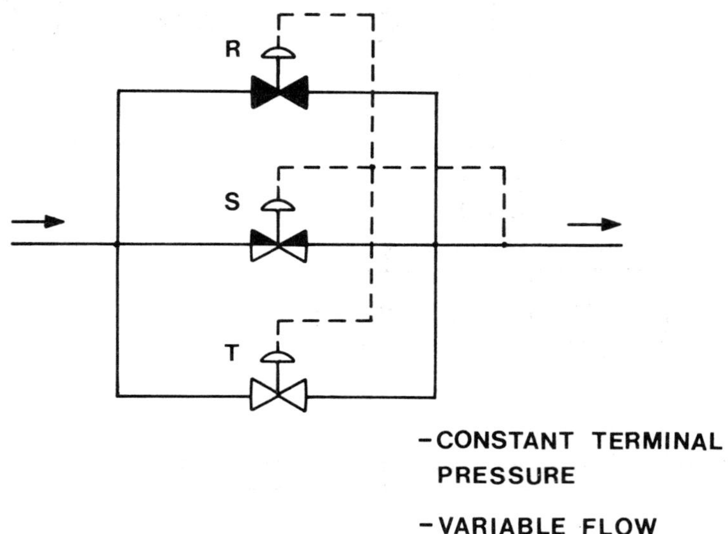

Fig. 2.   Simplified pressure regulating system

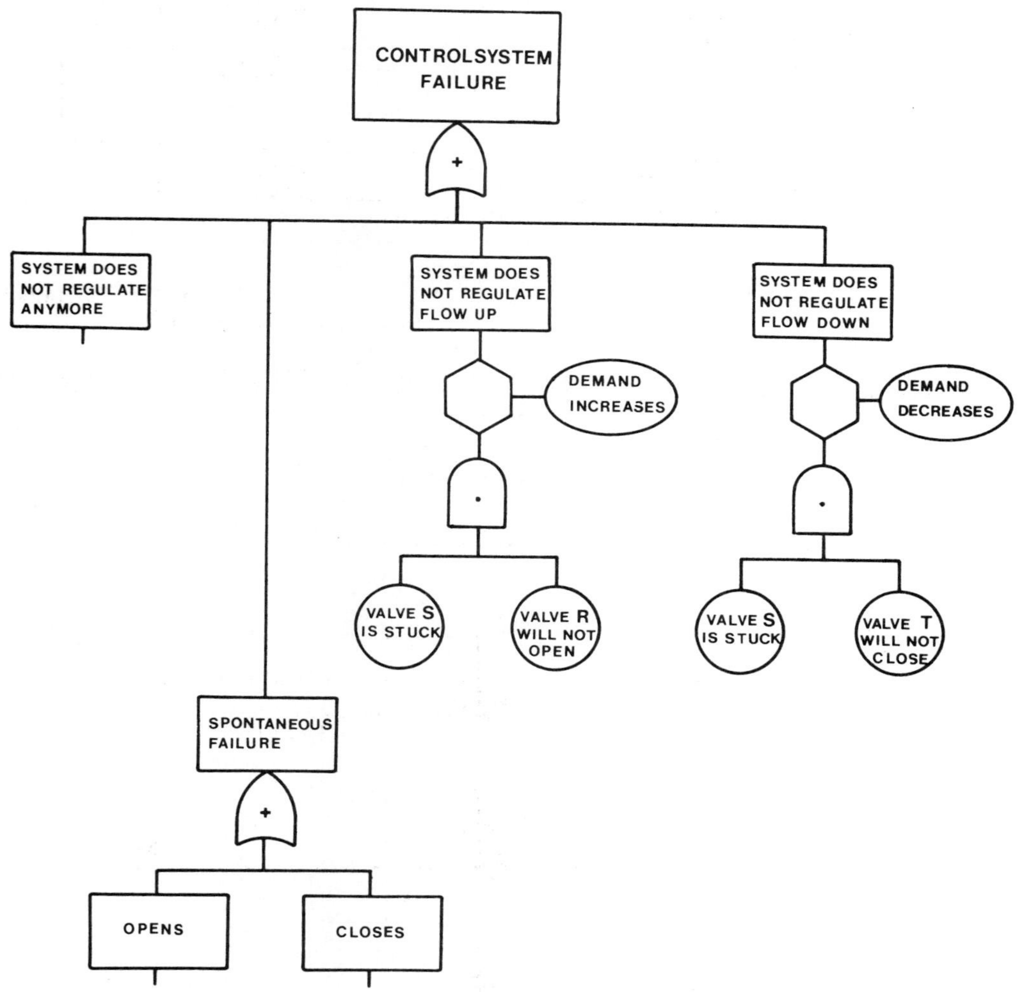

Fig. 3.   Upper part of the fault tree with top event 'Control system failure' for the simplified pressure regulating system

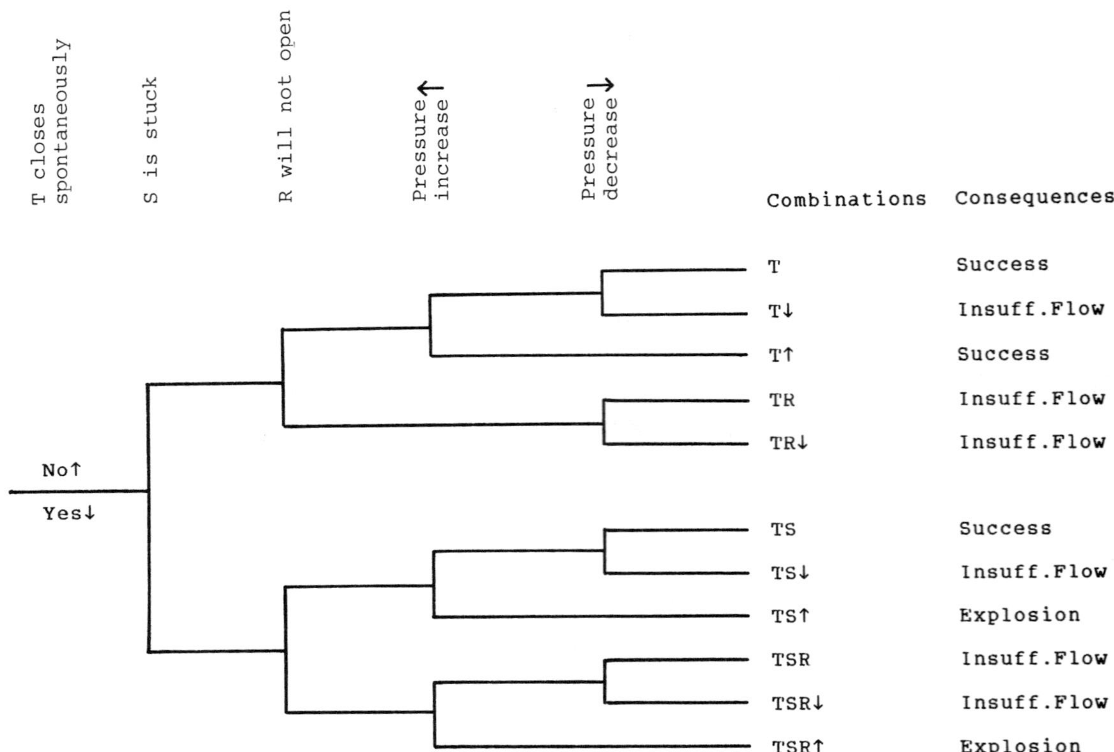

Fig. 4.   Event tree for the initiating event 'T closes spontaneously'.
A pressure increase (decrease) is caused by a demand decrease (increase)
in the pipeline network which the control system is serving.

Initiating event	Consequences	
	Insufficient flow	Explosion
T	T↓, TR (TR↓),(TS↓) (TSR),(TSR↓)	TS↑ (TSR↑)
R	...	...

Table 1   <u>The minimal combinations of events corresponding to
an initiating event and the resulting consequences</u>

INSTRUMENTATION SYSTEM MODELS FOR
COMPUTER-AIDED FAULT TREE ANALYSIS

A. Poucet* and C. Carletti**

*Systems Engineering and Reliability Division, Joint Research Centre, Ispra Establishment,
21020 Ispra (Varese), Italy
**Safety and Reliability Department, SNAM PROGETTI SpA, 20097 San Donato
Milanese, Italy

Abstract. Models are discussed ror analysing the reliability of instrumentation and
control loops in chemical process plants. The models are used in a computer code for
interactive fault tree construction and evaluation (CAPTS-Code). The paper first briefly
describes the CAFTS method and the concept of the modular component models used in that
method. The development of a library of such models for instrumentation systems is then
discussed. An example of the application of the models to a real system is given.

Keywords. System analysis; reliability; process control; control system analysis;
computer-aided circuit analysis.

INTRODUCTION

Fault tree analysis is a well known technique for
assessing the reliability and safety of complex
installations. Although there exist many computer
programs for logical and probabilistic analysis of
fault trees, the system modelling or construction
of the fault tree for a given system requires an
important effort.

Automatic fault tree construction codes have been
developed with the aim of reducing this effort
(Taylor, 1982; Lapp, 1977). However, a complete
automatisation of the fault tree construction process presents great difficulties both in controlling the process and finding an adequate and
flexible way of describing a system. Hence the use
of automatic methods has been limited to simple
straightforward cases.

The CAFTS method overcomes many of the drawbacks
because of its interactive approach which assists
rather than replaces the analyst. Using the CAFTS
method, the construction of a system fault tree is
performed in two phases (Poucet, 1981;1983;1985).

In a first phase, a high-level fault tree (Macro
Fault Tree MFT) is constructed. The construction
of a MFT for a selected undesired event (e.g. a
deviation of a process variable) is carried out in
an interactive procedure in which the analyst is
prompted to enter the system P&I diagram and other
relevant information in a guided way. CAFTS then
uses an "expert system" approach to build up the
fault tree progressively as the information is
entered. The undesired event is considered to be
a goal state whose causes are generated by back-
ward chaining of production rules. The production
rules are stored in a library and describe the
generic behaviour of components and subsystems.

In a second phase, the MFT is transformed into a
fully detailed fault tree by using "modular component models" (MCMs). MCMs are models which are
used to replace each event in the MFT by a mini-
fault tree which gives the detailed causes of the
event.

In this paper, we will focus on the second phase
and on the MCMs. The latter can be used in a more
conventional approach in which the MFT is constructed manually or results from a Hazard and

Operability Analysis. The MCMs can then be used to
generate mini-fault trees for selected high-level
events.

INSTRUMENTATION SYSTEMS CONSIDERED

This paper presents results of a study which was
performed in collaboration between JRC-Ispra (for
the methodological part) and SNAM PROGETTI (for
the instrumentation model part). The aim of the
study was to develop a library of general purpose
MCMs for instrumentation systems as found in
process plants. The instrumentation systems considered in the study were typical control loops for
the following process variables: level (L),
pressure (P), temperature (T) and flow (F).
The failure modes considered can be characterised
as "fails high" and "fails low", thus corresponding
to events such as "control loop fails causing high
temperature". The MCMs can be used to analyse such
events in its fine causes.

MODULAR COMPONENT MODELS (MCMs)

Before entering into the details of the instrumentation MCMs it is necessary to give some more ex-
planation on the MCM concept. In a detailed system
analysis, events such as "pump fails to start" or
"regulating valve fails", have normally to be
further decomposed into finer causes. One could say
that a mini-fault tree has to be constructed, for
example, for the regulating valve failure. The
mini-fault tree then would model the valve, its
supply, control, signalisation and possibly protection subsystems. The valve, including all subsystems related to its functioning is called
"macro-component".

The MCM concept is based on the observation that
- for different macro-components of the same type,
i.e. within a certain macro-component class - the
subsystem lay-out is relatively standard. Differences, if any, consist in the presence or not of
optional subsystems or in the fact that a small
number of alternatives may exist for some subsystems. For example, a regulating valve may or may
not have a protection subsystem, protecting, say,
against high flow; the valve logic may be of
"fails open" or "fails close" type; the protection
system may use an electrical or a pneumatical

transducer, etc. Small fault tree modules can be constructed for all possible subsystems. The mini-fault tree for a particular macro-component can then be assembled by linking the relevant modules according to the subsystems present in the macro-component. The MCM is used to perform this assembly is a computer-assisted way.

A MCM is needed for each macro-component class and for each failure mode considered in that class. The MCM contains:

- a questionnaire by which the user is prompted to give the options applicable to a real situation;
- a truth table which is used to identify the subsystem modules to be linked together for any combination of options;
- the modules for each subsystem;
- a small data base containing the reliability parameters of all primary events appearing in the MCM.

MCMs can be nested: i.e. a MCM can contain macro-events to be transformed by other MCMs. This allows one to build a MCM library with a hierarchical structure (from less to more detail).

INSTRUMENTATION MCMs

As already said, the instrumentation MCMs considered here concern control loops for L, P, T or F. Hence, the macro-events which can be expanded are:

{high/low} {level/pressure/temperature/flow}.

The first level causes of such a macro-event are, in the most general case, given by the mini-fault tree in Fig. 1. In a more specific case, some of the three indicated subsystems may or may not be present. Hence, MCMs can be constructed for the high-level causes.

The questionnaire for such a MCM could, for example, for "high level" be:

1   level regulation subsystem present? Yes(Y) or No(N)?

2   high-level signalisation subsystem present? Yes(Y) or No(N)?

3   high-level protection subsystem present? Yes(Y) or No(N)?

The corresponding truth table is:

Options	Structure function of mini-fault tree
YYY	$T = \alpha \cdot \beta \cdot \gamma$
YYN	$T = \alpha \cdot \beta$
YNY	$T = \alpha \cdot \gamma$
YNN	$T = \alpha$
NYY	$T = \beta \cdot \gamma$
NYN	$T = \beta$
NNY	$T = \gamma$
NNN	$T = \phi$

The events represented by the Greek letters are not primary events. They are to be developed further and, hence, are macro-events to be transformed by MCMs for the various high or low L, P, T, F regulation, signalisation and protection subsystems. This gives rise to 24 MCMs to be constructed.

The MCMs for regulation subsystems take into account:

- the type of regulation: standard, special;
- the type of actuator logic: fails open, fails closed, fails as is;
- the type of regulating valve: pneumatical, electrical;

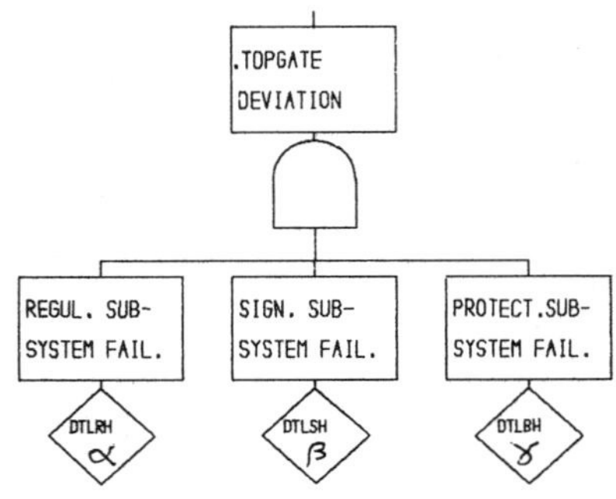

Fig. 1 Basic causes of deviation

- the type of signal transmission: pneumatical, electrical, pneumatical/electrical.

The MCMs for signalisation subsystems take into account:

- the type of signalisation: alarm, recorder, indicator;
- the type of measuring device: pneumatical, electrical, direct.

The MCMs for protection subsystems take into account:

- the possibility of manual protective action: time available or not;
- the type of sensoring device: pneumatical, electrical, direct;
- the type of actuating device: separate block valve, acting on regulating valve.

APPLICATION EXAMPLE

In Fig. 2 a simple example is given of a level control loop. The macro-event "high level" can first be expanded into a mini-fault tree by using the high-level MCM. Since all three regulation, signalisation and protection subsystems are present in this case, the resulting mini-fault tree is the same as in Fig. 1. This mini-fault tree is then further expanded by using the MCMs for level regulation, high-level signalisation and protection subsystems.

First, the actions for the regulation subsystem, such as type of regulation, actuator logic, etc., are entered and the regulation subsystem is expanded resulting in the tree shown in Fig. 3. After expansion of the signalisation and protection, the mini-fault tree is complete (see Fig. 4).

CONCLUSIONS

Since many control loops are present in process plant systems, the effort needed to set up the MCM library is paid off quickly. The interactive procedure eliminates repetitive work and reduces the effort of detailed analysis significantly. Furthermore, it decreases the likelihood of errors and provides a natural, engineering-oriented way of describing the control loops (questionnaires). It helps to obtain a constant level of quality throughout different analyses made by different people. The procedure can be used to expand system macro-fault trees as constructed by CAFTS. It can also be used directly to develop significant events as identified in an Operability Analysis.

REFERENCES

Lapp, S. and G. Powers (1977). Computer-aided syn-
    thesis of fault trees. *IEEE Transactions on
    Reliability*, Vol. R-26, 1.
Poucet, A. and P. De Meester (1981). Modular fault
    tree synthesis. *Reliability Engineering*, 2,
    65-76.
Poucet, A. (1983). Computer-aided fault tree syn-
    thesis. *Euratom report EUR-8707 EN*.
Poucet, A. (1985). CAFTS: Computer-aided fault tree
    analysis. *ANS/ENS Int. Top. Meet. on Probabi-
    listic Safety Methods and Applications*, San
    Francisco, February 24 - March 1, 1985,
    pp.115,1-115,10.
Taylor, J.R. (1982). An algorithm for fault-tree
    construction. *IEEE Transactions on Reliability*,
    Vol. R-31, 2.

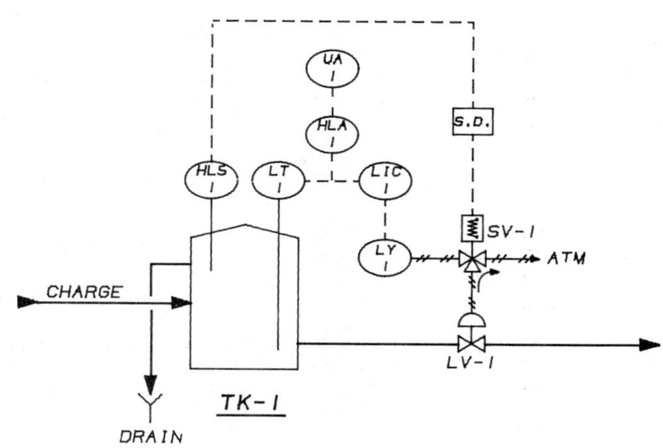

Fig. 2 Sample level control system

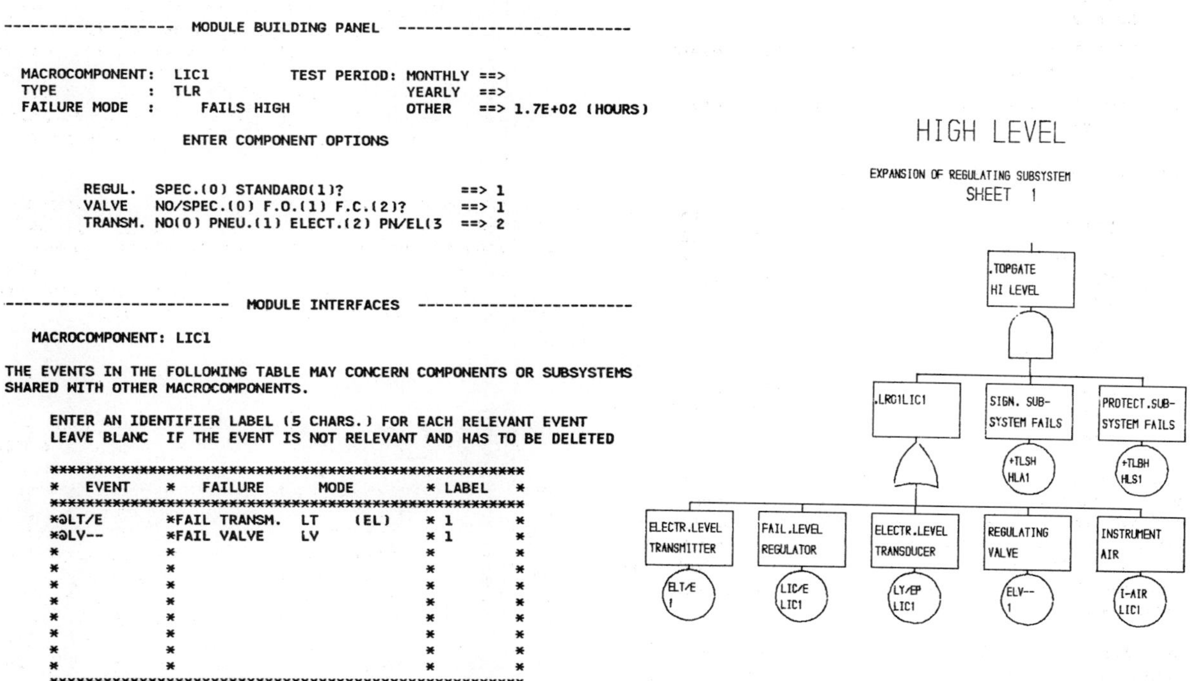

Fig. 3 Expansion of regulating subsystem: screens prompting for
information and resulting mini fault tree

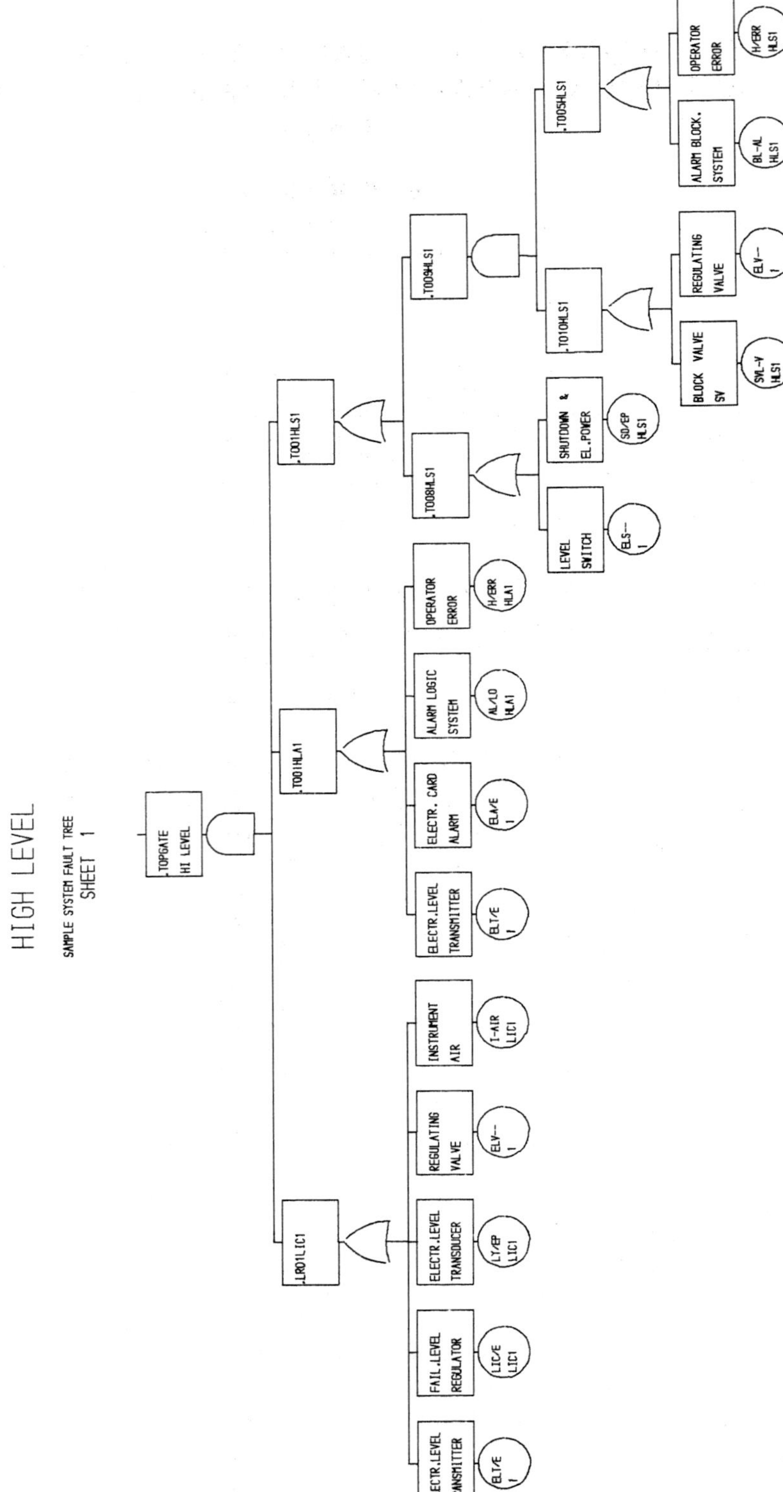

Fig. 4 Completely expanded fault tree

SENSITIVITY OF ANALYSIS OF RISK FROM CHEMICAL REACTOR EXPLOSION TO DATA USED

B. W. Robinson

Reliability Engineering Consultant, 573 London Road, Davenham, Northwich, Cheshire CW9 8LN, UK

Abstract.   This paper illustrates the various stages in analysing the development of the hazard from an explosion from the initiating event.   The analysis is broken down into seven stages from the identification of the probability of an explosion to the extent of the damage caused.   Each of these stages has uncertainties in the assumptions made and in the data used and it is not effective to aim at too high a degree of accuracy in one stage of the analysis if this cannot be matched in the others.

As an illustration brief consideration of the instrument system analysis demonstrates that

(a) the results are generally insensitive to most of the failure data used

(b) the analytical procedure identifies the instruments for which accurate data is required.

Keywords.   Reliability; sensitivity analysis; data; chemical reactor; explosion.

INTRODUCTION

Because of the range of disciplines involved in the overall examination of a potential hazard in a chemical process individual analysts are often only concerned with a small part of the problem. This can result in variations in the degree of detail included in the different parts of the analysis.   There is a tendency where the technology is well understood to develop the analysis in greater detail and in so doing to demand more and more sophisticated data on the equipment concerned.

This paper sets out seven stages (Fig. 1) into which the consideration of a typical reactor explosion hazard may be broken down.   Brief consideration is given to the problems and uncertainties at each stage in order that the balance of detail in the different stages of the analysis may be controlled.

The sensitivity of the analysis to particular data is illustrated by a simple analysis of the control and trip systems of a typical simplified reactor system.

STAGE 1 - RECOGNISE POTENTIAL EXPLOSION

Early in the life of a project when the process to be used to produce a particular product is being chosen several alternative processes may be considered.   In addition to the efficiency of the processes in producing the desired product the undesirable byproducts are also important.

The properties of the materials used in the reactions, formed as intermediaries or used as catalysts for example must be established under the range of operating conditions of interest.   From these considerations the most suitable process is chosen but this may well have potential hazards.

At this stage in the project when reliability engineers first become involved the process has been chosen.   It is essential to carry out a systematic examination of all of the reactants involved in the process and their possible hazardous combinations.

In general a fuel, an oxidant and an ignition source are required for an explosion.   At this stage we are looking for combinations of fuel and oxidant, sources of ignition will be considered later when more detailed design has been carried out.   Common fuels are gaseous ammonia, hydrogen and hydrocarbons such as ethylene.   The most common oxidant is of course air but oxygen is much more powerful and chlorine is not uncommon in the chemical industry.

Another type of explosion is the decomposition of unstable materials and such substances are used in their more controllable forms as explosives. Acetelyne is an extremely unstable gas widely used for welding but easily detonable in other circumstances.   Peroxides are found in many processes either as an intermediate or as a byproduct and accumulations can be dangerous especially when the plant is shut down and opened up for maintenance. Pyrophoric sulphur compounds may also accumulate as byproducts and be a hazard during shutdown or provide an ignition source during plant operation.

For an analysis to be comprehensive all hazardous combinations such as those referred to above must be identified.   Clearly some of the byproducts and occasional accumulations are difficult to predict.

STAGE 2 - MECHANISMS FOR ABNORMAL REACTOR CONDITIONS

In order that the potential hazards identified at Stage 1 develop there must be a mechanism for this to happen.   The materials required must meet at the appropriate pressures and temperatures in suitable physical conditions at the appropriate concentrations.

A reaction which may be used to illustrate several mechanisms is oxychlorination

$$2C_2H_4 + 4HCl + O_2 \quad = \quad 2C_2H_4Cl_2 + 2H_2O$$

The gaseous reactants, ethylene, hydrochloric acid and oxygen, are fed through a heated fluidised catalyst bed under flow ratio control.

1. HCl flow failure results in ethylene and oxygen mixture in the off gas.

2. C_2H_4 flow failure results in oxygen and EDC in off gas.

3. High oxygen effect as 2.

4. Low temperature - low reaction rates incomplete reaction therefore flammable exit gas.

The above type of deviations are relatively straightforward to identify but a systematic approach would also identify the more subtle side reactions which may of course take place in any part of the plant not just in the reactor.

STAGE 3 - GAS COMPOSITIONS

During normal running of the plant the gas compositions in the vapour spaces will be non-flammable. When a failure occurs such as described above the compostion begins to change. There are many factors which influence the rate of change and the final composition.

With simple reactions which take place rapidly there is little difficulty in predicting gas compositions. More complex reactions involving intermediate steps and solids and liquids have much more difficult reaction kinetics which would require considerable research for full understanding.

In the early days of project development considerable effort can be required to obtain sufficient understanding of the desired process. There is usually insufficient effort available to study in depth the reaction kinetics of deviations from normal some of which may be extremely hazardous.

It is clear therefore that efforts should be made early in the project to identify such hazardous deviations so that investigations and if necessary research, can be carried out to predict the gas compositions during potentially hazardous deviations.

The uncertainties in this stage could clearly be considerable.

STAGE 4 - EXPLOSIVE LIMITS

Combustion experts use Flammability Diagrams similar to Fig. 2 to describe the flammable limits of mixtures. Diagrams are available for many fuel gases and oxidising and inerting agents. Using this standard data diagrams can be constructed for mixtures. Apart from composition, pressure and temperature affect the shape of the diagrams.

If unusual gases or mixtures are encountered it may be necessary to carry our laboratory tests to identify the boundaries between the non-flammable, flammable and highly flammable zones. Tests may also be required to extrapolate existing data to higher, or lower, temperatures and pressures.

Clearly the boundaries between the three zones are not sharply defined and may more correctly be regarded as fuzzy lines.

Using these techniques Flammability Diagrams can be built up for the gas mixtures encountered in our reactor. The normal operating point of the reaction can be identified and the path of the changing gas composition following a deviation can be plotted. It can then be predicted whether the gas will become flammable or not.

STAGE 5 - FREQUENCY OF EXPLOSIVE MIXTURE

In order to illustrate this stage of the analysis in more detail a typical simplified chemical reactor Fig. 3 is considered.

The main feed reactant is flow controlled into the reactor where a constant flow of catalyst is added from a head tank. Loss of catalyst or high flow of feed results in a flammable off gas. The effect of loss of catalyst is gradual and can be detected by an analysis trip. High feed flow would cause too rapid a change for the analysis trip and so a high flow trip has been fitted.

The logic diagram for frequency of flammable off gas is shown in Fig. 4.

The two initiating events considered are loss of catalyst or flow control fails high. It can be seen from the logic diagram that the dominant equipment failure is the analysis trip and a factor of 3 change in this failure rate produces a factor of 2.5 change in the final frequency. A factor of 3 change in the flow trip or flow control failure rate changes the answer by only about 20% or a factor of 1.2. A change of 3 on the shutdown system including the valve changes the answer by only 50% which is perhaps surprising considering this common to both trips.

It is clear from the above that the data on the analysis trip is the most important and that the result is relatively insensitive to the other data.

STAGE 6 - PROBABILITY OF IGNITION

A considerable amount of data is available on energies of ignition required for particular substances. This is obtained from tests under well regulated conditions. Difficulties arise in predicting the energy levels of potential ignition sources in practical plant conditions.

It is often only possible to make a very crude estimate of the probability of ignition where the gas composition is expected to be say on the edge of the flammable region or well into the highly flammable. As a guide detonable mixtures take very little energy to set them off and for example weak ammonia air mixtures require a blow torch to cause them to burn with a soft blue flame.

The mechanisms and combination of circumstances leading to ignition is an area where considerable work needs to be done before a probability of ignition can with confidence be put on any but the simplest cases.

STAGE 7 - EXTENT OF DAMAGE

The final stage of the analysis is to consider the

damage or injuries caused by the explosion. For
an explosion in the gas space of a reactor the
external effects may vary from none at all for a
weak explosion to pieces of vessel hurled vast
distances for a large detonation.   More usually
weak attachments to the vessel would be blown off
or even just joints blown or small fittings
affected.

Assessment of either plant damage or the probability of injuries or fatalities is clearly far from
easy.

CONCLUSIONS

From consideration of the seven stages involved in
the complete analysis it can be seen that there are
significant problems at each stage which lead to
assumptions being made which cannot be wholly substantiated.   It is thus clearly not worthwhile to
over-elaborate the analysis in an area where the
knowledge is more complete because although this
may add a phoney air of authenticity to the analysis it will not have a real effect on the accuracy
of the prediction.

The more detailed consideration of the reactor
system demonstrated clearly that the result is
sensitive to only a small proportion of data.
Thus misplaced efforts to refine all of the data
used in an analysis are in general not cost
effective.

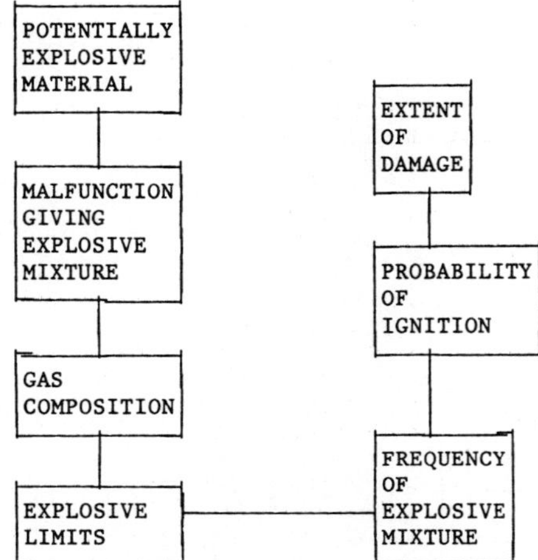

FIG. 1   SEVEN STAGES OF ANALYSIS

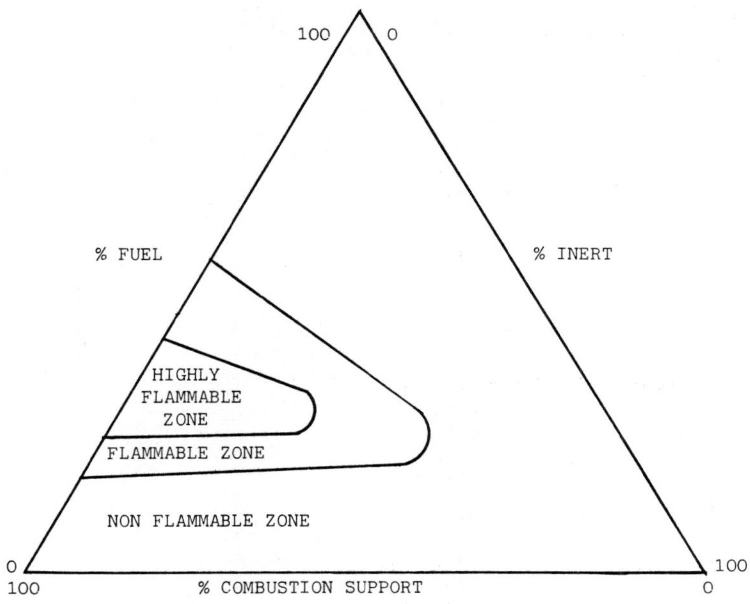

FIG. 2   TRIANGULAR FLAMMABILITY DIAGRAM

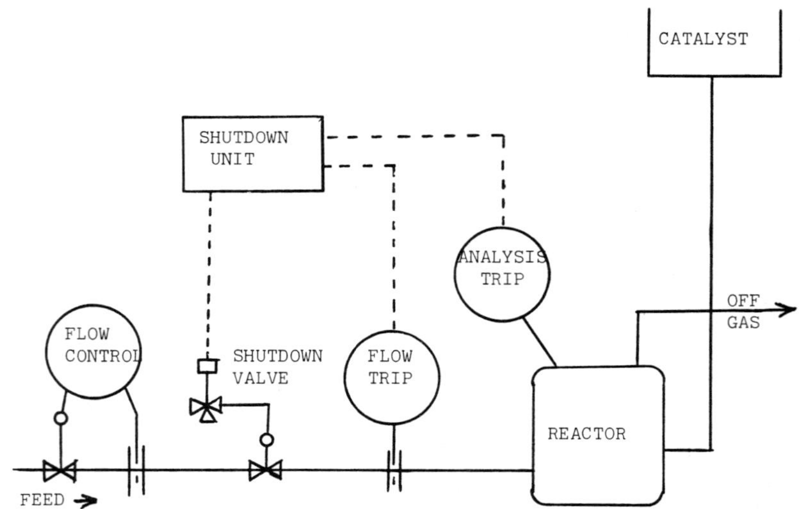

FIG. 3   PLANT DIAGRAM

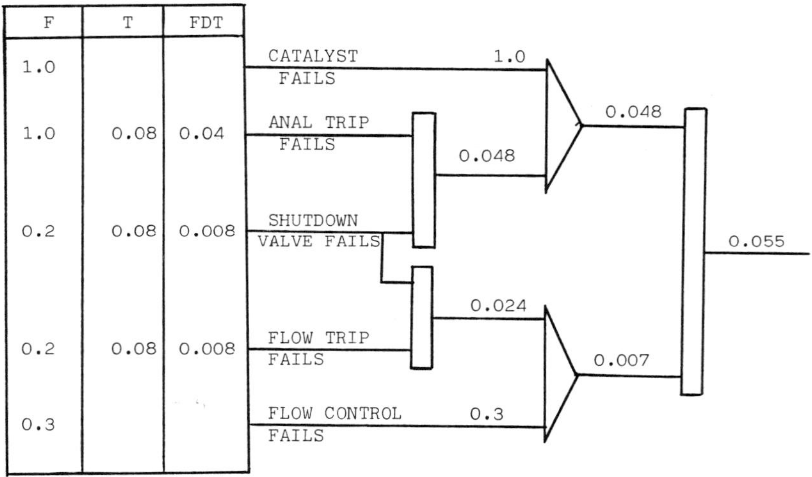

FIG. 4   LOGIC DIAGRAM

AN ALTERNATIVE TO PRESSURE SAFETY VALVES ON OFFSHORE PLATFORMS

P. Chamoux

Societe Nationale ELF Aquitaine (Production), 26 Avenue de Lilas, 64018 Pau, France
Present assignment: Elf Aquitaine Norge A/S, PO Box 168, 4001 Stavanger, Norway

Abstract
According to the usual practices in the oil industry, protection against overpressure
must be performed by two barriers. For gas fields the second barrier must be pressure
safety valves connected to a full-flow flare. On offshore installations, for large gas
flows, this flare can necessitate the erection of an additional platform.

On Frigg Field, an alternative system so-called OPPS (Over Pressure Protection System)
is being installed in order to cope with new conditions of production, the existing
flare platform being unable to handle additional third party gas flows pipelined from
other fields.

This OPPS has been designed through extensive reliability studies in order to meet a
higher level of safety than the conventional API 14 C based system.

Keywords
Control engineering computer applications, human factors, Oil technology, Reliability
theory, Safety systems.

INTRODUCTION

The Frigg Field, operated by ELF Aquitaine
Norge, located on the border line between
British and Norwegian waters, East of the
Shetlands. The gas produced from the two
drilling Platforms (CDP1 and DP2) is processed
and recompressed on two treatment platforms (TP1
and TCP2). This gas is then pipelined to Scotland
(St. Fergus) through a transportation system. This
system is also used for transportation of third
party gases coming from neighbouring gas fields
tied in to Frigg (fig. 1).

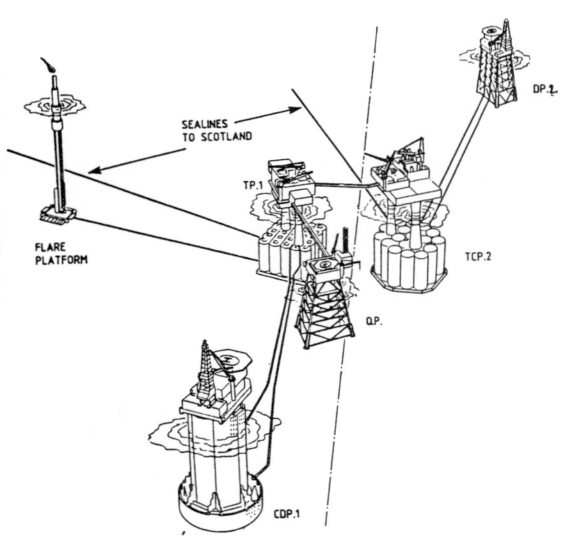

Fig. 1. General layout of Frigg Field

CONVENTIONAL PROTECTION AGAINST OVERPRESSURES

As regards the process and exportation equipment
on the treatment platforms, the maximum allowable
working pressures (MAWP) are:

- Process equipment   : 172 barg
- Exportation equipment
  (sales gas headers and : 153 barg

The static well head pressure being below 140 barg
no overpressure can originate from the Frigg
reservoir.

However, in case of closure of the ESDV located at
the outlet of the sales gas headers or in case of
obstruction of a sealine (hydrate plug or stuck
pig), the export turbo compressors can raise the
pressure in a header above the MAWP.

The Frigg platforms must comply with both
British and Norwegian regulations. In both
cases, the applicable standard is API R.P. 14 C
"Analysis, design, installation and testing of
basic surface safety systems on offshore
production platforms". In British waters, this is
a recommended practice but in Norwegian waters,
the safety devices must be selected according to
API 14 C.

The API 14 C states that the protection of the
headers against overpressure should be achieved
with 2 levels of protection:

- A primary protection provided by a Pressure
  Switch High (PSH) sensor able to shut off all
  input sources to the headers.

- A secondary protection provided by a pressure
  safety valve (PSV) linked to a flare system.

The more likely possibility of overpressure being
the accidental closure of the ESDV, primary
protection is actually performed by a valve.

position switch (ZS) in parallel with a PSH
able to trip the turbine driving the compressor
(fig. 2)

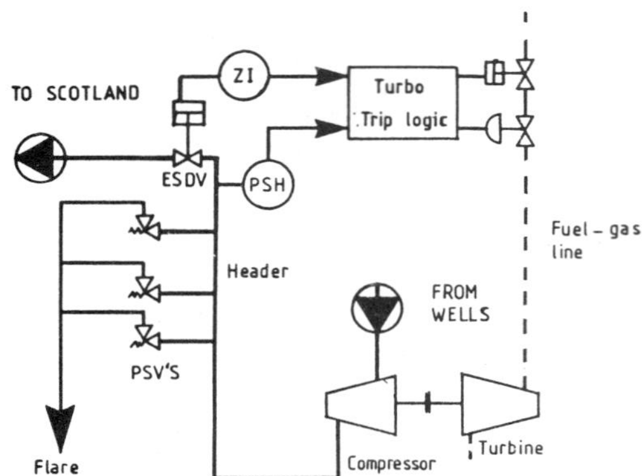

Fig. 2. Sketch of existing conventional system.

The relief system (PSV's) is able to handle the
full flowrate which is considerable (40 Million
of Standard cubic meters per day). This flow is
diverted to an articulated flare platform
installed 500 m away from the main platforms
(fig. 1).

NEW CONDITIONS OF OPERATION

So far, adjacent fields tied in to Frigg could
not cause an overpressure in the installation
since their reservoir pressure is lower than the
MAWP.

However, the gas coming from the North Alwyn
field in the British sector, for export to the UK
via the Frigg Pipeline system, is first compressed
on the N.A. Platform and could cause an
overpressure on Frigg.

Moreover, the arrival temperature of this gas
being low (5° C), in the event of flaring, the
temperature of the flare network is expected to
be around -80° C which exceeds the capabilities
of the existing flare (- 46° C).

The erection of a new flare platform has been
investigated and due to the severe environmental
conditions would require a 50 M $ worth structure,
able to withstand 120 knot winds and 30 meter high
waves in 100 meter water depth.

So, it has been decided to work out an
alternative system not making use of PSV's and
thus not needing a new flare platform.

BASIC PRINCIPLES FOR AN ALTERNATIVE SYSTEM

Since a dispensation was needed, it was decided
to design an alternative secondary protection
such that the reliability of the whole system
can be documented as greater than that of an API
14 C based system.

Moreover, the Norwegian regulatory authorities
recommend that the probability of occurence of a
socalled "excluded situation" should not exceed
10^{-4}/year.

So, it was decided to adopt this criterion for
the acceptance of a new system (Hazard rate
$<10^{-4}$/year).
In addition special attention should be paid to
the following points which are not easily
quantified in reliability assessments:

- Reduction of human interferences
- Avoidance of common mode failures
- Comprehensive testability.

The hazard rate HR is defined by:

    HR = FDT x r

Given,

FDT : Fractional dead time or probability of
missing a shut down upon demand
r   : rate of occurence of a dangerous situation
(demand rate)

The FDT is strongly dependant on the period
between tests particularly for redundant systems
Roughly it is proportional to the test interval
to the power of the order of redundancy (see
appendix 1).

So, as regards the secondary protection, with
conventional overpressure detection devices, the
interval between tests would be extremely short
in order to meet our criteria.

Such a frequency would lead to unacceptable
conditions of operation and numerous sources of .
human errors. Moreover, it would be difficult
to perform real proof tests especially as
regards impulse lines which can be easily
plugged by hydrates (solid gas/water compound).

So a solution has been evolved with the
following features:

- Automatic proof testing with short test
  intervals (daily) thus avoiding human errors.
- No removal or isolation of sensors
  during testing so as to keep the system fully
  available without production stoppages.

This has been achieved by:

 - the use of analog sensors (Pressure
transmitters) in 2 of 3 redundancy instead of
logic sensors (Pressure switches), with
continuous monitoring of signals;

- the periodic modulation of header pressure
  by overriding the pressure regulation at the
  outlet of compressors.

Continuous monitoring and comparison of the three
signals allows the elimination of problems
related to open or short circuits in wiring or
connections and of faults connected with drifting

Pressure modulation allows comprehensive checking
of sensors to be performed (see fig.5, including
impulse lines, without manual operation.

Moreover, the system is fully available during
testing. In fact, if an overpressure occurs, the
system is able to detect it and to trip the
compressors.

Obviously, this kind of device requires a logic
and control, which are more sophisticated than
relay based conventional systems, in order to
perform the voting, the comparison analysis and
the pressure modulation.

Industrial numerical systems such as Programmable
Controllers (P.L.C or P.C.) are able to perform
these tasks and appear to be suitable provided
that their implementation is cautious.

In previous studies, the following have been
assessed for various generic configurations of

P.L.C.'s.

- the probability of no shut down upon demand
  (FDT or unavailability).

- the spurious shut down rate.

The result of these calculations has been to
stress the importance of the following points:

- Good quality monitoring of input/output cards
  considerably reduces both the unavailability
  and the spurious shut down rate;

- Duplicating the power supply and C.P.U/memories
  avoids too numerous spurious shutdowns and
  enables operations to continue during a repair.

The more satisfactory system makes use of two
PLC's with unusual connection of output cards
(see fig. 3).

Each output is electrically supplied by a dual
bus which would require double failure to make
the system unavailable or to cause a spurious
shut down.

For each card, a shut-down is initiated by
opening both the ESDPC relay and the relevant
output(s).

Any spurious opening of these contacts is
automatically detected by the auxiliary bus
current monitoring.

Periodically, output contacts and ESDPC relays
are opened (not simultaneously of course) in
order to check short-circuited faults and stuck
contacts. A failure is, as before, detected by a
change in auxiliary bus current monitoring.

As regards the safety (no shut-down upon request)
the P.L.C's are in a 1 out of 2 configuration:

i.e. one P.L.C. only is necessary to initiate a
shut-down.

If a P.L.C. is detected to be faulty or if its
power supply is lost, the system is automatically
reconfigured in 1 out of 1 without shut-down via
RUN PC relays.

DESCRIPTION AND OPERATION OF THE ALTERNATIVE SYSTEM

A schematic of the system is given in fig. 4
Actually there are several headers to be
protected and several turbo compressors to be
tripped, which leads to numerous sensors and
actuators.

Obviously the total system is in principle "fail-
safe".

1. Primary Protection

The detection channel of the first barrier is, as
previously, composed of a limit switch on the
ESDV at the inlet of the sea line and one PSH
located on the header.

In case of alarm, a signal is sent to the trip
logic of the turbine which causes the closure of
the modulating and shut-off valves which are
located on the fuel gas line. Moreover, an
ESDV is closed on Alwyn gas line.

2. Secondary Protection

Three PT's are linked to three different input
cards on each of both PLC's. If 2 PT's out of 3
are in alarm the following actions are
initiated.

- Closure of two pneumatically operated valves
  (different types and separate air supplies),
  located on the fuel gas line of the turbine.

- Closure of two ESDV's on Alwyn gas line through
  separate output cards.

- Via a quadruple diversity tropospheric radio
  link a signal is automatically sent by the Frigg,
  PLC's to a similar system installed on Alwyn in
  order to stop the Alwyn export compressors.

Should the radio link be lost, the set point of
the Alwyn compressors trip logic is automatically
reduced to such a level so that no overpressure is
possible on Frigg.

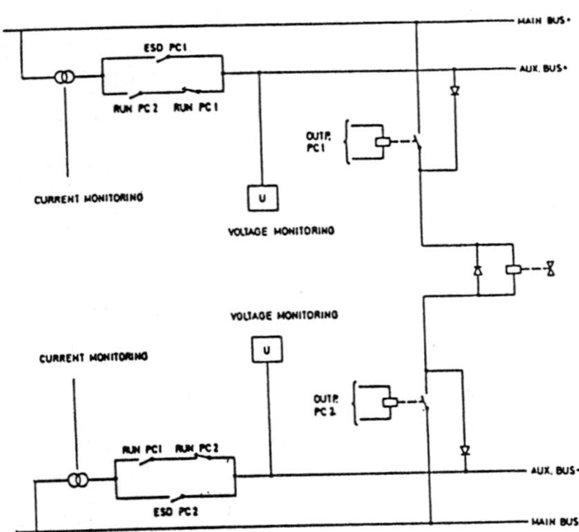

Fig. 3. Sketch of the connection of PLC's
Output cards.

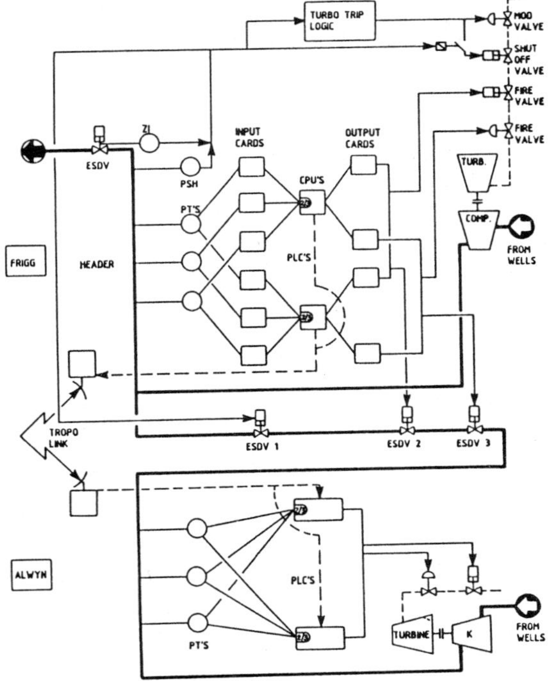

Fig. 4. Sketch of Alwyn/Frigg overpressure
protection system.

3. System Testing and Monitoring

The testing of the whole system is almost fully automatic. However, in order to keep the operators involved and to avoid a blind trust in the "machine", some test operations are kept manual, but under supervision of a third PLC specially dedicated to testing and monitoring. During testing, this PLC checks all the operations and when the test is over, supplies a detailed list of the faults, if any. It is to be noted that this PLC logs all the signals and the status of all the components of the system (sensors, relays, actuators and so on). Moreover it performs the daily pressure modulation in the headers.

The primary protection barrier is tested monthly.

The secondary protection barrier is tested daily except for the actuators, for obvious reasons.

Daily, the Testing and Monitoring PLC overrides the discharge pressure regulation of the turbo compressors in order to generate a pressure variation in the headers. This is to be detected by the PLC's. (cf. fig. 5)

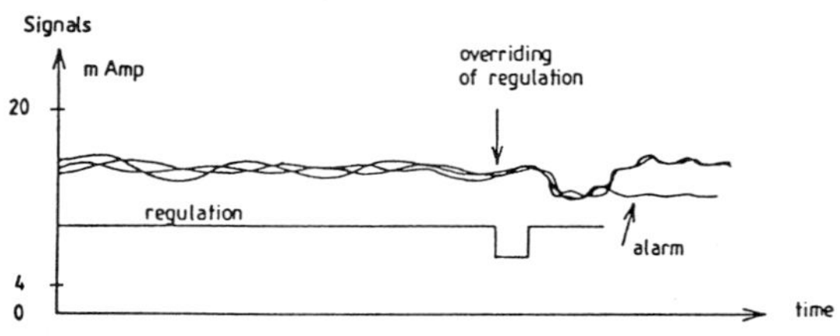

Fig. 5. Daily proof test of detection channel by pressure modulation.

Moreover, the input and output cards of the PLC's are checked by built-in auto-test; the first ones by feeding inputs by a increasing signal and the second ones by sequential opening of the outputs.

An example of manually performed test under PLC's supervision is shown in fig. 6. The PLC logs the setpoint of the PSH and checks that the operator has not left the isolation valve closed.

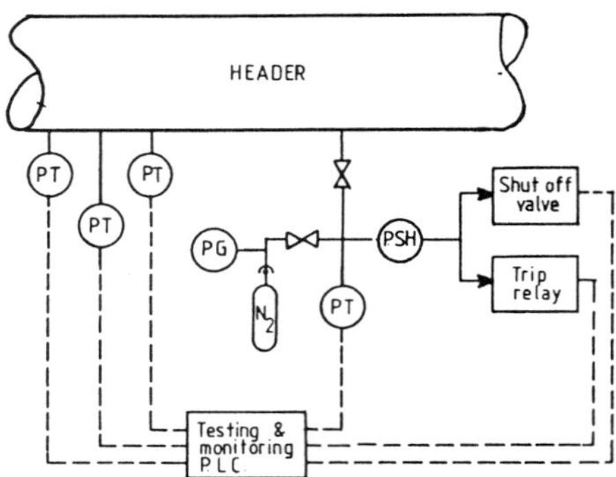

Fig. 6. Example of manual test under PLC's supervision.

It is to be noted that a third PLC is used for testing and monitoring in order to keep the task of the safety PLC's as simple as possible (2/3 voting and threshold) and thus avoid software bugs.

RELIABILITY ASSESSMENT

The hazard rates have been assessed for the two following events :

- the accidental closure of an outlet ESDV (estimated frequency 18/year)
- the plugging of a sea-line which is a rare occurence (estimated frequency 5 . 10^{-2}/year).

The related hazard rates and their 90 % confidence intervals are shown in fig. 7 for both the conventional API based system and for the alternative system.

The large confidence intervals are mainly due to the uncertainties in quantifying common mode failures. (See App. 2).

The fairly low figures related to the "plugging of a sea line" are explained by the very low probability of occurence of such an event.

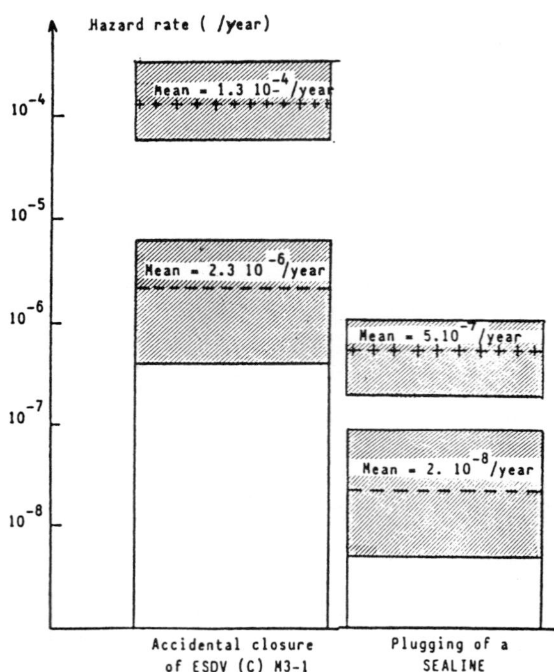

Fig. 7. Hazard rate (Hazard/year) For the conventionnal (+++) and alternative (---) systems (90 % confidence intervals)

It is to be noted that the possibility of human errors which are generally considered as very important in high safety systems has been carefully studied:

- No man-machine interface
- Continous monitoring of each component
- Inhibition possible only under PLC's supervision preventing a start-up under ab-normal conditions.

CONCLUSION

Such high reliability has been achieved by using in addition to short test intervals various diverse devices such as:

- Three different detection channels: limit
  switch, pressure switch, pressure transmitters
  in 2 out of 3.

- Three different signal processing units: a
  relay based electromechanical one and two
  extensively self tested programmable
  controllers.

- Four actuators to shut-off both gas flows.
  As regards the Frigg compressor fuel gas : 2
  gas operated aerospace type valves, 2 diverse
  air operated valves.
  As regards the Alwyn gas: 3 local hydraulically
  operated valves and a remote shut-off of Alwyn
  compressors through a tropospheric radio
  link.

These Reliability studies have been used to
document the applications for dispensation from
the regulatory and certifying authorities.
Moreover, at the end of the detailed engineering
phase this study has been brought up to date in
order to detect eventual deviations, but led to
similar results.

The British and Norwegian authorities gave a
conditional agreement for a period of
probation with both systems in operation in
order to validate by field data the figures used
in the reliability modelling.

REFERENCES

- GONDRAN M, Pages A (1980). _Fiabilite des
  systemes_, Eyrolles Paris, 323 P.

- CHAMOUX P. SCHMID O. (1983). PLC's into offshore shutdown systems. _IFAC Safecomp 83._
  Cambridge UK.

APPENDIX 1 - EXAMPLE OF FDT (UNAVAILABILITY)
CALCULATION FOR A DORMANT SYSTEM PERIODICALLY
TESTED:

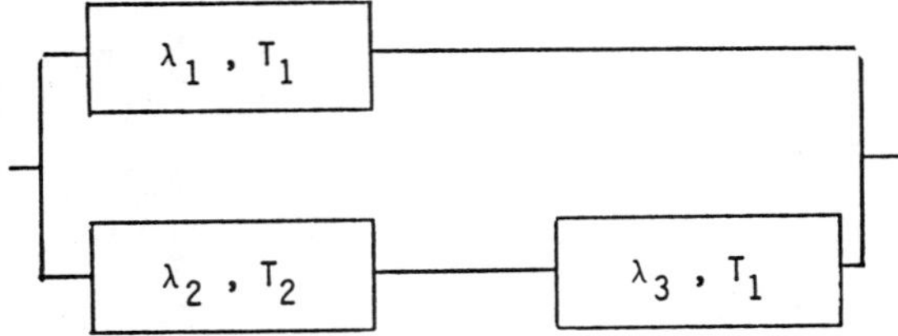

Fig. 8. Example of reliability block diagram.

T1: Interval between tests of block 1 and 3.

T2: Interval between tests of block 2.

$\lambda_1, \lambda_2, \lambda_3$: failure rates

T1   T2

$r = \dfrac{T1}{T2} - 1$, integer

The resulting overage unavailability will be :

$$\bar{A} = \frac{1}{T_1} \sum_{k=0}^{k=r} \int_{k T_2}^{(k+1) T_2} (\lambda_2(t - k T_2) + \lambda_3 t) \times \lambda_1 t \; dt$$

$$= \frac{1}{T_1} \sum_{k=0}^{k=r} \int_{k T_2}^{(k+1) T_2} \left((\lambda_1 \lambda_2 + \lambda_1 \lambda_3) \cdot t^2 - \lambda_1 \lambda_2 \; k T_2 \cdot t \right) dt$$

$$= \frac{1}{T_1} \sum_{k=0}^{k=r} \left[(\lambda_1 \lambda_2 + \lambda_1 \lambda_3) \, t^3/3 \right]_{k T_2}^{(k+1) T_2} - \left[\lambda_1 \lambda_2 \cdot k T_2 \cdot t^2/2 \right]_{k T_2}^{(k+1) T_2}$$

$$\bar{A} = T_2^2 \left(\frac{(T_1 - T_2)}{2 T_2} \cdot \left(2 \lambda_1 \lambda_3 \left(\frac{T_1 + T_2}{3 T_2} \right) + \frac{\lambda_1 \lambda_2}{2} \right) + \lambda_1 \left(\frac{\lambda_2 + \lambda_3}{3} \right) \right)$$

if $T_1 = T_2$

$$\bar{A} = \frac{1}{3} \cdot \lambda_1 \cdot (\lambda_2 + \lambda_3) \, T_1^2$$

APPENDIX 2 - COMMON MODE FAILURES

1. Definition
Often, redundant components of a system are not
stochastically independant.

Several type of so called "common mode failures"
(C.M.F.) can occur.

- Events or failures which cause the simultaneous
  failure of a set of components or which affect
  the probability of failure of some components.

- Statistical dependencies.

2. Quantitative processing

If redundant components have the same function
and technology, it is considered that the common
mode failure probability is in the range :

- at most, the probability of a single failure
  (upper bound).

- at least, the probability of the system failure
  (lower bound).

These bounds are considered as a 99.9 % confidence interval.

So, assuming a log normal distribution, we can
compute the mean and the 90 % error factor.

OPERATIONAL READINESS OF SAFETY SYSTEMS

G. W. E. Nieuwhof

Formerly, Atomic Energy of Canada Limited, Mississauga, Ontario, Canada (retired)
Present address: 25 McGilvray Crescent, Georgetown, Ontario, Canada L7G 1L7

__Abstract__. The Operational Readiness concept differs from the conventional Availability concept and must be modelled differently mathematically. Operational Readiness analyses are appropriate for systems which carry out missions as the demand for their operation arrives. These systems are normally in the dormant state. They operate when they are called upon to do so. Typical examples are emergency safety systems.

The conventional Availability concept applies only to systems which are subject to a continuous demand to operate and which only stop to carry out preventive or corrective maintenance actions. Typical examples are electric power generating systems, pumping systems, etc.

The Operational Readiness of a system is the probability that, at any time, the system is either operating satisfactorily or ready to operate on demand, when used under stated conditions. The total calendar time is the basis for computational Operational Readiness.

This paper develops a mathematical Operational Readiness model for safety systems which are regularly tested (and repaired if necessary) during its dormant state.

__Keywords__. Operational Readiness analysis; dormant safety systems; operation-on-demand systems; regularly tested drmant stand-by systems; reliability analysis.

INTRODUCTION

The Operational Readiness concept differs from the conventional Availability concept and must be modelled differently mathematically.

Operational Readiness analyses are appropriate for systems which carry out missions as the demand for operation arrives.

The Availability concept applies only to systems required to operate continuously, 'round-the-clock', such as electric generating equipment in power stations, which are either working (up) or being worked on (down) because of failure; that is, systems which are subject to a continuous demand to operate and which only stop to carry out preventive or corrective maintenance actions.

Definition

The Operational Readines of a system is the probability that, at any time, the system is either operating satisfactorily or ready to operate on demand when used under stated conditions.

The total calendar time is the basis for computational Operational Readiness analyses.

Typical exaples are:

- emergency diesel-electric generating systems;
- nuclear reactor shut-down systems;
- etc.

The mathematical model of Operational Readiness developed in this paper is based on a slightly modified version of Operational Readiness Model C of Igor Bazovsky, Sr. Reference 1),which has been extended to incorporate the regular testing of the system during its dormant period.

MATHEMATICAL MODEL OF OPERATIONAL READINESS WITHOUT REGULAR TESTING

The following assumptions have been made:

1. The mission duration time, t, is distributed with a probability density function $q(t)$;

2. Failures may develop during dormancy, i.e., after system checks out O.K. and before the next call arrives.

Operational Readiness, P_{OR}, may be expressed as

$$P_{OR} = R \cdot R_D + Q \cdot P_D \quad , \qquad (1)$$

where

R = the probability that the system returns from a mission without failure;

R_D = the probability that no dormant failures develop from the time the system returns in good condition and not requiring repair up to the next call arrival;

Q = the probability that the system returns in a state requiring repair, when the mission duration time is distributed according to $q(t)$;

P_D = the probability that the system, if requiring repair, will be fixed by the time $t_m < t_c$ and no dormant failures develop in the remaining time $t_c - t_m$.

t_c = call time;

t_m = time at which the system is repaired.

If we assume that all distributions are exponential with the following means:

M_1 = mean repair time of the system;

M_2 = mean time to next call arrival;

M_3 = mean time to failure occurrence during mission;

M_4 = mean mission duration time;

M_5 = mean time to dormancy failure;

then

$$P_{OR} = \frac{M_5}{M_2+M_5} \cdot \left[1 - \frac{M_4}{M_3+M_4} \cdot \frac{M_1}{M_1+M_2} \right] , \qquad (2)$$

For the derivation of (2) see Appendix I.

Numerical Example

Suppose we have an automatic starting emergency diesel-electric generating system (including sensing instrumentation, automatic starting controls and automatic voltage and frequency controls) as a stand-by electric power supply with the following failure data:

M_1 =   5 hours

M_2 = 40000   "

M_3 =   50   "   The system is not test-
                                    ed during its dormant
M_4 =   100   "   period.

M_5 =   8000   "

Using equation (2) we will get for the Operational Readiness of the system

$$P_{OR} = \frac{8000}{40000+8000} \cdot \left[1 - \frac{100}{50+100} \cdot \frac{5}{5+40000} \right] = 0.166$$

This result is not very good.

Suppose we have two of these 100% systems in parallel on stand-by, then

$$P_{OR} = 1 - (1-0.166).(1-0.166) = 0.304$$

This is an improvement, but is still not very good.

The Effect of the Average Call Time, M_2, on P_{OR} of the Single System of the Example

Intuitively we feel that when the Average Call Time, M_2, is shorter, the Operational Readiness, P_{OR}, will have a higher value, because the system will have less time to deteriorate.

To investigate this we have calculated the value of P_{OR} for various values of M_2 leaving all the other values unchanged. The calculation results are summarized in Table 1 and plotted in Figure 1.

TABLE 1 The Value of P_{OR} vs. Various Values of M_2

M_1	M_2	M_3	M_4	M_5	P_{OR}
5	5000	50	100	8000	0.615
5	10000	50	100	8000	0.444
5	15000	50	100	8000	0.348
5	20000	50	100	8000	0.286
5	25000	50	100	8000	0.242
5	30000	50	100	8000	0.211
5	35000	50	100	8000	0.186
5	40000	50	100	8000	0.166
5	45000	50	100	8000	0.151
5	50000	50	100	8000	0.138

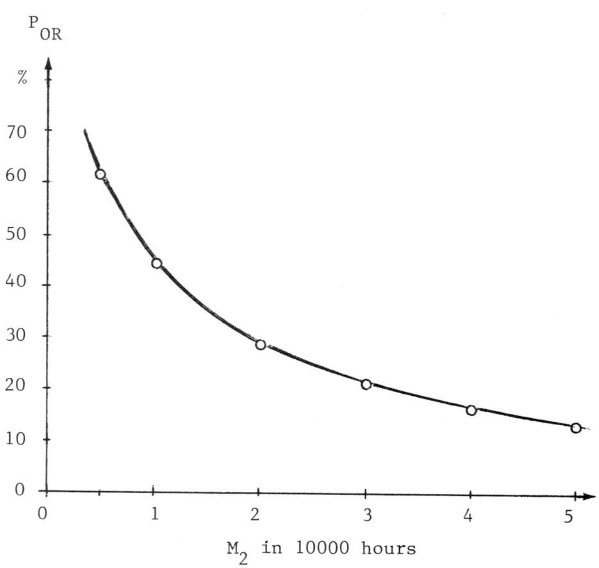

Fig. 1. The Effect of the Average Call Time (M_2) on the Operational Readiness (P_{OR}) of the System of the Example.

The Effect of Regular Testing of the System on the Operational Readiness

Suppose we test the system of the example at regular time intervals, say, after every T hours.

In Fig. 2 are illustrated:

- the time function of the probability that the system does not fail during its dormant period when tested after every T hours, $R_D(t_c)$;

- the time function of the probability of the call arrival, $g(t_c)$.

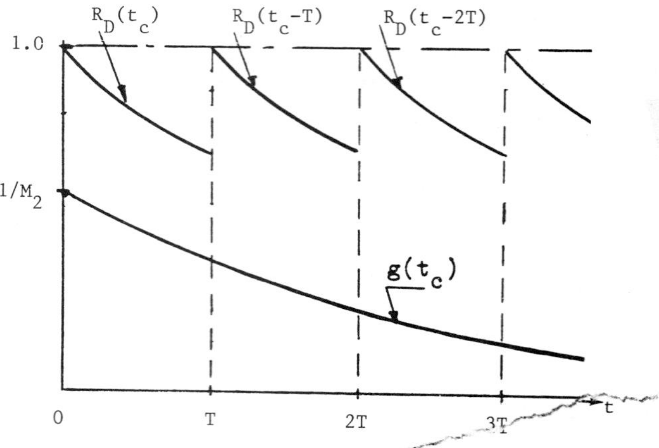

Fig. 2 $R_D(t_c)$ and $g(t_c)$ when the System is tested regularly during its dormant period.

$$R_D(t_c) = e^{-t_c/M_5} \quad \text{and} \quad g(t_c) = (1/M_2).e^{-t_c/M_2} , \qquad \ldots (3)$$

Let us consider the Nth interval, i.e., the period from NT to (N+1)T, where T is the length of the time interval.

The expected probability that the system does not fail during the interval NT to (N+1)T is

$$R_D(\text{expected})\Big|_{N}^{N+1} = \int_{NT}^{(N+1)T} g(t_c) \cdot R_D(t_c - NT)\, dt_c , \qquad (4)$$

$$= \int_{NT}^{(N+1)T} (1/M_2) e^{-t_c/M_2} \cdot e^{-(t_c-NT)/M_5}\, dt_c$$

Working this out will give (see Appendix II)

$$R_D(\text{expected})\Big|_{n}^{N+1} = \frac{M_5}{M_2+M_5} \cdot \left[1 - e^{-(1/M_2+1/M_5)T}\right] e^{-NT/M_2} \qquad \ldots(5)$$

Hence, the probability that no dormant failure develops from the time the system returns in good condition and not requiring repair up to the next call arrival, while the system is tested after every T hours, is

$$R_D' = \sum_{N=0}^{\infty} R_D(\text{expected})\Big|_{N}^{N+1} , \qquad (6)$$

hence

$$R_D' = \sum_{N=0}^{\infty} \frac{M_5}{M_2+M_5} \cdot \left[1 - e^{-(1/M_2+1/M_5)T}\right] e^{-NT/M_2}$$

Working this out will give (see Appendix II)

$$R_D' = \frac{M_5}{M_2+M_5} \cdot \left[\frac{1 - e^{-(1/M_2+1/M_5)T}}{1 - e^{-(1/M_2)T}}\right] , \qquad (7)$$

If we test the system of our example every 1000 hours, i.e., T=1000 hours, when in the dormant state, then for

M_5 = 8000 hours;
M_2 = 40000 hours,

$$R_D' = \frac{8000}{40000+8000} \cdot \left[\frac{1 - e^{-(1/8000+1/40000)\cdot 1000}}{1 - e^{-(1/40000)\cdot 1000}}\right]$$

$$R_D' = 0.9403$$

This is quite an improvement over the R_D when no testing is carried out, namely

$$R_D = \frac{M_5}{M_2+M_5} = 0.166$$

If we use R_D' for the value of R_D in the Operational Readiness equation (1), then

$$P_{OR}' = R \cdot R_D' + Q \cdot P_D' , \qquad (8)$$

where (see Appendix I)

$$P_D' = \int_{t_m=0}^{\infty} f(t_m) \left[\int_{t_c=t_m}^{\infty} g(t_c) \cdot R_D'\, dt_c\right] dt_m$$

$$= R_D' \cdot \int_{t_m=0}^{\infty} f(t_m) \left[\int_{t_c=t_m}^{\infty} (1/M_2) \cdot e^{-(1/M_2)t_c}\, dt_c\right] dt_m$$

$$= R_D' \cdot \int_{t_m=0}^{\infty} f(t_m) \left[(1/M_2) \cdot \left| -M_2 e^{-(1/M_2)t_c}\right|_{t_m}^{\infty}\right] dt_m$$

$$= R_D' \cdot \int_{t_m=0}^{\infty} f(t_m) \left[(1/M_2) \cdot M_2 e^{-(1/M_2)t_m}\right] dt_m$$

$$P_D' = R_D' \cdot \int_{t_m=0}^{\infty} (1/M_1) e^{-(1/M_1)t_m} \cdot e^{-(1/M_2)t_m}\, dt_m$$

$$= R_D' \cdot (1/M_1) \cdot \int_{t_m=0}^{\infty} e^{-(1/M_1+1/M_2)t_m}\, dt_m$$

$$= R_D' \cdot (1/M_1) \cdot \left| -\frac{M_1 M_2}{M_1+M_2} e^{-(1/M_1+1/M_2)t_m}\right|_{t_m=0}^{\infty}$$

$$= R_D' \cdot (1/M_1) \cdot \frac{M_1 M_2}{M_1+M_2}$$

$$P_D' = \frac{M_2}{M_1+M_2} \cdot R_D' , \qquad (9)$$

Substitution of this expression in equation (8) will give

$$P_{OR}' = R \cdot R_D' + Q \cdot \frac{M_2}{M_1+M_2} \cdot R_D'$$

$$= \left[R + Q \cdot \frac{M_2}{M_1+M_2}\right] \cdot R_D'$$

$$= \left[\frac{M_3}{M_3+M_4} + \frac{M_4}{M_3+M_4} \cdot \frac{M_2}{M_1+M_2}\right] \cdot R_D'$$

$$= \left[1 - \frac{M_4}{M_3+M_4} + \frac{M_4}{M_3+M_4} \cdot \frac{M_2}{M_1+M_2}\right] \cdot R_D'$$

$$= \left[1 - \frac{M_4}{M_3+M_4} \cdot \left\{1 - \frac{M_2}{M_1+M_2}\right\}\right] \cdot R_D'$$

$$= \left[1 - \frac{M_4}{M_3+M_4} \cdot \frac{M_1}{M_1+M_2}\right] \cdot R_D'$$

$$P_{OR}' = \left[1 - \frac{M_4}{M_3+M_4} \cdot \frac{M_1}{M_1+M_2}\right] \cdot \frac{M_5}{M_2+M_5} \left[\frac{1 - e^{-(1/M_2+1/M_5)T}}{1 - e^{-(1/M_2)T}}\right] , \qquad \ldots(10)$$

Applying this to our example (single system) with regular testing after every 1000 hours during its dormant period, will give

$$P_{OR}' = \left[1 - \frac{100}{50+100} \times \frac{5}{5+40000}\right] \times 0.9403$$

$$P_{OR}' = 0.9402$$

DISCUSSION OF THE OBTAINED RESULTS

If we compare the two equations:

(2) No testing during dormant period; and

(10) Regular testing during dormant period,

we notice that these equations differ only by one factor, namely:

$$L = \left[\frac{1 - e^{-(1/M_2+1/M_5)T}}{1 - e^{-(1/M_2)T}}\right] , \qquad (11)$$

L can be considered the Improvement Factor of Operational Readiness through regular testing of the system.

L is a function of T (= the regular time interval between tests), which in general can be varied easily. The values of M_2 and M_5 are more or less fixed (inherent to the system and its operation).

In our example M_2=40000 and M_5=8000 hours. The value of L has been calculated for various values of T. The results of these calculations have been plotted in figure 3.

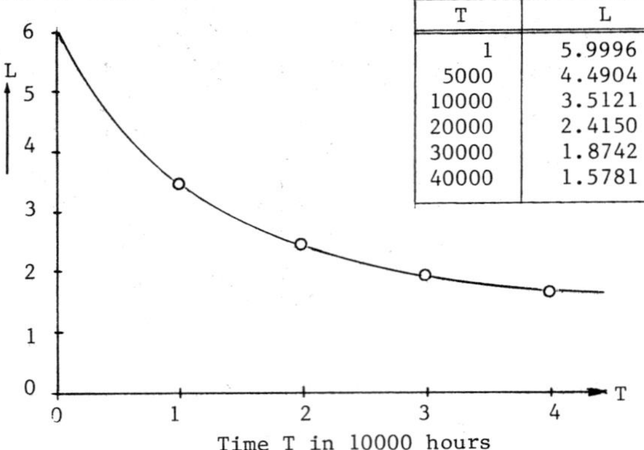

T	L
1	5.9996
5000	4.4904
10000	3.5121
20000	2.4150
30000	1.8742
40000	1.5781

Time T in 10000 hours

Fig. 3. L as a function of T for the example.

Note that T=0 hours has no physical meaning, however, if T approaches 0, then we have an interesting case for

$$L' = \left[\frac{1 - e^{-(1/M_2 + 1/M_5)T'}}{1 - e^{-(1/M_2)T'}} \right] , \text{ where } T' \text{ is very small.} \quad (12)$$

We know that $\quad 1 - e^{-\lambda t'} \approx \lambda t' , \quad (13)$

$$\text{if } \lambda t' \text{ is very small.}$$

Applying this in the above expression will give

$$L' \approx \frac{(1/M_2 + 1/M_5)T'}{(1/M_2)T'}$$

$$L' \approx \frac{M_2 + M_5}{M_5} , \quad \text{for } T' \text{ is very small.} \quad (14)$$

Note that, if T is very small, the product of the last two factors of equation (10) becomes

$$\frac{M_5}{M_2 + M_5} \cdot L' \approx \frac{M_5}{M_2 + M_5} \cdot \frac{M_2 + M_5}{M_5} = 1$$

Considering our example

$$\frac{M_5}{M_2 + M_5} = \frac{8000}{40000 + 8000} = \frac{1}{6}$$

If we test our example system once per month then T = 730 hours and L is

$$L = \left[\frac{1 - e^{-(1/40000 + 1/8000) \times 730}}{1 - e^{-(1/40000) \times 730}} \right] = 5.735$$

This is already quite close to its maximum value of 6.0.

The Operational Readiness of the example system, when tested regularly once per month, becomes

$$P'_{OR} = \left[1 - \frac{M_4}{M_3 + M_4} \cdot \frac{M_1}{M_1 + M_2} \right] \cdot \frac{M_5}{M_2 + M_5} \cdot L$$

$$= \left[1 - \frac{100}{50 + 100} \times \frac{5}{5 + 40000} \right] \times \frac{1}{6} \times 5.735$$

$$P'_{OR} = 0.9557$$

Let us investigate more closely the factor which is common to both the equations (2) and (10).

Let

$$K = \left[1 - \frac{M_4}{M_3 + M_4} \cdot \frac{M_1}{M_1 + M_2} \right] , \quad (15)$$

which can be written as

$$K = \left[1 - \frac{1}{M_3/M_4 + 1} \cdot \frac{1}{M_2/M_1 + 1} \right] = \left[1 - \frac{1}{A+1} \cdot \frac{1}{B+1} \right] , \quad (16)$$

where

$$A = M_3/M_4 = \frac{\text{Mean time to mission failure}}{\text{Mean mission time}}$$

$$B = M_2/M_1 = \frac{\text{Mean time to next call arrival}}{\text{Mean repair time of system}}$$

In general the value of ratio B is $\geqslant 1.0$. In our example

$$B = M_2/M_1 = 40000/5 = 8000.$$

This means that the factor $1/(B+1)$ is very small. The factor $1/(A+1)$ is always < 1 because $A > 0$. Hence, the product

$$\frac{1}{A+1} \cdot \frac{1}{B+1}$$

is relatively small and therefore K has in most systems a value close to 1.0.

To get a better insight into the relationship between A, B, and K, the value of K has been calculated for various values of A and B. The calculation results are summarized in Table 2 and plotted in figure 4.

TABLE 2   The Value of K for various Values of A and B

A	B	1/(A+1)	1/(B+1)	K
0.5	1	0.66667	0.50000	0.66667
	5	0.66667	0.16667	0.88889
	10	0.66667	0.09091	0.93939
	50	0.66667	0.01961	0.98693
	100	0.66667	0.00990	0.99340
	500	0.66667	0.00200	0.99867
	1000	0.66667	0.00100	0.99933
1.0	1	0.50000	0.50000	0.75000
	5	0.50000	0.16667	0.91667
	10	0.50000	0.09091	0.95455
	50	0.50000	0.01961	0.99020
	100	0.50000	0.00990	0.99505
	500	0.50000	0.00200	0.99900
	1000	0.50000	0.00100	0.99950
2.0	1	0.33333	0.50000	0.83333
	5	0.33333	0.16667	0.94444
	10	0.33333	0.09091	0.96970
	50	0.33333	0.01961	0.99346
	100	0.33333	0.00990	0.99670
	500	0.33333	0.00200	0.99933
	1000	0.33333	0.00100	0.99967
5.0	1	0.16667	0.50000	0.91667
	5	0.16667	0.16667	0.97222
	10	0.16667	0.09091	0.98485
	50	0.16667	0.01961	0.99673
	100	0.16667	0.00990	0.99835
	500	0.16667	0.00200	0.99967
	1000	0.16667	0.00100	0.99983

(continued)

TABLE 2 (cont'd)

A	B	1/(A+1)	1/(B+1)	K
10.0	1	0.09091	0.50000	0.95455
	5	0.09091	0.16667	0.98485
	10	0.09091	0.09091	0.99174
	50	0.09091	0.01961	0.99822
	100	0.09091	0.00990	0.99910
	500	0.09091	0.00200	0.99982
	1000	0.09091	0.00100	0.99991

Studying the plot in figure 4, it is obvious that if $A > 10$ and $B > 10$, then K is for all practical purposes 1.0 (see dotted surface in figure 4).

SUMMARY OF OBTAINED RESULTS

Let

M_1 = mean repair time of the system;

M_2 = mean time to next call arrival;

M_3 = mean time to failure occurrence during mission;

M_4 = mean mission time;

M_5 = mean time to dormancy failure.

1. The Operational Readiness of a system, which is <u>not tested during its dormant period</u>, P_{OR}, is

$$P_{OR} = \left[1 - \frac{M_4}{M_3+M_4} \cdot \frac{M_1}{M_1+M_2}\right] \cdot \frac{M_5}{M_2+M_5}$$

2. The Operational Readiness of a system, which is <u>regularly tested during its dormant period</u>, P'_{OR}, is

$$P'_{OR} = \left[1 - \frac{M_4}{M_3+M_4} \cdot \frac{M_1}{M_1+M_2}\right] \cdot \frac{M_5}{M_2+M_5} \left[\frac{1-e^{-(1/M_2+1/M_5)T}}{1-e^{-(1/M_2)T}}\right]$$

where

      T = the time between tests.

3. When $A = M_3/M_4$ is equal to or larger than 10.0 and $B = M_2/M_2$ is equal to or larger than 10.0, then for all practical purposes the Operational Readiness becomes;

$$P_{OR} \approx \frac{M_5}{M_2+M_5} \quad , \text{(no testing)}$$

$$P'_{OR} \approx \frac{M_5}{M_2+M_5} \cdot \left[\frac{1-e^{-(1/M_2+1/M_5)T}}{1-e^{-(1/M_2)T}}\right]$$

(with regular testing)

<u>Note:</u> If <u>T is very small</u>, then

$$P'_{OR} \text{ approaches } \frac{M_5}{M_2+M_5} \cdot \frac{M_2+M_5}{M_5} = 1.0$$

REFERENCE

Bazovsky; Igor Sr.; <u>Weapon Systems Operational Readiness</u>; Proceedings 1975 Annual Reliability and Maintainability Symposium, Washington, D.C.; Jan. 28, 29, and 30, 1975; Paper 1244/75RM031, pages 174 thru 178.

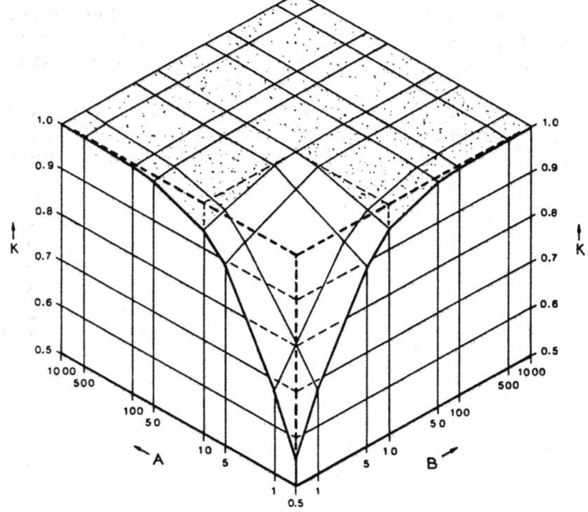

Fig. 4   K as a Function of A and B.

$$K = \left[1 - \frac{1}{A+1} \cdot \frac{1}{B+1}\right]$$

where $A = M_3/M_4$ and $B = M_2/M_1$

APPENDIX I   The Derivation of Equation (2)

Operational Readiness may be expressed as

$$P_{OR} = R.R_D + Q.P_D \tag{I-1}$$

See text for the definitions of the symbols.

The probability that the system returns from a mission without failure, when the mission time is distributed according to q(t), is

$$R = \int_0^{\infty} R(t) \cdot q(t)\, dt \quad , \tag{I-2}$$

where $R(t) = e^{-t/M_3}$,

and the probability that the system returns from a mission in a state requiring repair is

$$Q = 1 - R \quad , \tag{I-3}$$

Equation (I-2) can be worked out as follows

$$R = \int_0^{\infty} e^{-t/M_3} \cdot (1/M_4)e^{-t/M_4}\, dt$$

$$= (1/M_4) \cdot \int_0^{\infty} e^{-(1/M_3+1/M_4)t}\, dt$$

$$= (1/M_4) \cdot \left[-\frac{1}{(1/M_3+1/M_4)} e^{-(1/M_3+1/M_4)t} \right]_0^{\infty}$$

$$= (1/M_4) \cdot \left[0 + \frac{M_3 M_4}{M_3+M_4} \cdot 1\right]$$

$$R = \frac{M_3}{M_3+M_4} \quad , \tag{I-4}$$

and

$$Q = 1 - R = \frac{M_4}{M_3+M_4} \quad , \tag{I-5}$$

APPENDIX I (cont'd)

Computation of P_D

Let

$f(t_m)$ = the p.d.f. of the maintenance time;

$g(t_c)$ = the p.d.f. of the time arrival of the next call, counted from the instant the system returned from last mission in a state requiring repair,

then

$$P_D = \int_{t_m=0}^{\infty} f(t_m) \left[\int_{t_c=t_m}^{\infty} g(t_c) \cdot R_D(t_c - t_m)\, dt_c \right] dt_m \quad ,\qquad \ldots\ldots (I\text{-}6)$$

The integral within $\left[\ldots\ldots\right]$ is the probability that the call arrives at time t_c, after the variable t_m, required to complete maintenance, __AND__ that the system does not fail between t_m and t_c.

$$f(t_m) = (1/M_1)e^{-(1/M_1)t_m} \quad , \qquad (I\text{-}7)$$

$$g(t_c) = (1/M_2)e^{-(1/M_2)t_c} \quad , \qquad (I\text{-}8)$$

$$R_D(t_c - t_m) = e^{-(1/M_5)(t_c - t_m)} \quad , \qquad (I\text{-}9)$$

Substitution of these expressions in equation (I-6) gives for the expression within the square brackets

$$\left[\ldots\ldots\right] = \int_{t_c=t_m}^{\infty} (1/M_2)e^{-(1/M_2)t_c} \cdot e^{-(1/M_5)(t_c - t_m)}\, dt_c$$

$$= (1/M_2) \int_{t_c=t_m}^{\infty} e^{-(1/M_2 + 1/M_5)t_c} \cdot e^{+(1/M_5)t_m}\, dt_c$$

$$= (1/M_2)e^{+(1/M_5)t_m} \left. \cdot \left[-\frac{M_2 M_5}{M_2 + M_5} e^{-(1/M_2 + 1/M_5)t_c} \right] \right|_{t_m}^{\infty}$$

$$= (1/M_2)e^{(1/M_5)t_m} \cdot \frac{M_2 M_5}{M_2 + M_5} e^{-(1/M_2 + 1/M_5)t_m}$$

$$= \frac{M_5}{M_2 + M_5} e^{-(1/M_2)t_m} \quad ,$$

hence

$$P_D = \int_{t_m=0}^{\infty} (1/M_1)e^{-(1/M_1)t_m} \left[\frac{M_5}{M_2 + M_5} e^{-(1/M_2)t_m} \right] dt_m$$

$$= \frac{1}{M_1} \cdot \frac{M_5}{M_2 + M_5} \cdot \int_{t_m=0}^{\infty} e^{-(1/M_1 + 1/M_2)t_m}\, dt_m$$

$$= \frac{1}{M_1} \cdot \frac{M_5}{M_2 + M_5} \cdot \left. \left[-\frac{M_1 M_2}{M_1 + M_2} e^{-(1/M_1 + 1/M_2)t_m} \right] \right|_{t_m=0}^{\infty}$$

$$P_D = \frac{M_2}{M_1 + M_2} \cdot \frac{M_5}{M_2 + M_5} \quad , \qquad (I\text{-}10)$$

Computation of R_D

$$R_D = \int_{0}^{\infty} R_D(t_c) \cdot g(t_c)\, dt_c \quad , \qquad (I\text{-}11)$$

$$= \int_{0}^{\infty} e^{-(1/M_5)t_c} \cdot (1/M_2)e^{-(1/M_2)t_c}\, dt_c$$

$$= (1/M_2) \int_{0}^{\infty} e^{-(1/M_2 + 1/M_5)t_c}\, dt_c$$

$$R_D = \frac{M_5}{M_2 + M_5} \quad , \qquad (I\text{-}12)$$

Equation (1) for Operational Readiness can now be written as

$$P_{OR} = \frac{M_3}{M_3 + M_4} \cdot \frac{M_5}{M_2 + M_5} + \frac{M_4}{M_3 + M_4} \cdot \frac{M_2}{M_1 + M_2} \cdot \frac{M_5}{M_2 + M_5}$$

This can be written, after some manipulations, as

$$P_{OR} = \frac{M_5}{M_2 + M_5} \cdot \left[1 - \frac{M_4}{M_3 + M_4} \cdot \frac{M_1}{M_1 + M_2} \right]$$

APPENDIX II   The Derivation of Equation (7)

$$R_D(\text{expected}) \Big\uparrow_N^{N+1} = \int_{NT}^{(N+1)T} (1/M_2)e^{-t_c/M_2} \cdot e^{-(t_c - NT)/M5}\, dt_c$$

(see text)

$$= (1/M_2)e^{NT/M_5} \cdot \int_{NT}^{(N+1)T} e^{-(1/M_2 + 1/M_5)t_c}\, dt_c$$

$$= \frac{1}{M_2} e^{NT/M_5} \cdot \frac{M_2 M_5}{M_2 + M_5} \cdot \left. \left[-e^{-(1/M_2 + 1/M_5)t_c} \right] \right|_{NT}^{(N+1)T}$$

$$= \frac{M_5}{M_2 + M_5} \cdot e^{NT/M_5} \cdot \left[-e^{-(\frac{1}{M_2} + \frac{1}{M_5})(N+1)T} + e^{-(\frac{1}{M_2} + \frac{1}{M_5})NT} \right]$$

$$= \frac{M_5}{M_2 + M_5} e^{\frac{NT}{M_5}} \cdot e^{-(\frac{1}{M_2} + \frac{1}{M_5})NT} \cdot \left[1 - e^{-(\frac{1}{M_2} + \frac{1}{M_5})T} \right]$$

$$= \frac{M_5}{M_2 + M_5} \cdot \left[1 - e^{-(1/M_2 + 1/M_5)T} \right] \cdot e^{-NT/M_2}$$

$$R_D' = \sum_{N=0}^{\infty} R_D(\text{expected}) \Big\uparrow_N^{N+1}$$

$$= \sum_{N=0}^{\infty} \frac{M_5}{M_2 + M_5} \cdot \left[1 - e^{-(1/M_2 + 1/M_5)T} \right] \cdot e^{-NT/M_2}$$

$$= \frac{M_5}{M_2 + M_5} \cdot \left[1 - e^{-(1/M_2 + 1/M_5)T} \right] \sum_{N=0}^{\infty} e^{-NT/M_2}$$

$$\sum_{N=0}^{\infty} e^{-NT/M_2} = \sum_{N=0}^{\infty} \left[e^{-(T/M_2)} \right]^N = \sum_{N=0}^{\infty} r^N \quad ,$$

which is a geometric series of which the first term is 1, the common ratio is

$$r = e^{-T/M_2} \quad ,$$

and the number of terms is infinite.

Since r^2 is less than unity,

$$\sum_{N=0}^{\infty} e^{-NT/M_2} = \sum_{N=0}^{\infty} r^N = \frac{1}{1-r} = \frac{1}{1 - e^{-T/M_2}} \quad ,$$

hence

$$R_D' = \frac{M_5}{M_2 + M_5} \cdot \left[\frac{1 - e^{-(1/M_2 + 1/M_5)T}}{1 - e^{-(1/M_2)T}} \right] \quad , \ldots\ldots (7)$$

____

DISCUSSION

Session: Safety Systems

Paper: Operational Readiness of Safety Systems.

Questions: What percentage of the "operational"
failure rate should we take for the "dormant"
failure rate?
Before the presented model can be applied the
various "mean times" $(M_1...M_5)$ must be determined.
How can these "mean times" be obtained?

Author's Reply: If no actual field data on "dormant"
failure rates or "mean times" are available
engineering judgement should be used. The type
of equipment, the working environment, the maintenance, spare parts availability, etc, should be
taken into account.
Various data sources are available both in the
U.S. and Europe.
When analysing a system always mention the source
of data or method used to obtain the data applied
in the analysis.
The purpose of the presentation was to introduce
the concept: "Operational Readiness"; simplifica-
tions were used in order not to complicate the
models.

EXPERIENCE WITH INTEGRATED CONTROL SYSTEMS

M. Roodhuyzen

*Systems Support Division, Shell Nederland Raffinaderij BV,
Rotterdam, The Netherlands*

Abstract. The reliability, as experienced by the user, of the
integrated control systems (ICS systems) from three different
manufacturers is discussed. An evaluation of the recorded faults
in the systems is presented. They show the modern ICS systems to
be quite reliable.

Keywords. Integrated plant control; reliability; maintenance.

INTRODUCTION

Since 1981 we have at our refinery and chemical
plant complex experience with the modern integrated
control systems (ICS systems). At this moment we
have 7 Honeywell TDC3000 Basic systems, 2 Foxboro
Spectrum systems and 2 Fischer & Porter DCI4000
systems. Failures of the systems are put into a
failure recording program implemented on a personal
computer. Each failure is assigned a severity
category: A, B or C, according to the average
severity of the such a failure to plant operations.

Category A failures pose no threat to the plant.
There may be some loss of information. Repair is no
urgent matter. An example is a failure of a single
control loop.

Category B failures still pose no threat to the
installation. There is loss of information. Part
of the system may have switched over to its back-up
unit so there may be loss of redundancy. An example
is a failure of a control box, the back-up having
taken over.

Category C: Any failure which could cause the plant
to go down. Part of the system may not be available
(for example a control unit has flunked). Much
information may be lost. An example is a failure
of a control box and the back-up failing to take
over.

RESULTS

In fig. 1 the numbers of failures for the various
systems at our site are plotted.

Systems A, B and C are Honeywell TDC3000 Basic
systems for continous control applications.

Systems D, E, and F are Honeywell TDC3000 Basic
systems for batch control applications.

Systems H and I are Foxboro Spectrum systems for
continous control applications.

Systems J and K are Fischer & Porter DCI4000
systems for sequence control applications.

In most plots the early part of the bathtub curve
can be distinguished: during the first half year of
operation the number of faults per month decreases
and settles to a more or less stable figure. This
effect is more pronounced with Foxboro and Fischer
& Porter than with Honeywell. We do hope (and
expect) we will not experience the end of the
bathtub curve when it starts to rise again!

The oscillating nature of the plot of system J is
caused by the fact that the main plant controlled
by the system only operates in the winter months.

The distribution of the failures over the 3
severity categories is:

	cat A	cat B	cat C
Honeywell	62 %	33 %	5 %
Foxboro	28 %	71 %	0.6 %
Fischer & Porter	29 %	64 %	6 %

Foxboro and Fischer & Porter have a high proportion
cat B faults. In the Foxboro systems cat B faults
were mainly problems with one UCM, of which the
main unit often failed and who's function was then
taken over by its back-up, and by FOX300 problems
caused by errors in the standard Foxboro software.
In the Fischer & Porter systems most cat B faults
were failures ("blockages") of one of the two
COP's.

It is tempting to compare the systems of the 3
suppliers. But because of the very different
nature of the plants, the different years of
installation, the maturity of the product, the age
of the installed systems this is not possible.
Another factor which makes a comparison difficult
is the fact that the failures for the various
systems are entered into the Failure Recording
Program by different persons. There are bound to
differences in classification of the faults by the
various persons. We do not have the feeling that
one supplier is much better or much worse than the
other two. Keeping this in mind we can put the
figures of the first full year of operation of
systems A + B (1982 Honeywell) and of system H
(1985 Foxboro) next to each other. The plants
controlled by these systems are of similiar
complexity, both are continuous processes with
about 400 control loops.

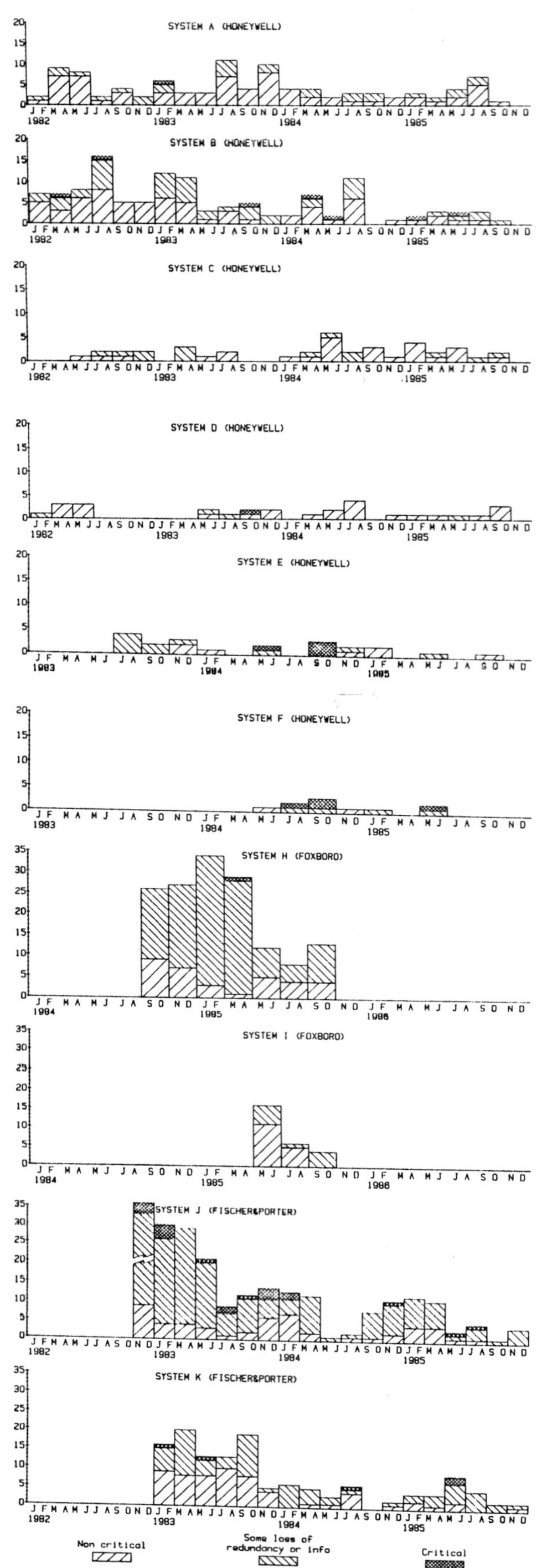

Fig. 1. Number of faults

The number of failures in the 3 categories are:

	cat A	cat B	cat C
Systems A+B, 1982 (Honeywell)	40	21	2
System H, 1985 (Foxboro)	17	80	1

The only conclusion that can be drawn is that the distribution over cat A and B is different for Honeywell and Foxboro. Cat. A faults are minor problems and cat. B faults are occasions in which some redundancy is lost (the back-up has succesfully taken over). Both types of problems do not really interfere with the normal operation of the plant.

C-CATEGORY FAILURES

The most interesting part of the records are the C-category failures. These are the most serious type of failures which potentially can cause the plant to shut down.

In the Honeywell systems we had a few occasions that the reserve controller did not correctly take over from the main unit. It also happened a few times that the communication between a controller and the operator console was lost. In the very beginning it happened once that the configuration of the controller was lost when the main unit was taken back into operation.

In the Foxboro systems we had one occasion in which, because of a fault in the buffer card, the the back-up of a unit control module (UCM) did not take over from the failing main unit.

In the Fischer & Porter systems we had problems with the central operator's panels (COP's) being blocked. Both COP's of a system being blocked simultaneously is a C-category fault. This happened several times. We had several occasions in which the back-up of a digital back-up unit (DBU) failed to take over, mainly because of faults in parts which are not automatically checked by the system.

SPECIAL PROBLEMS

Many of our systems have or have had a particular reoccuring type of problem throughout the system or a particular device constantly giving (maybe different) problems which is responsible for a great number of the recorded faults, and therefor has a distinct effect on the plot of the faults in the system. These problems may not be very critical but because they keep reoccuring they do undermine the trust in the system.

For Honeywell these were problems with the DPS20 printers and the Conrac monitors. Fig. 2 shows the effect of the removal of the printer faults from the log.

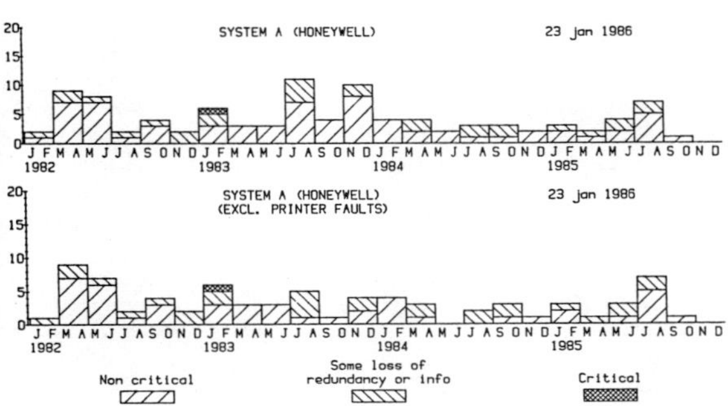

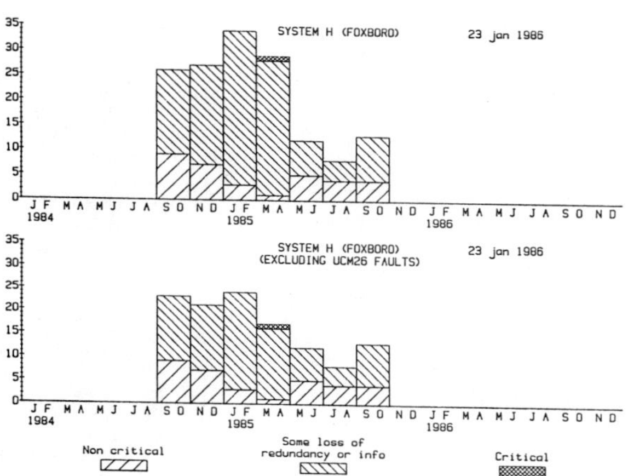

Fig. 2.   Number of failures in system A,
          including and excluding printer failures

In the Foxboro systems we had problems with the
memory chips. We also have a universal control
module (UCM) which regularly showed errors.
Replacement of cards did not help, so in the end
the complete nest with the main and the back-up UCM
had to be replaced. Fig. 3 illustrates the effect
of the problems with this UCM.

Fig. 3.   Number of failures in system H,
          including and excluding UCM26 failures

In the Fischer & Porter systems we had many
occasions in which the central operator's panel
(COP) was blocked. In one system we had various
problems with all digital back-up units (DBU's),
the number of which decreased significantly after
replacement of one particular DBU. The effects
of these problems on the plots is shown in fig. 4.

COMPARISON WITH MTBF FIGURES

The only item of which we have gathered sufficient
data to be able to compare with the mtbf figures
provided by the supplier is the basic controller of
Honeywell TDC3000 Basic.

The MTBF figure published by Honeywell are
(theoretical worst case conditions)
   1.5 year for a single basic controller
   78 year for a basic controller backed by a UAC,
      and both main and UAC failing (assuming 1 UAC
      per 8 basic controllers and MTTR=8 hours)

We have experienced following number of faults in
the basic controllers of systems A and B:

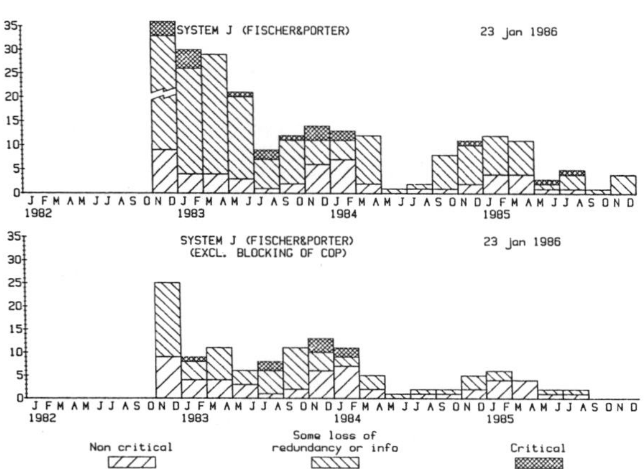

Fig. 4. Number of failures in system J,
         including and excluding COP failures

	System A	System B
Number of basic controllers	32	38

Number of faults	A	B	C	A	B	C
1982	3	1	-	1	5	1
1983	4	6	1	1	9	1
1984	1	5	-	-	7	1
1985	-	-	-	-	3	2

For the calculation of the MTBF figure of a single
basic controller we have to add the cat. A, B and C
failures:

   System A: 21 failures in 4 years
   System B: 31 failures in 4 years

The MTBF for a single basic controller is

   System A: 32 × 4 / 21 = 6 years
   System B: 38 × 4 / 31 = 5 years

For the calculation of the MTBF figure for the
basic controllers in a system with UAC's, only the
C-category failures have to be counted. These are
the cases in which the reserve controller did not
succesfully take over from the main controller.

   System A: 1 C-cat. failure in 4 years
   System B: 5 C-cat. failures in 4 years

The MTBF figure are:

   System A: 32 × 4 / 1 = 128 years
   System B: 38 × 4 / 5 = 30 years

The somewhat low figure for system B is mainly
caused by the problem of the basic controller
loosing its highway address.

ENGINEERING

The way a system is engineered for a particular
plant has a great influence on the availability of
the system to the final user. Some guidelines are:

All important IO signals are put into the redundant
control units and not in the non-redundant IO
units.

In a continuous process plant the various important
control loops are distributed over the various
control units, such that in case of a total loss of
a control unit it may still be possible to keep the
plant stable for some time. In the case of a batch
process plant with parallel reactors the aim is to
keep the signals of one reactor together in one
control unit, so that in case of a total failure of
a control unit never more than one reactor is
involved.

At our site ICS systems are powered by battery
backed vital power supplies. This is of importance
for batch processes, as all information on a
running batch is lost after a power failure.

Our aim is to keep the configurations simple and to
rely as far as possible on standard equipment of
the ICS supplier. Experience has shown that
special keyboards, display panels etc. can be a
source of problems and they do complicate
maintenance.

In case of an emergency, with danger to personnel
and or equipment it may be necessary to shut down
the plant. The trip systems have to be very
reliable and are therefor realised mostly by means
of separate hardwired systems.

SOFTWARE

In the ICS systems for continuous control
applications the user can hardly make any
configuration mistakes which could cause a
dangerous situation to occur. The standard software
supplied by the manufacturer can be assumed to be
nearly bugfree. As it is fixed in PROM memory it
cannot be altered by the user.

The systems for batch control applications are a
different matter. These require extensive
programming in a sequence control language for a
particular application, with all the risks of
errors (But because of the extensive standard
facilities supplied by the system, the risk will be
smaller than with a system based on a free
programmable computer). By certain programming
techniques the risk to the plant can be reduced
further.

One technique is the use of a separate interlock
program. The primary reason for this program is to
prevent certain valves to be opened in the case of
manual control. But in automatic control this
program is also active. In each cycle of the
control unit a part of each of the sequence
programs is performed. At the end of the cycle the
actual output to the field is done. The interlock
program is executed just before the actual output
is performed. It prevents illegal combinations of
signals to be output to the field. An illegal
combination could be for example the simultaneous
opening of more than one valve to a reactor. So the
interlock program prevents programming errors in
the sequence programs to have a damaging effect in
the field.

We also keep the program sections which handle the
abnormal situations quite separate from the normal
sequence. We distinguish 3 of these "abnormals".
The "hold" sequence and the "shutdown" sequence
stops the process, the process can be continued by
the operator, after the faulty device in the plant
has been repaired or the feedback signal has been
put in "ignore" by the operator. The emergency
shutdown sequence is the most drastic action to
bring the plant into a safe state. In most cases it
means the batch is lost. We lay great weight on the
handling of the abnormal situations. Some 40% of
all programming effort and memory space in a
control unit is devoted to the abnormals and
interlocks.

DISCUSSION

In general the ICS systems are quite reliable.
Mechanical devices such as printers cause many
problems. Also software is a very difficult item to
get correct, as illustrated by the problems Fischer
& Porter has had with the COP (central operator's
panel) software and Foxboro with the FOX300
software.

In virtually all cases the back-up unit took over
correctly from a failed main control unit. Where
this did not happen, it was mostly because of some
small part of the system which is not automatically
checked. This could be a buffer board between the
main and its back-up or the logic with the highway
address setting. Regular preventive maintenance is
required to prevent these type of faults.

Compared with systems based on free-programmable
computer systems the ICS systems are very reliable.
This is illustrated by the case of a particular
plant in which a redundant computer system was
replaced by an ICS system. With the redundant
computer system we had 6 C-category faults in 2
years and after its replacement in mid 1984 only
one. Correcting faults in an ICS system system is
simpler than correcting faults in free-programmable
computer systems. So from a reliability and
maintenance cost point of view ICS systems are a
very good proposition.

AIMS, TASKS AND METHOD OF ON-LINE DIAGNOSTIC OF INDUSTRIAL CONTROL SYSTEMS

J. M. Kościelny

*Institute of Industrial Automatic Control, Warsaw Technical University,
Warsaw, Poland*

Abstract. The paper describes aims and tasks of on-line diagnostics in industrial automatic control systems. General classification of diagnostic problems has been presented. Detection and localization of failures occuring in measuring and control equipment and technological devices has been discussed in greater detail. Different kinds of models of diagnostic systems and certain method of description of diagnostic object have been presented. Diagnostic checkings and rules of their choice have also been given. A method of the checkings' realization during the system's operation has been described. An example of diagnosis of tank level control system has been presented and also properties of the presented diagnostic method have been discussed.

Keywords. Alarm systems; automatic testing; on-line diagnostics; failure detection and localization; computer control.

INTRODUCTION

Appropriate control and progress of the industrial process is one of the most important tasks of the computer-controlled automatic control system. Failures of technological elements and automatic control devices are the main reason of damage states (Jublanc, 1975). The failures are unavoidable because of great concentration of the elements and devices within the system, their limited reliabilities and because of long period of operation of such systems. Damage states cause serious economic losses and also could be dangerous for people or instruments. The most important from the system's operation security point of view are these periods of time, during which existing failures are not known. This can be described by the system's operation protection factor which can be defined as follows (Krüger, 1977):

$$P_s = \frac{T_\lambda + T_N}{T_\lambda + T_D + T_N}, \qquad (1)$$

where T_λ, T_N and T_D are mean values of times of operation without failures, renewal and diagnostics times respectively.

Automatic realization of diagnostic actions during the system's operation allows significantly to reduce time of a failure detection and localization in comparison with the diagnostics being performed by the system's operator. This results in significant increasing of many reliability parameters of the system and also increasing of economic effects. Moreover, exact and quick diagnostic information enables for appropriate protection to be performed within the system in order to avoid or to reduce the failure's effects before they could influence the system's operation in an dangerous way. Thanks to this one can obtain the effect of tolerance of certain failures.

Failure-tolerant systems (Kopetz, 1983) are being realized more and more often in modern decentralized automatic control systems. It has been achieved by application of different kinds of redundancy and also by realization of the auto-diagnostics and mutual testing by particular intelligent units of the automatic control system (Bonn, 1980; Schäfer, 1983; Schmidt, 1980). Thanks to this the losses connected with failures of elements belonging to this group can be reduced. Now the losses related to failures of technological devices and also measurement and control devices compose the main part of the losses resulting from the instrumentation failures. Because of that the problems of on-line diagnostics of these devices have very important practical significance. The general classification of aims and methods of technical diagnostics in automatic control systems has been shown in Table 1.

TABLE 1 Classification of aims and methods of diagnostics in automatic control systems

No	Criterion of the division	Distinguished elements

1	The diagnosis object	-The diagnostics of instrumentation:
		  -digital devices
		  -analogue devices
		-The diagnostics of software

2	The method of realization	-With interruption of the system's operation (off-line) -Without interruption of the system's operation (on-line)
3	The aim of the action	-Failure detection -Failure localization -Failure forecasting -Failure origin finding
4	Methods of realization	-Hardware methods -Software methods -Software and hardware methods

MODELS OF THE DIAGNOSIS OBJECTS

Set of devices external towards the computer-controlled automatic control system will be considered as the diagnosis object. It is being assumed that the diagnosis of elements composing the digital control system is being realized by the system. For the description of such a division, Schwager (1983) have used following formulations: "diagnose steuerungs-externe Fehler" and "diagnose steuerungsinterne Fehler". The diagnostics problems are also known in automatic control as "Alarm Analysis" (Plamping, 1983). Exemplary object of the diagnosis has been shown in Fig. 1.

sidered states is usually limited to the state of full efficiency and the states with failures of particular single elements of the system. In this case the a_{ji} element of the table answers to the d_j result of the checking in the state with failure of the i element of the system. The table of checkings can be generalized assuming that the considered set of reliability states concerns not only to the states with failures but also to states with other errors which should be recognized, such as operator´s errors, shortage of raw materials at inputs of technological assemblies etc.

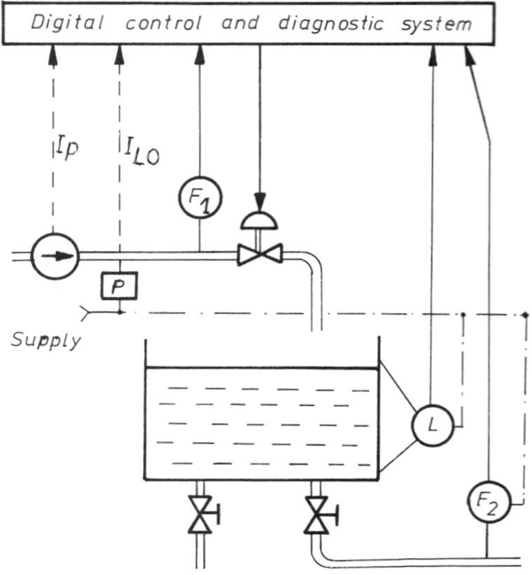

Fig. 1.   An example of the diagnosis object.

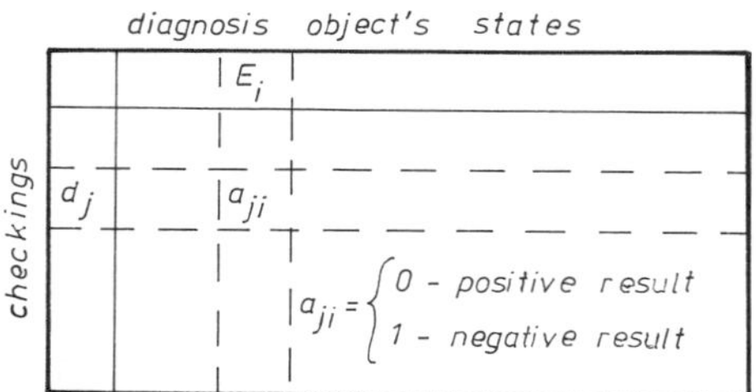

Fig. 2.   Table of states.

The logical models´ advantage rely on their construction simplicity and on that they grasp the hardware structure of the diagnosis object. Their disadvantage however rely on that they overlook the diagnosis object´s dynamics.

In papers concerning diagnostics of external devices, different methods of the object´s formal description are being applied. Logical model ist the least complicated of them. It is being applied by Plamping (1983) in a following way. The result of a diagnosis is defined as a logical function of the error symptoms (e.g. checking results), such as exceeding of absolute limits or relative limits (related to set value), admissible change rates or meter´s measuring ranges exceedings by the process measuring variables. For example, the failure of the control valve or the controller in the object presented in Fig. 1 is ascertained, when following exceeding states occur: $LO(L) \wedge LO(F1) \wedge LO(F2) \wedge HI(P1)$, where P1 denotes the inflow pressure (not shown in Fig. 1). "The failure causality model" presented by Sata (1979) belongs to the group of logic models as well.

Table of states (Parchomenko, 1976) shown in Fig. 2 is also a kind of logical model which is commonly applied in technical diagnostics of complex objects. Rows of the table answer to particular checkings (measurable failure symptoms) and columns answer to reliability states of the set of system´s elements. If each of the elements can exist in one of two states: efficiency or failure, then the number of possible reliability states $n=2^m$, where m is the number of devices. The set of con-

Analitical models are the second group of models commonly applied in diagnostics of external devices. These models have been exhaustively described by Isermann (1984) and Willsky (1976). The models are much more complete but simultaneously much more complicated description of the diagnosis object. They are applied mainly for failure detection purposes (Isermann, 1980), but there are being performed tests for construction of such analytic models suitable also for the failure localization (Singh, 1983; Willsky, 1976).

Topological models labeled as graphs, for example by Baldeweg (1980,1982), are the third method applied for description of the diagnosis object. These models can constitute still more complete and more precise description of the diagnosis object than the logical models, because they can take the object´s dynamics, among others, into considerations. They often describe the way which the diagnosis is being performed, i.e. the sequence of particular checkings. Their disadvantage however relies on difficulty of simple and fast modification of such models when changes of set of checkings and set of considered elements (the system´s structure) occur.

SUGGESTED WAY OF THE DIAGNOSIS OBJECT DESCRIPTION

It seems that the most sensible for practical applications in complex industrial plants supervisory systems would be the following way of the diagnosis object description:
- Adopting of a logical model for the diagnosis object´s general structure description. For this purpose one should define the set of errors $E = \{e_i\}$, occurence of which should

be detected and localized and also the set of diagnostic checkings $D = \{d_j\}$. While the set of checkings defining, to each of the checkings d_j one should attribute the subset of errors $E_j = \{e_i(d_j)\}$, occurence of which gives the negative result of the checking. Such a diagnosis object description corresponds with the table of checkings if one can assume that only single errors (failures) occur.

- Applying of analytical models and other simplified ways of the diagnosis object description for construction of particular diagnostic checkings belonging to the set D. The results of all of the checkings are being treated always in a two-stage way (negative-positive).

- Supplementing of the logical model with the additional subset of time τ_i parameters in order to take into considerations the dynamic properties of the object and also the dynamics of the diagnostic actions. The period of time τ_i attributed to the checking d_j is being defined as maximal time interval from the moment of the error occurence to the moment when there appear symptoms being recorded by the checking and indicating the presence of this error. Not taking the τ_i parameters into considerations can lead to a false diagnosis. It can be explained on a following example: failure of an element is being detected after different time intervals by different checkings encompassing the same element (even after supposing they are being realized simultaneously). Therefore in any given moment of time after the failure occured, one of the checking may bring the negative result while the result of the other checking could be still positive (but after the τ_i time the other one would also indicate that the failure occured). Not considering the τ_i times could bring therefore false conclusion about the failure reason.

CHARACTERISTIC OF APPLIED CHECKINGS

Checking of an element or a set of elements in a general case consists in controlling of their operation when the model describing operation of the diagnosis object in state of efficiency is known. Such a model can take into considerations the absolute values of input and output variables, their increases or in the least complicated version only directions of the signals' changes which are being compared then with real directions of the process variables. Therefore in many cases models of objects being used for diagnostic purposes are less complicated than models applied for control. For practical use there are often applied simple checkings, not requiring any mathematical models of controlled elements and based only on control of certain parameters or certain known properties of these elements.

The basic group of checkings being applied in automatic control systems relies on control of exceedings of limiting values by the process analogue variables which are being measured in a direct or indirect way and also on control of values of binary variables. Limiting of diagnostic actions to only realizations of these checkings however does not allow quickly and accurately to determine the place where the failure occured. Therefore a number of other checkings performed during the system's operation is needed. Here are some examples of such checkings:
- measuring analogue signals' credibility of

checking by:
a) control of their admissible ranges and their change rates,
b) comparison of different measurements of the same quantity (Williams, 1974),
c) comparison of interrelated process variables basing on the model or on simple relations occuring between them. For example, in a sugar factory's evaporator, simple physics laws require that particular temperature values beginning from the evaporator's bottom and ending at its top fulfill the relation $t_k \leq t_{k+1}$ (where t_k denote the temperature at the k level which is lower than the k+1 level). These relations enable simple detections of inaccurately operating measuring circuits,
d) detection of events which should not appear with correct operation of measuring circuits. As an example one could mention control of stability of viscous medium level being measured by a bubble probe. Long lasting absence of any change of its signal testifies that the probe got stuck;
- checking of the actuators; in this case it consists in measuring of the feedback signal from the actuator's positioner.

Applied checkings can be divided into two groups. To the first of them one can count these checkings for which both following statements are correct:
1. Negative result of the checking d_j testifies to occurence of an error out of the set $E_j(d_j)$;
2. Positive result of the checking d_j testifies that none of the errors out of the set $E_j(d_j)$ had occured.

Legitimacy of these two statements proves that checkings belonging to the first group can be utilize both to detection and to localization of the errors. To the second group there belong these checkings, for which only first statement is correct. These checkings can not be used in the process of error localization. As an example of such checkings one can count checkings detecting events which should not appear in the state of efficiency (e.g. checkings listed in d) of this chapter).

AN EXAMPLE OF THE SET OF CHECKINGS CHOICE

Choice of the set of checkings will be discussed with help of an example. Fig. 1 presents a tank, inflow of which is forced by a pump and outflow is free (by gravity forces). The liquid level is being digitally controlled by acting on a valve placed on a inflow pipe. Absence of the control signal means that the valve is closed. Three analogue measuring signals are connected to the digital control and diagnostic system. They are: L - level signal, F1 and F2 - flows. There are also two binary signals connected: I_{LO} - supply pressure signal and I_p - the pump operation signal. The set E of considered errors consists of:
1 - valve failure,
2 - pump failure,
3 - level measuring circuit L failure,
4 - flow measuring circuit F1 failure,
5 - flow measuring circuit F2 failure,
6 - absence of input medium or absence of medium pressure,
7 - valve Z2 opened,
8 - valve Z1 closed,
9 - meters supplying pressure unacceptably low,

10 - supply pressure extreme indicator failure,
11 - failure of a relay or measuring circuit
    of the pump operation signal,
12 - incorrect set point.
In this set, failure of the digital controller has been omitted because localization of this failure should be performed by internal devices of the digital system.
Following set D of checkings has been taken for realization:

d_1(1,2,3,6,7,8,9,12) - checking basing on control of the admissible range of the level changes: $LO(L) < L < HI(L)$,

d_2(3,4,5,7,9) - checking of the signals L,F1,F2 conformability changes based on a model: $\Delta L = K(\Delta F1 - \Delta F2)$,

d_3(3,5,8,9) - checking of the conformability of the signals F2 and L changes: $L\uparrow \Rightarrow F2\uparrow \wedge L\downarrow \Rightarrow F2\downarrow$,

d_4(1,2,4,6,9) - changing of the conformability of the signal F1 and the control signal u changes: $u\uparrow \Rightarrow F1\uparrow \wedge u\downarrow \Rightarrow F1\downarrow$,

d_5(2,11) - checking of the pump operation state,

d_6(9,10) - checking of the meters supply pressure,

d_7(3) - checking of credibility of the level measuring signal basing on control of technically admissible change range and admissible change rates: $L_{min} < L < L_{max} \wedge \Delta L/T < \dot{L}_{max}$,

d_8(1,2,3,6,7,9,12) - checking if the level lower limiting value has been exceeded $LO(L)$,

d_9(1,3,8) - checking if the level higher limiting value has been exceeded - $HI(L)$

Numbers of errors being detected by particular checkings have been placed in brackets.

In order to compute available accuracy (differentiation) of the errors, the table of states has been presented (Fig. 3). The table does not include however the checkings useful only to error detection, i.e. checkings d_8 and d_9. Basing on this table one can determine sets of undistinguishable errors when the set of errors D has been assumed. These are such errors, for which rows of the table of states are identical. It is however possible to evaluate additional checkings usefulnesses for increasing the diagnosis accuracy and for minimalization of the set D. Table 2 shows that in this example only checkings 1 and 6 are undistinguishable. All the other errors can be located precisely. By defining the diagnosis accuracy δ as a ratio of a number of distinguishable set of errors m_L to the number of all of the errors m_o one can compute that for the example illustrated by the Table 2, the diagnosis accuracy is equal $\delta = 11/12$.

D\E	0	1	2	3	4	5	6	7	8	9	10	11	12
d_1		1	1	1			1	1	1	1			1
d_2				1	1	1		1		1			
d_3				1		1			1	1			
d_4		1	1		1		1			1			
d_5			1									1	
d_6										1	1		
d_7				1									

Fig. 3.   Table of states for the object presented at Fig. 1.

One can also notice that elimination of the checking d_7 out of the set D does not decrease the diagnosis accuracy. Usefulness of existence of redundant checkings results however from the need of obtaining high diagnosis credibility by its verification. This problem will be discussed in the next chapter.

METHOD OF THE CHECKING'S RESULTS ANALYSIS

A set of diagnostic checkings $D = \{d_j\}$ is being realized in order to detect failures and other errors which appear during the system's operation. Software checking procedures like all other procedures of the process variables conversion are independent modules, out of which sequences of data conversions are being realized. The diagnostic checkings are therefore being realized in a normal way with frequency of sampling of particular process variables. It should be stressed that all of the checkings are being treated in the same way, not taking into consideration their usefulnesses to the error localization (for example checkings d_8 and d_9 mentioned above). Negative result of any one of the checkings (usually after it has been confirmed twice) triggers action of the error localization procedure.

Error localization begins with knowledge of the error set E_s being detected by the checking d_s, result of which was negative. This set composes therefore the primary set of possible errors $E_o^o = E_s$. The set E_o^o is being reduced by realization of successive checkings, performed since this time in a special way, i.e. independently from normal frequencies of the variables sampling. If the result of any successive checking d_j is positive, the set is being determined in following way:

$$E_o^k = E_o^{k-1} - \left[E_o^{k-1} \wedge E_j(d_j) \right] \qquad (2)$$

If the result of the checking is negative one can assume that the event which is the reason of the error belongs to the set $E_j(d_j)$. Therefore

$$E_o^k = E_o^{k-1} \wedge E_j(d_j) \qquad (3)$$

In any moment of time $t > t_o$ only these checkings can be realized, which fulfill the requirement:

$$t - t_o > \tau_j \qquad (4)$$

It is necessary in order to avoid the diagnosis errors connected to nonidentical times of the error symptoms appearances in different checking lines. The choice of checkings out of the set of available checkings can be performed in a dynamic way according to a criterion being assumed. For example it can be the criterion of halving of the set of possible errors E^k. The checking quality factor according to this criterion has the shape

$$n_j(d_j) = \left| 1 - 2\frac{m_j}{m_o} \right| \qquad (5)$$

where m_j denotes the size of the sets $E_j(d_j) \wedge E_o^j$ and m_o denotes the size of the set E_o^j. The choice of a checking is therefore being performed according to the rule

$$\hat{n}(d) = \min_{d_j \in D} n_j(d_j) \qquad (6)$$

Also other, more complex criteria can be applied. They take into consideration error probabilities and error "weights" according to losses which they cause (Kuźniecow, 1972; Parchomenko, 1981). The localization process stops when for all checkings belonging to the set of available checkings, the following condition is fulfilled:

$$E_j(d_j) \wedge E_o^k = \emptyset \vee E_j(d_j) \wedge E_o^k = E_o^k \quad (7)$$

As the result of the error localization process finishing, minimal set of errors $\hat{E}_o$ (encompassing the occuring error) is defined.

In order to ensure high credibility of the diagnosis, it is being verified in an automatic way. For verification purposes there are being chosen first of all these checkings, which have not been chosen for diagnosis and which have the ability for detection of events belonging to the set $\hat{E}_o$. If such checkings do not exist, one can apply action of checkings already used in previous error localization process. When the diagnosis is being confirmed, there occurs automatic suspension of availabilities of all checkings encompassing errors belonging to the set $\hat{E}_o$ for the time needed for the object´s renewal, because these checkings´ results are a priori known in this time.

Examples of error localization for the object being analyzed have been presented in Fig. 4. An assumption has been made, which checkings would detect an error in the first order in particular diagnostic actions. An error can usualy be detected by more than one of the checkings. In general case one can not unmistakably tell, which of these checkings will detect the error in the first order. It depends, among other, on the moment of time of the error occurence. Fig. 4 shows the sequence of the checking realization with the criterion (6) being applied and with assumption, that examining of the condition (4) does not alter the sequence of the checkings´ realization. In cases when the choice factor values are identical, the checking having the lowest number is being performed. Fig. 4a shows the sequence of checkings realized after occurence of transducer F2 failure detected by the checking d_2. The diagnosis indicating this error causes suspension of action of the checkings d_2 and d_3. The sequence of checkings in case of occurence of the error 7 as the next one (valve Z2 opened) has been presented in Fig. 4c and Fig. 4d. When the error 7 occured but the transducer F2 is efficient (all of the checkings are available) the diagnosis accuracy is higher (Fig. 4b). This example illustrates the possibility of correct diagnosis in cases of failures appearing one by one. It also indicates that automatically realized diagnostic actions can adapt to the system´s structure changes and the set of checkings´ changes. Comparisons of examples shown in Figs. 4e and 4f and also in Figs. 4c and 4d indicate the advisability of the checkings d_8 and d_9 being applied for failure detection. They determine the lesser set of possible errors than more general checkings (d_1 for example), what simplifies and quickens the process of error localization and sometimes increases the diagnosis accuracy.

DESCRIPTION OF THE METHOD´S PROPERTIES

- The main advantage of, the method relies on its universality, enabling its application in the

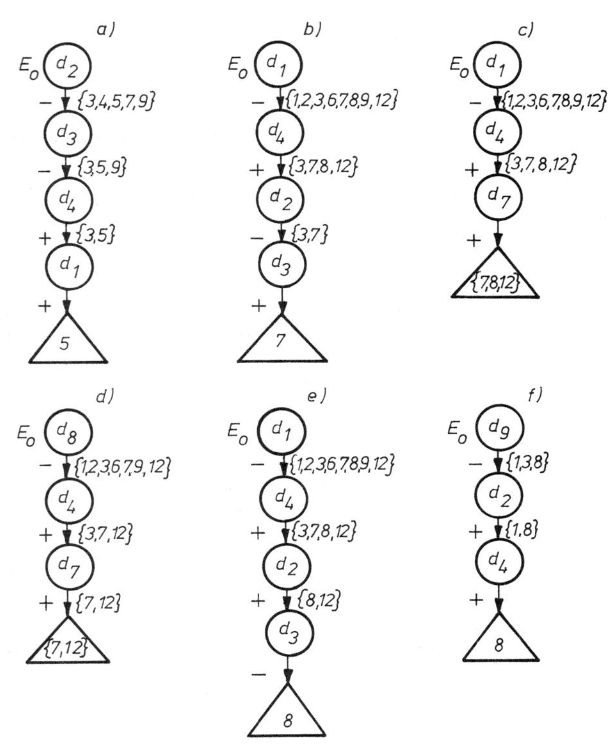

- checkings   - diagnosis
"+" - positiwe result of the checking
"-" - negative result of the checking

Fig. 4.   Sequence of checkings in different diagnosis cases:

a) Sequence of checkings after the error 5 has been detected by d_2,
b) Sequence of checkings after the error 7 has been detected by d_1,
c) Sequence of checkings after the error 7 has been detected by d_1 after localization of the error 5 and suspension of d_2 and d_3,
d) Sequence of checkings after the error 7 has been detected by d_8 after localization of the error 5 and suspension of d_2 and d_3,
e) Sequence of checkings after the error 8 has been detected by d_1,
f) Sequence of checkings after the error 8 has been detected by d_9.

diagnostics of different objects. Only the set of the checking algorithm should be individually determined.
- The diagnosis accuracy depends on the set of checkings $D = \{d_j\}$ chosen for realization. It is advisable to obtain the error distinguishability down to the single error. The choice of the set of checkings can be realized with the help of the table of checkings at the designing stage.
- Realization of the error localization according to a strategy which determines rules of dynamic choice of checkings enables automatic adaptation of diagnostic actions to the system´s structure changes and the set of available checkings´ changes. This advantage does not exist in diagnostic methods determined at the designing stage. Such methods have been presented by Baldeweg (1980, 1982) and

Bieliński (1977). The criterion of the choice of checkings however is less important practically because ef high speed of realization of the checkings by a computer (Kościelny, 1985).
- The error localization is being performed under condition that only single errors occur. This process is being properly realized also in cases of greater numbers of errors under condition that any new error occurs only after the old one has been localized. The diagnosis accuracy in these cases can be lower because the suspended checkings are being eliminated from the set D. False diagnosis however is almost impossible because of small possibility of simultaneous errors appearing in the very short period of time needed in modern systems for the diagnosis realization and because of the diagnosis verification being performed in an automatic way.
- Established diagnosis allows for quick action to be performed in order to restore the state of efficiency and in many cases also to initiate the appropriate protection algorythm in an automatic way (Kościelny, 1983). For the object presented in Fig. 1 for example the protection after the failure of the level measuring circuit has been localized relies on the change of the level control algorythm (Fig. 5a) for the algorythm of the flow follow-up control (Fig. 5b). This change is being realized by the software without any hardware elements connection changes. This replacing algorythm can usually ensure sufficiently good control till the level measuring circuit's full efficiency will have been restored.

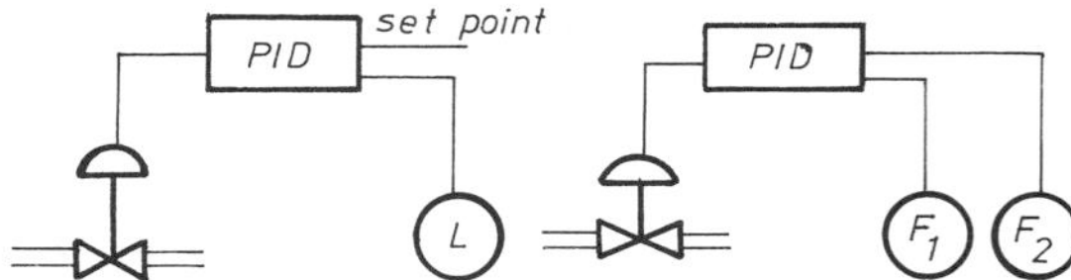

Fig. 5.   a) algorythm of the level control in the state of efficiency
          b) algorythm of control in the state with the failure of the level measuring transducer.

REFERENCES

Baldeweg, F., and F.H. Gassmann (1980). Formale Beschreibung von Diagnose und Therapiesteuerung in einen gestörten diskontinuierlich - diskreten Basissystem. Messen-Steuern-Regeln, 10, 575-579.
Baldeweg, F., and U. Fiedler (1982). Dialog - orientierte Analyse von Ereignisgraphen - ein Verfahren zur Diagnose komplexer technologischer Anlagen. Messen-Steuern-Regeln, 3, 126-128.
Bieliński, H. and J. Ligorowska-Pięta (1977). System diagnostyki przekroczeń dopuszczalnych wartości zmiennych procesowych w złożonym chemicznym procesie technologicznym. Pomiary Automatyka Kontrola, 8, 29-31.
Bonn, G., M. Patz, and F. Saenger (1980). Grundprinzipien und Betriebserfahrungen mit Fehlererkennung und - anzeige bei fehler - toleranten Prozessrechnersystemen mit funktionsbetei-
ligter Redundanz. Fachberichte Messen-Steuern-Regeln, 5, 329-352. Springer-Verlag.
Isermann, R. (1980). Methoden zur Fehlererkenung für die Überwachung technischer Prozesse. Regelungstechnische Praxis, 9 and 10, 321-325 and 363-368.
Jublanc, P. (1975). Les passages à charge nulle non programmés des tranches thermiques à Elektricité de France. Revue Generale Thermigue, 158, 107.
Kopetz, H. and M. Syrbe (1983). Konzepte zur Realisierung hochzuverlässiger Automatisierungssysteme. Fachberichte Messen-Steuern-Regeln, 10, 746-758, Springer-Verlag.
Kościelny, J. M. (1985). Metoda oceny zbioru testów i struktury diagnostycznej komputerowych układów automatyki. Archiwum Automatyki i Telemechaniki, 1, 91-104.
Kościelny, J. M. and P. Wasiewicz (1983). On-line Diagnostik - und Sicherungsfunktionen des Steuerungssystems eines verfahrentechnischen Prozesses. Wissenschaftliche Berichte der Technischen Hochschule Leipzig, 3, 75-79.
Krüger, G. and J. Nehmer (1977). Metoden zur Steigerung der Zuverlässigkeit von PR-Systemen. Fachberichte Messen-Steuern-Regeln, 1, 509-539. Springer-Verlag.
Kuźniecow, P. I., L. A. Pezelincew, and W. S. Gajdienko (1972). Optymalizacja diagnostyki systemów. PWN, Warszawa.
Parchomenko, P. P. (1976). Osnowy techniceskoj diagnostiki. Energija, Moskwa.
Parchomenko, P.P. and E.C. Sogomonjan (1981). Osnowy techniceskoj diagnostiki. Energoizdat, Moskwa.
Plamping, K. and P.K. Andow (1983). The design of process alarm systems. Trans Inst MC, 5, No 3, 161-168.
Sata, T., S. Takata, M. Sato and K. Suzuki, (1979). Monitoring and diagnosis system of maschine tools. Information control problems in manufacturing technology. Procedings of the second IFAC/IFIP symposium, Stuttgart, Germany, Oxford, England, XI8319, 73-82.
Schäfer, M., and F. Saenger (1983). Verfügbarkeit und Fehlerdiagnose bei verteilten Mikrorechner - Automatisierungssystemen. Fachberichte Messen-Steuern-Regeln, 10, 777-791. Springer-Verlag.
Schmidt, G., and W.Sendler (1980). Redundanzkonzepte in modernen Prozessautomatisierungssystemen. Regelungstechnische Praxis, 9, 310-313.
Schwager, J. (1983). Diagnose steuerungsexterner Fehler am Fertigungseinrichtungen. ISW48 Universitet Stuttgart. Springer-Verlag.
Singh, M. G., M.F. Hassan, Y.L. Chen, D.S. Li, and Q.R. Pan (1983). New approach to failure detection in large-scale systems. IEE Procedings, 130, Pt.D. No 5, 243-249.
Williams, J.R. (1974). Reliability in a process control system. Computer system reliability. Meidenhead: Infotech. Inf., 373-388.
Willsky, A.S. (1976). A survey of design methods for failure detection in dynamic systems. Automatica, 12, 601-611.

RELIABILITY GROWTH PROGRAM ENSURES HIGH AVAILABILITY FOR NEXT GENERATION INDUSTRIAL INSTRUMENTATION SYSTEMS

R. C. Crombe

Taylor Instrument, Combustion Engineering, Rochester, NY, USA

Abstract. Customers of high-technology industrial process control instrumentation systems are demanding that these systems provide enhanced functionality, features, and flexibility. Examples are: an integrated system with a distributed architecture, user-friendly configuration and reconfiguration control and monitoring from an operator's console, supervisory and optimizing control through the use of Computer Interfacing, trend and other historical recording, built-in diagnostics and error message handling, and other Management Information Systems capabilities.

Customers are also demanding that these systems give improved performance, including reliability, maintainability, and availability (RMA). To ensure that these RMA performance characteristics are met during the design, development, manufacture, and deployment (use) of the equipment requires Reliability, Quality Control, and Quality Assurance Programs be put in place. The Design and Development Program must include redundancy and back-up operating modes for any critical functions, methods of operating around any single, critical hardware failure (this may include software methods) and, in cases of critical control loops, the back-up operating modes also need to include continuous automatic control even for the unusual case of the operator losing all communication to the control process instrumentation. This also would need to allow repair by replacement of the local primary Controllers while operating on-line in a back-up mode. These capabilities plus built-in diagnostics will improve maintainability, reduce downtimes, and thereby improve availability.

Improving reliability will then reduce required maintenance actions, which will also improve availability. The entire Reliability Program is aimed at assuring and improving reliability. The Reliability Program includes the following major parts:

>   Component Qualifications
>   IC Screening to the Equivalent of MIL-STD-883
>   Quality Level B-2
>   Component Design Derating Requirements
>   Design Reviews, including Stress and Application
>   Analyses
>   Hardware and Software Reliability Growth Testing

The Reliability Growth Testing consists of:

>   Board Level Testing
>   Subsystem Testing
>   Software Integration Testing
>   System Functional and Hardware Testing

This paper describes the Reliability Growth Testing that has taken place for the Taylor™ MOD 300™ System product line, the advantages that are being experienced from that testing for that 3rd-generation microprocessor-based control instrumentation equipment, and also some of the improvement results. In the manufacturing area the expected results of getting further Reliability Growth (reducing infant mortality failures from getting into the field) by applying an all-equipment Environmental Stress Screening (consisting of a powered and monitored temperature cycling test) is also described.

Keywords. Reliability Growth; Reliability Assurance Program; High Reliability; High Availability; Redundancy; Environmental Stress Screening; Reliability Predictions; Field Results.

RISC-J

INTRODUCTION

Customers of high-technology industrial process control instrumentation and information management systems are demanding that these systems provide enhanced functionality, features, and flexibility. They are also demanding that these systems give improved performance, including high reliability, easily performed maintainability, and high availability.

To achieve all of these goals requires applying concerted R&QA (Reliability and Quality Assurance) efforts throughout the design, development, manufacture, and field use of the equipment. This may also require the use of redundant success paths to still allow proper critical functions to be performed even when some of the equipment is not functioning. The R&QA efforts include applying a comprehensive Reliability Assurance Program, which is described in this paper.

In the manufacturing QA (Quality Assurance) efforts, an all-equipment ESS (Environmental Stress Screening) is applied, which also adds to the reliability growth (for reducing infant mortality and latent workmanship failures from getting into the field).

This paper describes how applying our Reliability Assurance Program has enhanced reliability and resulted in reliability growth for our MOD 3™ System and MOD 30™ System products. It then goes on to describe how the Reliability Assurance Program is being applied to our MOD 300 System products with expected and preliminary results.

BRIEF MOD 300 SYSTEM DESCRIPTION

A basic MOD 300 System designed and configured for high reliability and availability consists of the following subsystems: Controller (with built-in redundancy), Redundant Data Processors, and Redundant Consoles, all linked together by a Redundant DCN (Distributed Communications Network).

The general philosophy for achieving high reliability and availability is that the design shall be proven to be robust (achieved by applying the Reliability Assurance Program) and, in addition, any single failure shall not cause a critical system function to be lost. Less critical functions, such as advanced supervisory, optimization control, or monitoring Bulk I/O (inputs and outputs) may not need redundancy to meet their reliability or availability requirements.

Having a distributed architecture improves reliability and availability by isolating potential failures to controllable areas (e.g. eliminates common cause failures from resulting in loss of another redundant functioning success path).

Hardware standardization (there are only 13 major hardware modules used throughout the system) improves maintainability and reduces spare provisioning costs. Commonality of system-wide software services (there are common operating systems, data base managers, diagnostics, and communications services) reduces training and maintenance costs.

Comprehensive power-up and runtime diagnostics improves maintainability and reduces downtime, and thereby improves reliability and availability.

RELIABILITY ASSURANCE PROGRAM

Our Reliability Assurance Program is based on the cost-effective application (i.e., applying the most effort on the most critical parts, applications, functions, and applications where it can have the most effect on enhancing or ensuring Reliability, Maintainability, and Availability) of the following:

1.   Reliability Program Requirements (Ref. 4) and Reliability Program (Ref. 5)

A Reliability Program is defined as: Reliability Engineering tasks aimed at preventing, detecting, and correcting design deficiencies, weak parts and workmanship defects, and providing reliability-related information.

2.   Standard Definition of Reliability Terms (Ref. 15)

3.   Technical Reviews and Audits (Ref. 9)

4.   Reliability Modeling and Prediction Tasks and Methods (Ref. 6), Reliability Block Diagrams and Reliability Allocation (Ref. 5), and Mathematical Models and Failure Rate Prediction (Ref. 3).

5.   FMECA: Failure Modes, Effects, and Criticality Analysis (Ref. 10)

FMECA's primary purpose is the early detection of all critical failure possibilities so their effects can be minimized or eliminated through system configuration change or design corrections. Criticality is ranked as the combined influence of severity of failure and probability of occurrence.

6.   Parts Control Program (Refs. 4 and 8), Design Derating Requirements (Ref. 14) and Controlled Quality (Reliability) Level Purchased Parts (Ref. 16)

7.   Reliability Critical Items (Ref. 5) are to be controlled and tested in a cost-effective manner. Such items may be identified with:

• Limited life
• High or unknown failure rates
• Specials with selected parameters
• Failures critical to system function
• High complexity or used in large quantities
• A sole source

8.   ESS: Environmental Stress Screening (Refs. 5 and 12) to precipitate early failures and other weaknesses out of the equipment.

MOD 300 Equipment is subjected to a dynamically operated and monitored Temperature Cycling ESS, one of the most cost-effective screens.

9.   Reliability Growth Test Program (Refs. 5 and 2)
TAAF: Test, Analyze, And Fix
FRACAS: Failure Reporting, Analysis, and Corrective Action System

10.   FTA: Fault Tree Analysis (Ref. 13)

Applying probabilities and failure rates to the FTA
events can make it quantitative and directly analogous
to Reliability Block Diagrams and Mathematical
Models when carefully and consistently applying the
pertinent aspects of probabilities of either failure or
success.

MOD 3 AND 30 FIELD RESULTS, RELIABILITY GROWTH AND COMPARISON TO MIL-HDBK-217 PREDICTIONS

Taylor MOD 3 Process Control Instrumentation
Systems were designed, developed, and had reliability
and other enhancements designed into them in the
mid-to-late 1970's. Over 21 system-years of field data
were collected and analyzed on them and reported in
a paper presented at the Third National Reliability
Conference in Birmingham, U.K., April-May 1981
(Ref. 1). Those experiences taught us several important lessons in reliability and reliability growth. First,
we needed to purchase our ICs (integrated circuits) to
an enhanced quality level. We chose MIL-STD-883
Level B-2 (Ref. 7) for its cost-effectiveness: it gives a
major reliability improvement (a factor of 3 to 10 times
improvement, depending on the initial quality of the
parts), with only a nominal purchase price adder. And
second, we found that the combination of adding a
Reliability Assurance Program (some of the details of
that program are given in Ref. 1), adding a 72 hour,
50 deg. C System Burn-In, and a follow-up design
enhancement program for the discovered reliability
and redundancy deficiencies, gave us a mature and
robust design. After applying these enhancements and
improvements, systems with similar complexities (normalized to the same complexity values) showed an
average of 3 to 4 times improved reliability and
availability for simplex systems and about 9 to 16 times
improvement for redundant systems. For additional
detailed data, see Ref. 1 and Fig. 1. Figure 1 also
shows that while both immature and mature systems
achieve field reliabilities of 3 times that predicted by
MIL-HDBK-217, the mature system immediately exceeds the predicted value in the first month of
customer operation.

CE/Taylor MOD 30 Process Control Instruments were
designed, developed, and introduced in the early
1980's. These instruments also went through two
phases of predictions, but for different reasons. The
MIL-HDBK-217 prediction didn't change, but our ex-
pectations did. Some new components, representing
what were previously referred to as Reliability Critical
Items, are being used in these instruments. Before our
TE&Q (Testing, Evaluation, and Qualification) of these
new components (principally Vacuum Fluorescent
Displays, Lithium Primary Batteries, Membrane
Keyboard Switches, Thermal Print Heads, Chart Drive
Motors and Internally Designed Instrument Power Supply), we expected they would have relatively high
F.R.s (failure rates) or they would have relatively
serious wear-out failure mechanisms (which, in the
case of critical failure effects, could require scheduled
preventive maintenance procedures to keep their effective F.R.s low). Because of our extensive TE&Q
(some of which have lasted 4 or 5 years and are still
going on) and because of other reliability growth
practices, we made a second reliability prediction. As

part of these practices we have selected vendors of
more reliable products, we have made suggestions to
manufacturers of methods to improve the quality and
reliability of their product, and, in the case of the
Vacuum Fluorescent Displays, extensive experimental
tests derived optimum operating points of anode plate
voltages and filament power, which takes advantage of
their built-in characteristics and significantly reduces
wear-out failure mechanisms and, therefore, greatly
prolongs their life.

The second MOD 30 System reliability prediction then
took advantage of the above improvements and also
used the same type of reliability growth which we had
experienced in our MOD 3 System product line.
Doing this showed that CE/Taylor expected the MOD
30 Instruments to be about 10 times more reliable than
predicted by MIL-HDBK-217 after they have been
field operated for 1 year. MOD 30 System field
operating experience shows that they are about twice
as good as our second prediction (about 20 times better than the MIL-HDBK-217 prediction) and getting
better all the time (nearly 39 times better than the
MIL-HDBK-217 prediction at 21 months of field operation). Figure 2 shows this information graphically.

Another trend, besides reliability growth, is noticed
with respect to the correlation of field experienced
reliability and MIL-HDBK-217 predictions. MIL-
HDBK-217 shows almost a direct relationship between
F.R. and IC chip complexity (e.g. for Dynamic RAMs,
MIL-HDBK-217 shows F.R.s increasing by a factor of
about 3 for every increase of a factor of 4 in complexity in going from 1K to 4K to 16K and to 64K bit
devices). Manufacturer's data, CE/Taylor test data,
and CE/Taylor field experience shows that this chip
complexity has practically no affect on F.R. (Figures
3, 4, and 5 show more detail on this).

Even RAC (Reliability Analysis Center) publication
MDR-21, Microcircuit Device Reliability Trend
Analysis, July 9, 1985 (Ref. 11), in making "General
Observations of VLSI Failure Rates," makes statements
to the effect of:

1.   There is no apparent trend to associate failure rate
with complexity for microprocessors.
2.   Memory reliability shows little dependence on
complexity.
3.   Devices with similar complexities fail independently of one another.

Along with this, it is interesting to note that RAC is a
part of RADC (Rome Air Development Command),
which controls publication of MIL-HDBK-217.

MOD 300 SYSTEM CORRELATIONS ON RELIABILITY GROWTH, PREDICTIONS, AND TEST RESULTS

Since the MOD 300 System project has the same type
of Reliability Assurance Program applied to it as the
MOD 30 System product, very similar reliability
growth is expected.

CE/Taylor has more than 10 systems undergoing
Development, Demonstration, and Reliability Tests, on
which we are able to gather reliability and reliability
growth results and compare the results with MIL-

HDBK-217 predictions. With an average of greater than 6 months operating time on these systems, we have already experienced improvement factors of 1.19 to 7.04 times the MIL-HDBK-217 predictions on various subsystem units, with an average value of 2.84 times. We expect considerable continued reliability growth because of continuing feedback of corrective actions. Just one example is that we have experienced 3 failures in Removable Hard Disk Drives, have referred this problem back to the manufacturer, who has identified and resolved the problem. In all of this testing there has not been one failure that would take down a system with the proper redundant configuration for critical functions.

CONCLUSION

Principally, because of rigorously applying a Reliability Assurance Program, including Reliability Growth Testing and feedback of previous experience and other corrective actions, CE/Taylor has been able to experience considerable reliability, maintainability, and, therefore, availability improvements for three generations of microprocessor-based process control instrumentation and management information systems.

This paper shows that by following these actions, CE/Taylor was able to effect reliability-related improvements of 3 to 4 times our first and second field experiences as compared to MIL-HDBK-217 predictions for the first generation MOD 3 System simplex systems (for redundant configurations, this would result in system reliability-related improvements of 9 to 16 times). For our CE/Taylor second generation MOD 30 Instruments, these actions resulted in reliability-related improvements of about 20 (and greater) times the MIL-HDBK-217 predictions.

CE/Taylor has been continuously applying the same type of Reliability Assurance Program for several years. Preliminary results on MOD 300 Systems are already showing improvements of 1.19 to 7.04 times (average of 2.84 times), and CE/Taylor is expecting to eventually achieve the same kinds of improvements as for our two previous generations of products.

1. Crombe, R. C. and R. A. Merrill (1981). Industrial process control instrumentation system's reliability assurance program, reliability growth and field results. Quality Assurance, 7, 55-58.
2. MIL-HDBK-189 (1981) Reliability Growth Management
3. MIL-HDBK-217D (1982) Reliability Prediction of Electronic Equipment Modified by Notice 1 (1983)
4. MIL-STD-454H (1982) Standard General Requirements for Electronic Equipment
5. MIL-STD-485B (1980) Reliability Program for Systems and Equipment Development and Production
6. MIL-STD-765B (1981) Reliability Modeling and Prediction (Tasks and Methods)
7. MIL-STD-883C (1983) Test Methods and Procedures for Microelectronics
8. MIL-STD-965 (1977) Parts Control Program Modified by: Notice 1 (1978), Notice 2 (1981), Notice 3 (1983)
9. MIL-STD-1521A (1976) Technical Reviews and Audits for Systems, Equipments and Computer Programs
10. MIL-STD-1629A (1980) Procedures for Performing a Failure Mode, Effects and Criticality Analysis
11. RAC MDR-21 (1985) Microcircuit Device Reliability Trend Analysis
12. RADC-TR-82-87 (1982) Stress Screening of Electronic Hardware
13. RADC-TR-83-72 (1983) The Evolution and Practical Applications of Failure Modes and Effects Analyses
14. Taylor Engineering Department Design & Drafting Manual Instructions Section 25.1.4.16 (1979) Design Derating Requirements for Electronic Component Parts
15. Taylor Standards Section 107-8 (1980) Standard Definitions of Reliability Terms
16. Taylor Standards Section 208-1 (1982) Specification TK1: Specifications for 100% Screening Tests for Digital and Linear Integrated Circuits (IC's)

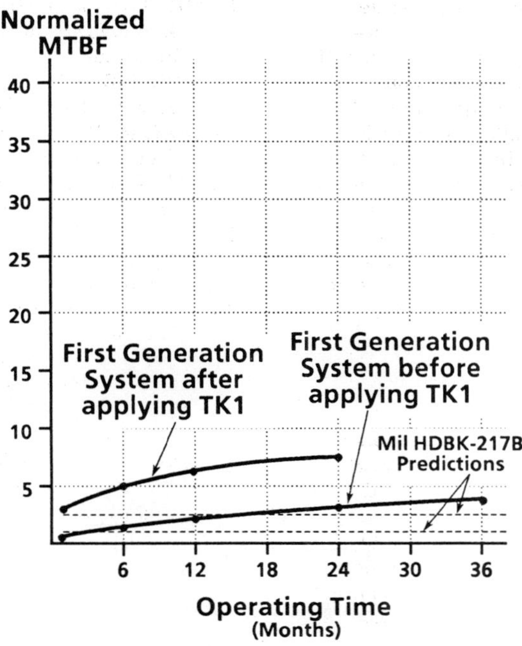

Fig. 1.   Field experience vs. predicted reliability for
Taylor's first generation microprocessor-based
system (MOD 3 Systems)

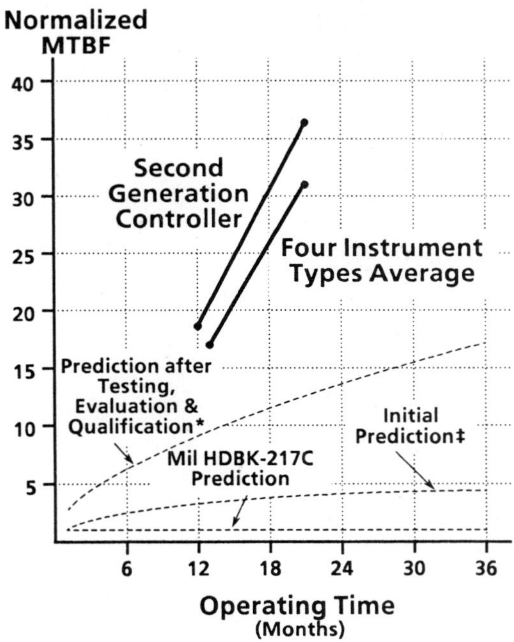

*Trend is from field experience on previous equipments
‡Including suspected wearout mechanisms

Fig. 2.   Field experiences vs. predicted reliabilities for
Taylor's second generation microprocessor-
based instrumentation systems (MOD 30
Systems)

R. C. Crombe

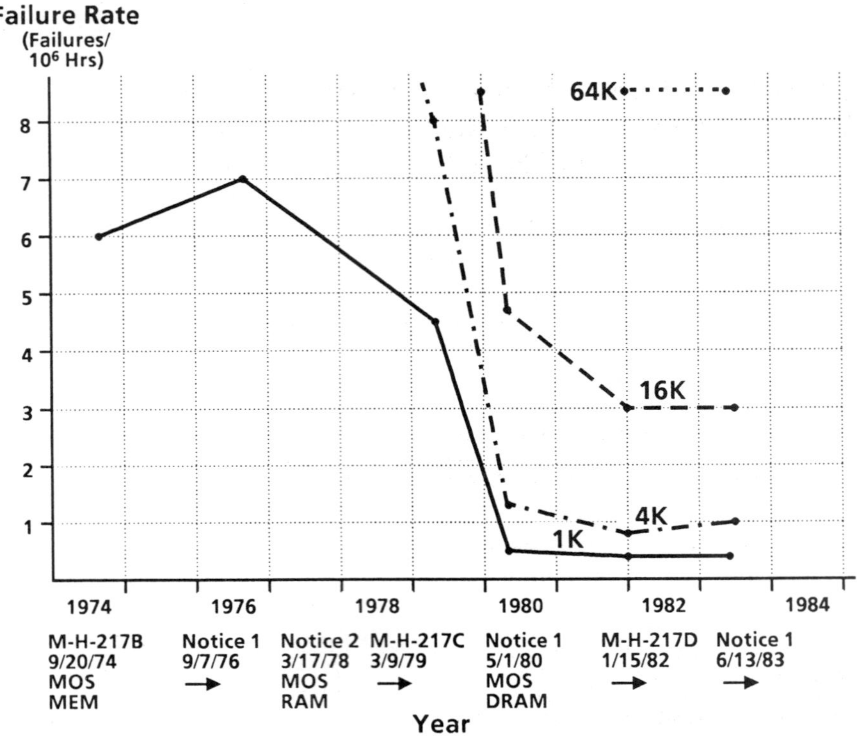

Fig. 3.   M-H-217 predicted dram failure rates

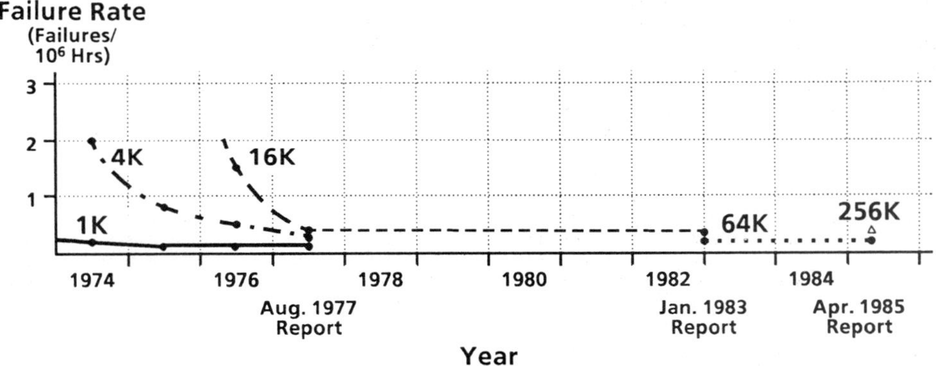

Fig. 4.   Major manufacturer's dram failure rates

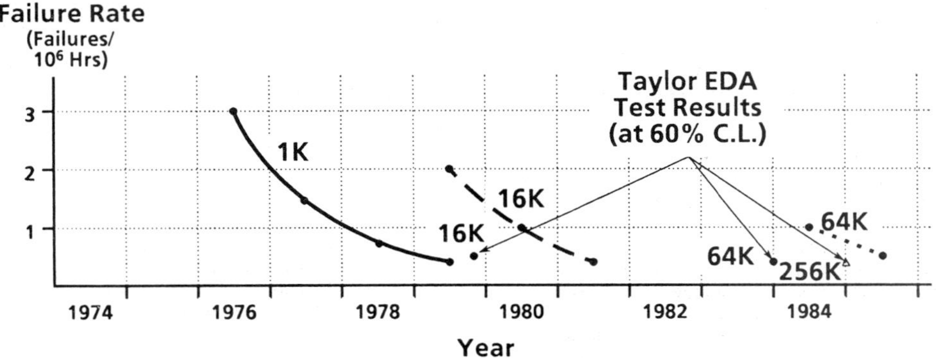

Fig. 5.   Taylor test and field experienced dram failure
rates derived from equipment averages

APPLICATION OF AN INNOVATIVE PROCESS DIAGNOSTICS ALGORITHM TO TUBE LEAK DETECTION IN A HEAT EXCHANGER

K. S. Vasudeva*, A. Cubukcu, K. A. Loparo**, M. R. Buchner****
and R. Yoel**

**Bailey Controls Company, Wickliffe, Ohio, USA*
***Department of Systems Engineering, Case Western Reserve University, Cleveland,
Ohio, USA*

Abstract. This paper describes an experimental investigation of a multiple model nonlinear filtering approach to leak detection in a laboratory heat exchanger. Although the work is still in its embryonic stage the results obtained are promising.

Keywords. identification, fault detection, filtering, process control, sensors

INTRODUCTION

In the current environment of increased capital equipment and energy costs, a primary goal for industrial process control systems is to ensure the reliable and economical operation of plant and equipment with minimal adverse impact on people. Normally it is necessary that this goal be achieved despite the random occurrence of contingency events such as the malfunction and failure of system components. When such events occur, either the system itself or the operator must take prompt and appropriate action to mitigate any adverse effects on the safety and productivity of the plant. Such action is important if the costly and catastrophic effects of plant malfunctions on the process are to be avoided. For catastrophic events, the operator would, in general, know that something had gone wrong, even though it may be too late to take any corrective action. There is, however, another class of failures that may initially have an insignificant effect but if left undetected, could lead to catastrophic failures. The discussion in this paper is concerned with the detection of such failures; and in particular to the detection of leaks in an experimental heat exchanger.

Due to the complexity of most industrial processes, all the process variables are not directly observable by the operator, therefore, instrumentation must be relied upon to gather information related to the plant's operating status. It is, therefore, important that the information presented to the operator be an accurate representation of the plant's actual status. The responsibility of establishing the validity of the data collected by the plant data acquisition system is a formidable task for an operator, especially since prompt corrective action is generally necessary to avert the possibility of a disaster. There is, thus, a need to develop automatic diagnostic techniques that continually monitor the plant's operation and inform the operator prior to the occurrence of a failure.

Several different techniques for process diagnostics have been proposed in the literature. Excellent survey papers on the subject have been provided by Iserman and Willsky. The major classes of failure detection methods may be categorized as follows:

Innovations Based Techniques

The innovations based failure detection techniques are in essence based on the correlating filter optimality with failure detection. If abnormalities in the system exist, changes in the statistical properties of the innovations process are expected to occur. Therefore, by performing statistical tests on the filter innovations it is possible to determine whether or not a failure in the system has occurred. Some common forms of statistical tests of the innovations process used to detect system failures are tests for whiteness, zero mean, larger than expected variance, and chi-squared tests. Refer to references by Mehra and Peschon; Willsky, Deyst, and Crawford; and Kerr for details.

Analytic Redundancy-Dedicated Observer Systems

The analytic redundancy or dedicated observer approach to failure detection is similar to a hardware redundancy approach except that an analytic estimate of the system output is compared to the actual output. For a multiple output system, one dedicated observer is constructed for each of the process outputs. Each of the observers process all of the system data and calculate an estimate of one of the system outputs. An error signal is generated which can be processed by a logical algorithm to determine if a sensor failure has occurred. Typically, if the actual output differs from the estimated output of a particular sensor by more than a predetermined threshold level, then a failure in that instrument is declared. For details see references by Clark, Fosth, and Walton; Clark; Clark and Cambell; Clark and Hertell; Deckert, Fisher, Laning, and Ray; Frank and Keller; Kitamura; Ray, Geiger, Desai, and Deyst; and Tylee.

Generalized Likelihood Ratio

The generalized likelihood ratio (GLR) approach to the failure detection problem is an extension of the simple innovations based detection techniques described earlier. The first step in the GLR approach is to enumerate all of the anticipated failures. Secondly, the effect of each failure on the innovations process and the failure detection filter must be determined. Assuming a linear system and filter, the general form of the innovations processes corresponding to the failed modes will be:

$$\nu_i(t) = G_i(t) + \bar{\nu}(t) \qquad (1)$$

where

$\bar{\nu}(t)$ = innovations process of the unfailed system,

$G_i(t)$ = "failure signature matrix",

$\nu_i(t)$ = innovations process assuming failure i has occurred.

If there are n possible failures that are anticipated in a system, then at each point in time n + 1 hypotheses must be examined to verify the existence of a failure.

Let H_0 = hypothesis no failure occurred,

H_i = hypothesis a failure of type i occurred,

i = 1, 2, ..., n.

For each anticipated failure a log-likelihood ratio, $\ell_i(t)$, may be calculated which gives a measure of the probability of hypothesis H_i being true relative to the hypothesis that no failure occurred, H_0. By comparing each of the log-likelihood ratios to some predetermined threshold level ε,

1) if $\ell_i(t) > \varepsilon$, then a failure of type i is declared,

2) if $\ell_i(t) < \varepsilon$ for all i, then no failure is declared.

A detailed description of the GLR method of failure detection has been provided by Tylee and Willsky and Jones.

Model-Based Equation Error Methods

The equation error methods essentially relate to material balance relationships, e.g., conservation laws. The basic idea is to use hypothesis testing techniques on the equation error to determine if a particular fault condition has occurred. The fault is localized by using a threshold on the statistical tests for the equation error vector. This information is then used to construct a fault incidence matrix. This technique, like other essentially linear estimation based methods, is sensitive to modeling errors. Refer to some recent work, Shutty and Sundbar, and the references therein for more details.

Nonlinear Filtering Applied to Systems with Random Structure

The nonlinear filtering approach to failure detection is based on the modeling of failure events as a random jump process. The approach requires that a system model be constructed for each of the possible failure modes. These models are used to construct a nonlinear detection filter whose output is the joint conditional probability law of the state/jump parameter process. Thus, rather than producing a failure/no failure type decision, the nonlinear (detection) filter calculates the conditional probability that the actual system is operating in each of the N possible modes.

The types of systems with random structure that are of interest in this study are those of the following form:

$$x(t) = A_j x(t) + B_j u(t) + V_j v(t)$$
$$z(t) = H_j x(t) + W w(t) \qquad (2)$$

where x(t), u(t), z(t) are the state, input, and output vectors respectively, the index parameter j is governed by a time homogeneous finite state Markov process, J(t), which takes values in a finite set {1, 2, 3, ..., N} and v(t), w(t) are uncorrelated Gaussian white noise processes. Each of the N possible modes of operation are distinguished by a set of five matrices, {A_j, B_j, H_j, V_j, and W}, each of which is assumed to be known.

Let $p_j(t)$ = prob{J(t) = j|z(s),$t_0 \leq s \leq t$} denote the conditional probability that the system is operating in mode j at time t given all of the output data up to time t. Also, let P(t) = $[p_1(t),p_2(t),...,p_N(t)]^T$ denote the vector of conditional probabilities, one for each possible mode of operation. Then the failure detection problem described earlier is equivalent to determining the vector P(t). A filtering scheme based on reparameterization of a Kalman filter is proposed by Davis and Loparo, Roth, and Eckert, characterize the optimal filter in terms of known finite dimensional filtering procedures. For this paper an approximate filter which assumes that the transition rates for the process J(t) are zero is used. That is, we simply have a parameter uncertainty problem and the filter has a finite dimensional recursive realization. This is a reasonable approximation if the detection time of the filter is small compared to the mean time between failure events.

THE PROCESS AND EXPERIMENTAL APPARATUS

A single pass shell and tube heat exchanger process, at Case Western Reserve University, was used for the experiments. A schematic of the process is shown in Fig. 1. The NETWORK 90 distributed digital control system manufactured by Bailey Controls was used for data acquisition and implementation of the nonlinear filtering and detection algorithm as well as for the control of the inlet water flow. The heat exchanger process was instrumented with industrial thermocouples to monitor the outlet temperature of the water and the temperature of the steam. Additional thermocouples were installed along the water tube to calibrate a lumped parameter model of the process. A differential pressure transmitter was installed at the inlet side of the heat exchanger to obtain data on the inlet water flow and provide flow information to the NETWORK 90 Control System so that the inlet water valve could be regulated to control the external disturbances in the inlet water flow.

A distributed parameter mathematical model of the process was developed from first principles by Polichronis. This distributed model of the process was simulated and the results were compared with actual experimental data to verify the correctness of the model structure and the underlying assumptions made in its development. The distributed model was of the form

$$\frac{\partial T_f}{\partial t} + \mu_f \frac{\partial T_f}{\partial z} = B(T_s - T_f) \qquad (3)$$

where T_f is the fluid temperature, T_s is the steam temperature, and B is an aggregate heat transfer coefficient between the fluid and the steam, and μ_f is the fluid velocity. The distributed parameter model was reduced to a lumped parameter form by dividing the heat exchanger into a tribe of N segments by transforming the spatial co-ordinate z into discrete segments. This yields a system of ordinary differential equations of the form:

$$\frac{dT_n(t)}{dt} + \mu_{f_n} \frac{T_n(t)T_{n-1}(t)}{\Delta z} = B(T_s - T_n(t)) \qquad (4)$$

where n = 1,2,...,N refers to the number of segments in the lumped parameter approximation. A detailed simulation study was carried out by Polichronis and Cubucku. The results obtained during our experiments confirmed the validity of the model and indicated that a seven segment approximation was adequate for our purposes. The lumped parameter model (4) described the operation of the process during normal operation, i.e., the no leak situation, a leak in the inner tube of the heat exchanger is considered as a failure. The mathematical analysis developed in by Polichronis indicates that the water pressure is sufficiently greater than the steam pressure so that the leak flow is from the tube to the shell. The leak affects the velocity and pressure distribution of the water flow downstream from the point at which the leak occurs. This, of course, affects the overall heat transfer coefficient for the process; the parameter B in the lumped model (4). For this analysis the changes in the coefficient B are neglected since we were only concerned with leaks flows in the range of 6-15% of nominal water flow. This adds an error to the detection filter and provides some information on the robustness of the proposed detection algorithm.

A leak model based on energy balance and continuity relationships was developed and this model was used to determine the effect of a leak on the normal system model structure and parameter values. Simulation tests were conducted to validate the qualitative behavior of the leak model for a distributed and lumped parameter representation of the heat exchanger process.

As indicated in Fig 1, the experimental heat exchanger process was modified to include a leak in the inner water tube at the second and sixth segments. The leak tubes were brought out of the shell through the end caps and fitted with valves to control the leak flow rate. During the experiments the leak flow was determined from real-time data.

An extensive set of simulation results were obtained and reported by Cubukcu.

EXPERIMENTAL PROCEDURE AND RESULTS

The experimental procedure consisted of an extensive set of tests to evaluate the performance of the nonlinear filtering diagnostic algorithm. Two representative experimental tests are discussed in this section. For a more detailed description refer to Yoel.

The first test corresponds to a nominal input water flow rate of two gallons per minute (2 GPM) and a leak of 6.51% of nominal water flow in the second segment of the process. The heat exchanger process was brought to steady-state operating conditions before the experiment was begun.

The real-time decision statistics are generated by a set of parallel estimators which feed-forward into a nonlinear post-processing filter as shown in Fig. 2. The algorithm is based on a **single temperature measurement**, the exit water temperature and the decision logic component was a simple threshold detector operating on the decision statistics.

At the 4th data sample time a leak of 6.51% of the inlet flow was introduced into the process. The real-time data was collected from the process using the Bailey NETWORK 90 system and is shown in Figs. 3 through 5. In all of these figures the arrow indicates the time at which the leak was introduced. Fig. 3 shows the outlet water temperature, on a scale from 0-200°F, versus time. There is no obvious signature introduced by the leak. Fig. 4 shows the outlet water temperature, on a scale from 130-140°F, versus time. The water temperature measurement is quite noisy and it would be very difficult for an operator to make a decision on the basis of this data. Fig. 5 shows a plot of a normalized decision statistic indicating the no-leak situation. The abrupt changes in the decision statistic are a result of feedback from the threshold decision logic to the algorithm. If a decision statistic crosses the 0.8 threshold, a flag is set indicating the status of the process and the algorithm is re-initialized. For this test the algorithm was initialized three times, each indicating a no-leak situation. The arrow indicates the time at which the leak was introduced. Since the decision statistics are normalized and constrained to add to one, it is only necessary to monitor one statistic and a leak decision corresponds to the no leak decision statistic crossing the 0.2 threshold. After the leak is introduced the algorithm responds, and after approximately 15 samples the decision statistic indicates the leak situation. This continues throughout the experiment.

It is noteworthy that the decision logic implemented in this experiment is the simplest possible. The threshold, 0.8, indicating detection was arbitrarily selected. Even in this simple threshold detection scheme, there are a variety of trade-offs between detection time, false detections, and missed detections which need to be quantified in terms of the threshold limit, process disturbances, and measurement noise statistics. Considerable work remains to be done in addressing these, and other issues related to the development of a robust decision logic system.

The second set of experiments were undertaken to determine how the algorithm would perform in the following situation: A leak is introduced at the 17th data sample time and the leak is turned off at the 47th sample time. The flow rate for this experiment was 2.5 GPM and the leak flow was 6.5% of nominal inlet water flow. Fig. 6 shows the outlet water temperature on a scale from 0-200°F and Fig. 7 shows the outlet water temperature on a scale from 130-140°F. The first arrow indicates when the leak was introduced and the second arrow indicates when the leak was turned off. For this experiment, there were substantial inlet water flow disturbances which caused the outlet temperature to be noisy. Fig. 8 shows the performance of the algorithm in terms of the no leak decision statistic.

CONCLUSION

This paper discussed an application of a multiple model nonlinear filtering technique for the detection of leaks in a laboratory heat exchanger. A simple decision logic algorithm (threshold crossing) was used to process the filter outputs to indicate the status of the process, i.e., the leak or no leak situation.

The results of the study indicate the feasibility of the proposed diagnostic technique to detect small leaks in an experimental heat exchange process. The study also indicates that the technique is robust in the midst of significant flow disturbances on the inlet water flow, using only simple models for processing the data, but is sensitive to the aggregate heat transfer parameter (B) used in the detection algorithm. Further work is required to investigate the use of on-line parameter identification for model tuning, i.e., adjustment of the B parameter to improve the performance of the algorithm. Also, significant work needs to be undertaken in the area of selecting an appropriate decision logic algorithm which takes into account the various objectives, such as minimum detection time and minimum false alarms.

ACKNOWLEDGEMENTS

The work described in this paper was supported by Bailey Controls Company and the State of Ohio Department of Development.

REFERENCES

Cubukcu, A. (December 1985). "Leak Detection in a Heat Exchanger System: A Nonlinear Filtering Approach", U.S. Thesis, Department of Systems Engineering, Case Western Reserve University, Cleveland, Ohio.

Clark, R.N., D.C. Fosth, and V.M. Walton (July, 1975). "Detecting Instrument Malfunctions in Control Systems", _IEEE Transactions on Aerospace and Electronic Systems_, AES-11, pp. 465-473.

Clark, R.N. (May 1978). "Instrument Fault Detection", _IEEE Transactions on Aerospace and Electronic Systems_, AES-14, pp. 456-465.

Clark, R.N. and B. Cambell (January, 1982). "Instrument Fault Detection in a Pressurized Water Reactor Pressurizer", _Nuclear Technology_, Vol. 56-pp. 23-32.

Clark, R.N. and J. E. Hertel (May 1982). "Instrument Failure Detection in Partially Observable Systems", _IEEE Transactions on Aerospace and Electronic Systems_, AES-18, pp. 310-317.

Davis, M.H.A. (April 1975). "The Application of Nonlinear Filtering to Fault Detection in Linear Systems", _IEEE Transactions on Automatic Control_, AC-20, pp. 257-259.

Deckert, J.C. and others (June 14-16, 1982). "A Signal Validation Methodology for Nuclear Power Plants", _Proceedings of American Control Conference_, pp. 113-120, Arlington, Virginia.

Frank, P.M. and L. Keller (July 1980). "Sensitivity Discriminating Observer Design for Instrument Failure Detection", _IEEE Transactions on Aerospace and Electronic Systems_, AES-16, pp. 460-476.

Iserman, R. (1984). "Process Fault Detection Based on Modeling and Estimation Methods - A Survey", _Automatica_, Vol. 20, pp. 387-404.

Kerr, T.H. (July 1982). "False Alarm and Correct Detection Probabilities Over a Time Interval for Restricted Classes of Failure Detection Algorithms", _IEEE Transactions on Information Theory_, IT-28, pp. 619-631.

Kitamura, M. (1980). "Detection of Sensor Failures in Nuclear Plants Using Analytic Redundancy", _Transactions of the American Nuclear Society_, Vol. 34, p. 581.

Loparo, K.A., Z. S. Roth, and S. J. Eckert, "Nonlinear Filtering for Systems with Random Structure", submitted for publication.

Mehra, R.K. and J. Peschon (1971). "An Innovations Approach to Fault Detection and Diagnosis in Dynamic Systems", _Automatica_, Vol. 7, pp. 637-640.

Polichronis, R.P. (August 1985). "Dynamic Model, Failure Analysis and Simulation of a Double Pipe Heat Exchanger", M.S. Thesis, Department of Systems Engineering, Case Western Reserve University, Cleveland, Ohio.

Ray, A. and others (June 14-16, 1982). "Computerized Fault Diagnostics in a Nuclear Reactor Via Analytic Redundancy", _Proceedings of American Control Conference_, pp. 784-791, Arlington, Virginia.

Shutty, J. (January 1986). "A Multilevel Approach to Fault Detection", M.S. Thesis, Department of Systems Engineering, Case Western Reserve University, Cleveland, Ohio.

Sundbar, A.C. (January 1986). "Process Fault Detection using Augmented System Model Approach", M.S. Thesis, Department of Systems Engineering, Case Western Reserve University, Cleveland, Ohio.

Tylee, J. L. (March 1982). "A Generalized Likelihood Ratio Approach to Detecting and Identifying Failures in Pressurizer Instrumentation", _Nuclear Technology_, Vol. 56, pp. 484-492.

Tylee, J.L. (July 1982). "Real Time Instrument Failure Detection in the LOFT Pressurizer Using Functional Redundancy", report prepared for the U.S. Department of Energy under DOE Contract No. DE-AC-07-761D01570.

Tylee, J.L. (March 1983). "On-Line Failure Detection in Nuclear Power Plant Instrumentation", _IEEE Transactions on Automatic Control_, AC-28, pp. 406-415.

Willsky, A.S., J. J. Deyst, and B.S. Crawford (June 19-21, 1974). "Adaptive Filtering and Self-Test Methods for Failure Detection and Compensation", _Proceedings of the 1974 JACC_, Austin, Texas.

Willsky, A.S. (1976). "A Survey of Design Methods for Failure Detection in Dynamic Systems", _Automatica_, Vol. 12, pp. 601-611.

Willsky, A.S. and H. L. Jones (February, 1976). "A Generalized Likelihood Ratio Approach to the Detection and Estimation of Jumps in Linear Systems", _IEEE Transactions on Automatic Control_, AC-21, pp. 108-112.

Yoel, R. (December 1985). "Real-Time Implementation of a Nonlinear Filtering Process Diagnostic System for Leak Detection in a Heat Exchanger", MS Project Report, Department of Systems Engineering, Case Western Reserve University, Cleveland, Ohio.

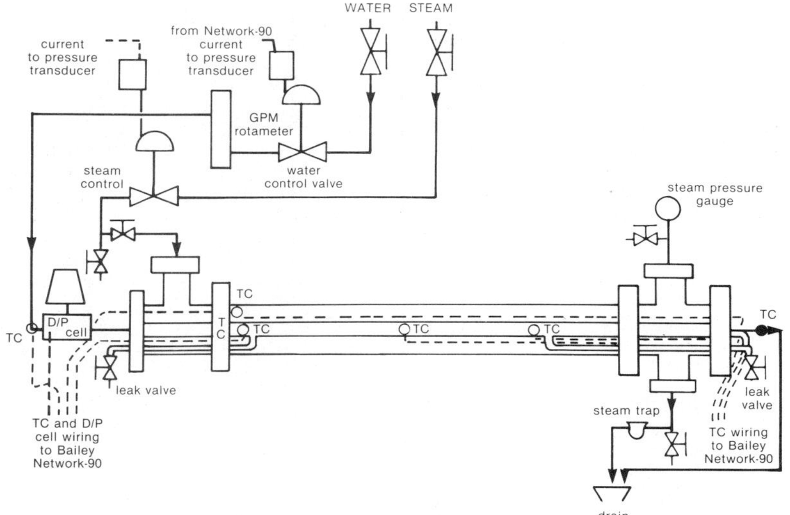

Fig. 1. Heat exchanger process

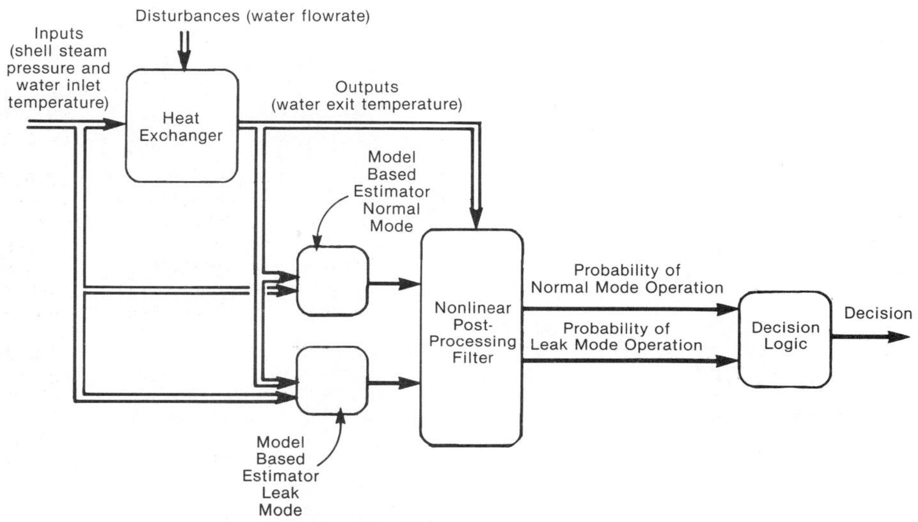

**Fig. 2. Process diagnostic system applied to a heat exchanger for leak
detection**

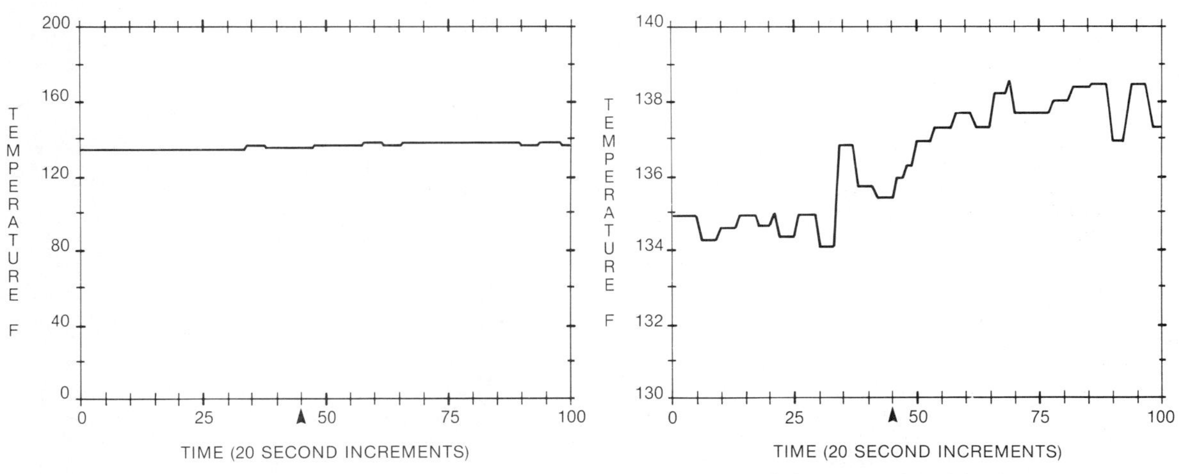

Fig. 3. Outlet temperature against time

Fig. 4. Outlet temperature (expanded scale) against time

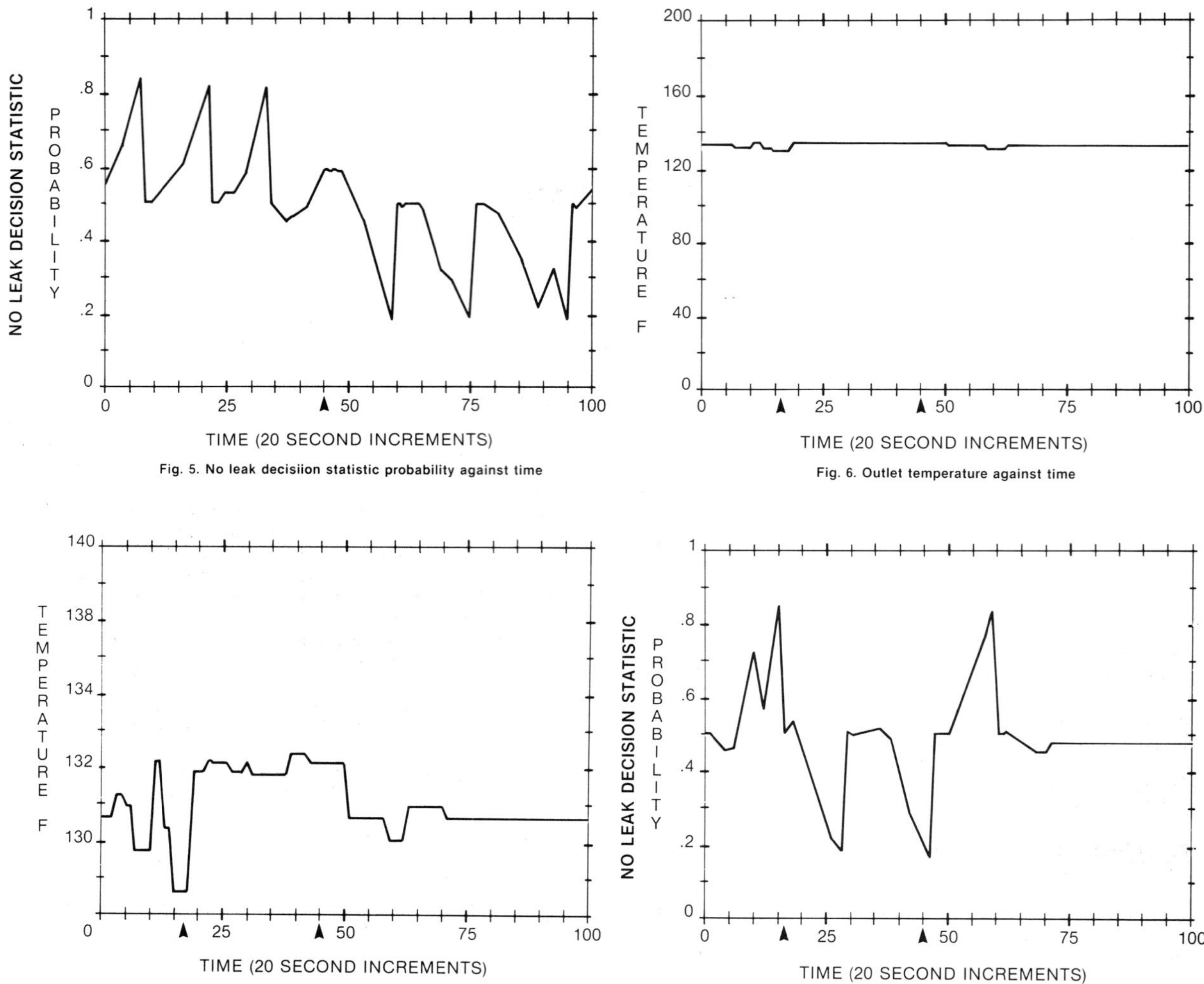

Fig. 5. No leak decisiion statistic probability against time

Fig. 6. Outlet temperature against time

Fig. 7. Outlet temperature (expanded scale) against time

Fig. 8. No leak decision statistic probability against time

DISCUSSION

Session: Field data and maintenance

Paper: Experience with integrated control systems.

Question: Many systems had a particular reoccurring
type of problem which had a significant influence
on the failure frequency graph. Does this mean,
that repairs were not done successfully?

Author's reply: Not necessarily, one of the systems
e.g. had many unrelated problems with a printer,
another showed software problems which took several
years before the real cause was found and resolved.

Question: Did you experience outputs to change
randomly because of a fault?

Author's reply: No, in case of a total failure the
outputs went to a pre-determined state.

Question: Do you include emergency shut-down functions
in the systems?

Author's reply: We use independent hardwired trip
systems. In the integrated control systems emergency
shut-down sequences are incorporated.

Paper: Reliability growth program ensures high
availability for next generation industrial instrumentation systems.

Question: In the paper is stated that complexity
of standard VLSI circuits don't influence the failure
rate.
- Do you use semi-custom or custom VLSI-circuits
  in your products?
- Such circuits enhance reliability of pc boards,
  but what about the reliability of those VLSI
  circuits and their influence on the reliability
  of complete systems?

Author's reply: We don't use custom or semi-custom
VLSI-circuits in our MOD-300 products. We do use
them on our new 400T and 500T series of pressure
and differential pressure ceramic capacitive-sensor
transmitters. These transmitters also use hybrid
type PC boards using all surface-mounted components
and these also use operational laser trimming.
- These types of VLSI circuits are still experiencing
  learning curve and reliability growth improvements,
  but are expected to eventually achieve equivalent
  reliability to other standard VLSI-circuits.
  Therefore, eventually their use should improve
  reliability for the PC-boards and for the complete
  system (this includes improvement from a reduced
  number of connections).

Question: Can you expand on how maintainability is
improved by standardization?

Author's reply: Standardization makes spares more
readily available, makes for easy and simple repair
and, thereby, reduces downtime.

Question: How do you control reliability for changes
in "design masks" or manufacturing country?

Author's reply: We control Quality and Reliability
of incoming complex IC's by our TK1-Specification,
which is based on MIL-STD-883, quality level B-2.
TK1 has incoming acceptance criteria on either
a sampling or 100% test basis. We also control
the change in mask version, major packaging changes
or country of manufacture by requiring the
manufacturer to notify us of these major changes.
We then require Qualification and Reliability types
of information from the manufacturer and we may
also require a requalification in-house.

Question: Why do you limit your Component Reliability
test results to a 60% confidence level?

Author's reply: This is a single step in our cost-
effective reliability assurance program. In
addition to a complete Qualification test of the
component parts, we assure proper de-rating,
design reviews and testing under actual pc-board
and sub-assembly operating conditions.

TREND PRESENTATION AND HUMAN'S PREDICTABILITY

T. N. White* and P. van der Meijden**

**Ergonomics Group, Twente University of Technology, Enschede, The Netherlands*
***Department of Applied Mathematics, Twente University of Technology, Enschede,
The Netherlands*

Abstract. Various time series with different physical sizes of amplitude- and
time-scales are experimentally investigated. The task is to extrapolate the
discrete points of the trend by making a prediction of one data point in future.
Objective measures of the prediction accuracy are analysed under the various
experimental conditions. These measures are compared with subjective ratings of
the perceived difficulty to predict under the investigated conditions. Results
show that the larger the physical size of the trend presentation becomes, the
higher prediction accuracy is found. In contrast to these results, the higher
the prediction accuracy becomes, the more difficulties one perceives to predict.
This discrepancy can have a large impact on safety. It leads to the conclusion
that using face validity only is hazardous in the decision phase of formatting
trend information on VDT's.
Applying a line between the succesive data points in a trend as a visual aid is
dependent of the 'number of data points'; 10 data points the effect is
possitive, 15 data points is neutral and 5 data points the effect is negative on
the prediction performance. Again subjective ratings are far from identical to
the objective performance measurements. As suggestion, one should always apply
the objective performance data and preferably use both types of measurements in
designing VDT presentations.

Keywords: trend information, human prediction, VDT's, display formatting.

INTRODUCTION

The supervision and control of large, multivariable and complex systems is increasingly performed
by a small number of operators. The tools applied
in those cases are modern instrumentation systems
connected with huge system computers. As a consequence, the conventional displays such as indicators, face plates and hard-wired or chart recorders are removed from the control room and replaced
by VDT-screens. The layout on those screens can in
principle be changed rapidly and groupings of
information can be altered without rebuilding the
complete control panels. This flexibility has at
least one disadvantage, i.e. large quantities of
information cannot be displayed parallel but has
to be presented serially. This argument is regularly counterbalanced by the remark that grouping
of related information solves this problem to an
important extend. Besides it is argued that
modern computer driven instrumentation systems
allow the ability to make very accurate post-
mortem analyses: analyses to locate the cause of
the occurance of severe system malfunctions and
accidents.

The discussion about the sense of presenting
historical information to the operator is far from
being closed. Due to safety regulations in some
countries the hard-wired recorders are still
installed. Besides, trendrecorders are frequently
accepted by the fact that many operators say to
rely on the information being presented by those
recorders.

Although a plea is made for using hard-wired
trendrecorders based on safety regulations and
emotional considerations, the tendency to implement historical information on VDT-screens is
gaining preference.

The discussion of this paper is primarily focussed
on the basis of two questions. When trends are
displayed on VDT's how should they be layouted and
what physical size should these trends have in
order to be a decision aid for the operator to
supervise the system.

One of the main reasons to display trend information to operators is that operators are thought to
be better able to extrapolate on the basis of
historical and present information to future
values of variables. Due to trend presentation
anticipating control can thus principally made.
Moreover, trendpresentation can be used to facilitate early detection of malfunctioning in the
control system.

The method applied here is to introduce different
forms of trend presentation, fundamentally based
on changing time and amplitude scale of the
historical information, and gather experimental
data about the difference in prediction accuracy
between the presentation forms. These objective
data are then statistically tested and compared
with the subjective ratings of the testees about
their perceived performance in doing the prediction task.

This research is performed because practically no
information can be found in scientific literature
about layouting trend presentation. When information is present in literature it is mostly beyond
the scope of this purpose. Exceptions are Rouse
and Enstrom, 1976; Van Bussel, 1980; Van Heusden,
1980; Bösser and Melchior, 1984; White and Van
Heusden, 1984; Van Schaik and Rijnsdorp, 1985; and
White and Van Schaik, 1985.

DESCRIPTION OF THE EXPERIMENTS

Definitions and theoretical aspects

A trend is defined as a representation of a certain (proces) variable, as a function of time. In
a majority of cases the horizontal axis is allocated for time information, where the vertical axis
is applied for amplitude information. Contrary to
many hard-wired recorders the horizontal axis is
the time-axis in the experiments discussed later.
The choice to represent time on the x-axis and the
most recent information on the right hand side is
made on the findings of Koch, Edman and Guenther
(1982).
The signal shown in the trend is a second-order
autoregressive time series. This implies that the
value of the amplitude of the signal on time
instant i, x_i, is derived from the equation:

$$x_i = a_1 x_{i-1} + a_2 x_{i-2} + r_i \qquad (1)$$

The term r_i is a gausssian noise. The parameters
a_1 and a_2 are fixed and have the values 1.30
and -0.65 respectivily. With these values a rather
representative output signal of a trendrecorder is
generated.
As said, a trend can be represented in a variety
of ways. The number of data points, the shape of
the data points, the presence of connected lines
between two successive data points (polynomial,
straight), one or more trends sharing the same
zero-line, trends next to each other or above each
other and last but not least the scaling of time
and amplitude axis can be topic of investigation
and optimization. The last two aspects are
responsible for the shape of the trend, expanded
in vertical or horizontal direction.
The position of a data point on the screen can be
expressed in the number of pixels (the smallest
addressable unit of the screen) distant of both
the axes. Moreover, the lenght of the axes can be
expressed in number of pixels.
The display screen applied in the experiments has
a vertical resolution of 192 pixels and a horizontal resolution of 280 pixels. The screen is part
of the standard configuration of an Apple IIe
which is used to run the experiments.
To determine the prediction or extrapolation error
made by the testees the predicted value is sub-
tracted from the value to be predicted. In order
to eliminate the effect of the unit of length of
the amplitude axis, the absolute value of the
found quantity is divided by the unit of
amplitude. The relative error score is thereby
obtained. To quantify the total prediction accuracy after n prediction trials, the relative errors
are squared and summed. The socalled RMS-error
score (Rooted Mean of Squares) is calculated by
applying the following equation:

$$RMS = [1/n \sum_{i=1}^{n} (\text{relative error i})^2]^{1/2} \qquad (2)$$

The RMS-value expresses the prediction inaccuracy
made by the subject when performing the prediction
task in one time series. So the higher the RMS-
value becomes the less accurate the time series is
predicted.
From a pure theoretical point of view, only one
noise realization should be applied to compare the
different trend layouts. From a practical point of
view one noise is apt to be recognised very soon
by the subjects. The learning curve of recognising
the series is then detemined instead of quantifying the differences in trend layout.
Therefore different time series with comparable
statistical properties such as variances of the
noise term in the signal, variances in the total
signal and theoretical predictability (TP) are

generated. The theoretical predictability is a
quantity which expresses the difficulty of the
predictions in each time serie:

$$TP = [1 - \frac{S^2 \text{ noise}}{S^2 \text{ total}}] * 100\% \qquad (3)$$

In the experiment a TP= $\pm$ 81% is applied. For each
series the optimal predictions can be determined
by means of an optimal predictor algorithm based
on ARMA-modelling, see for instance Van Heusden
(1980). Applying all optimal predictions within
one series, a RMS-error score can be calculated.
Besides the statistical constraints mentioned
above, another statistical constraint is important
when using different time series in order to
prefend signal recognition. The optimal RMS-error
scores has to be comparable when applying different series. The optimal RMS-error scores are never
completely identical. The subjects RMS-error score
is therefore related to the optimal RMS-error
score of that particular time series. The final
prediction accuracy measure the Performance Factor
(PF), is defined as:

$$PF = \frac{RMS \text{ optimal}}{RMS \text{ subject}} \qquad (4)$$

- The RMS optimal is the RMS-error score of
  the optimal predictor for a certain time serie.
- The RMS subject is the RMS-error score of a
  subject for that particular time series
  obtained as a result of one of the experimental
  trials.

The larger the PF-score is, the better the
predictions of that subject for that serie. The
Performance Factor is consequently serie
dependent, although much smaller then the original
RMS-error score of the subject. Using a complete
experimental design a fair comparison of
differences between subjects and more important
between trendlayouts can be made without measuring
just artefacts.

Hypotheses and dependent variables

The hypotheses carry along two basic aspects:
 -the size of trend presentations will
  influence the prediction accuracy output,
 -the ratio of time- and amplitude-scaling
  influences the prediction accuracy.
 -connection of the data points by a line
  will influence the prediction error.

The following postulations are experimentally
tested:
o trend presentation on a large physical scale
  results in smaller prediction errors than trends
  having small screen formats,
o the ratio between amplitude and time scale
  around the value 1.0 (45 degrees of angle) show
  higher prediction accuracy than more diverging
  values.
o the connection of two successive data points by
  means of a straight line results in a better
  prediction performance.

Experimental conditions

The experiments are a continuation and expansion
of those described by White and Van Heusden
(1984). Therefore some variables are kept
constant, which are:
 -the number of trends on the VDT, i.e. 1,
 -the length of the time-axis, i.e. 249
  pixels,

-the length of the amplitude axis, i.e. 192
 (the screen maximum in vertical direction),
-the kind of symbol to indicate the several
 data points in the trend, i.e. a square of three
 by three pixels,

In the experiment ten distinctive time series are
applied with approximately identical statistics.
The next variables are non-fixed:
 -the unit on the time axis,
 -the unit on the amplitude axis,
 -the number of data point shown to the
  subject (the rate of historical information),
 -the presence or absense of a line between
  the successive data points.

Given the resolution of the screen, (192 * 280
pixels) and the values of the 10 time series,
lying between +5 and -5, the following limitations
are put to the unit of time and the unit of
amplitude:
         5 <= unit of time <= 30
         4 <= unit of amplitude <= 16
Due to the physical limitation of the screen 57
different trend presentations can be used in the
experiments; 26 trends with 5 data point as
history, 20 trends with 10 data points, and 11
trends with 20 data points of visible information.

The task to be performed

The subject will be seated behind a green VDT-
screen and is supposed to extrapolate the 5, 10 or
20 data points of the trend by indicating the
vertical position of the next point. The indication is made visible on the screen by moving a
dedicated joystick up or down in the activated (y-
axis) direction. When the pointer is placed on the
predicted position the prediction is aknowledged
by pushing one of the two buttons on the joystick
box. All data points in the trend will then shift
one position to the left while the actual to be
predicted point appears at the right hand side of
the trend. The oldest data point at the left hand
side disappears as a consequence. The subject gets
the opportunity to compare the rate of correctness
of his prediction with the actual value before
moving the pointer to indicate the next prediction. A counter shows how many of the predictions
has still to be made in that particular series,
see figure 1.
Having completed the predictions the subject is
requested to answer three questions about the
perceived difficulty, the prediction strategy, and
give an indication about the concentration during
the prediction activity. After the complete
session the subjects filled in a leaflet with
questions about age, sex, education, VDT/text-
editing/computer programming experience, the
introduction given, the impression about the
experiment, the recognition of one or more signals
and if yes where and when and the perceived
benefit, and a question about critical remarks.
The subjects get a small payment for the
participation.

EXPERIMENT 1

The 19 subjects/participants started the
experimental session with reading a paper printed
instruction and one or, if requested, two trials
to get exercised.
Eight different ratios of amplitude- and time-axis
are applied: 1:6, 1:5, 1:4, 1:3, 1:2, 1:1, 2:1,
and 3:1. Ratios of 1:6, 1:5 and 1:4 are concurrent
with angles between 10-20 degrees; ratio 1:3 = 20-
30 degrees; ratio 1:2 = 30-40 degrees; ratio 1:1 =
40-60 degrees; ratio 2:1 = 60:70 degrees and ratio
3:1 = 70-80 degrees.

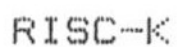
RISC-K

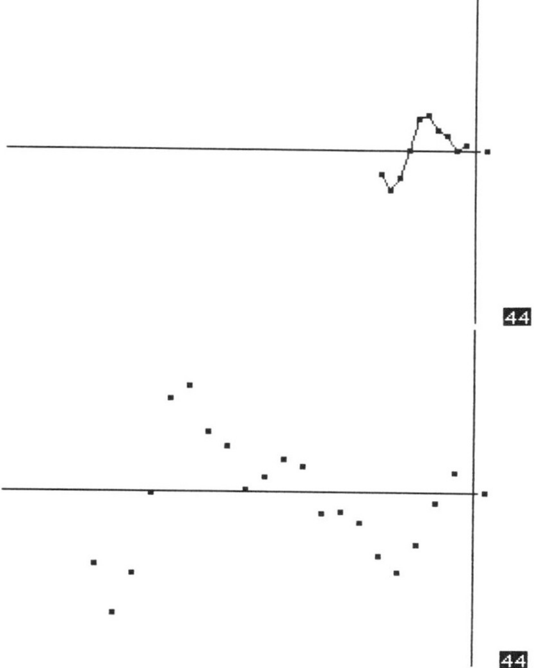

Figure 1: Layout of two trend presentations

In the first experiment no line will be drawn
between the data points.
Each subject makes 50 predictions in 15 series.
Computing the PF-value, the first 5 predictions
are skipped to compensate for accustomation
effects by the subject from one trend presentation
to another and to compensate for starting fluctua-
tions of the optimal predictor.
The 15 time series are ordered in such a way that
each subject has an equal number of 5-, 10- and
20-data point series.
All statistical testing is executed with SPSS
(Statistical Package for the Social Sciences).

NUMERICAL RESULTS OF EXPERIMENT 1

Introduction

A first test is made to check the existence of a
gaussian distribution by means of the Kolmogorov-
Smirnov test. If the result is possitive the
comparison of mean differences can be made by a T-
test. The result of the test has indicated that
T-testing is allowed. The level of accepting
statistical differences or significance is for p
<= .05.

The effect of number of data points

To check whether or not the number of historical
data points influences the prediction performance
two test conditions can be distinguished:
(1) with all 57 trend presentations involved,
(2) with trend presentations having identical
units (11 units for each number of data points).
A comparison between 20 and 10 data points under
condition (1) and (2) results in a significant
difference in favour of 10 data points. A comparison between 10 and 5 points is of a significant
difference under condition (1) and almost significant in favour of 10 data points under condition
(2). Under condition (1) and (2) 5 points are
showing better performance than 20 points which is
near to significant, see Table 1.

Table 1   Effect of data points on PF

Condition (1)

Results test with all 57 trend presentations	two-sided probability P
10 points better than 5 points	.005
10 points better than 20 points	.000
5 points better than 20 points	.072

Condition (2)

Results of test with the 11 identical units trends	two sided probability P
10 points better than 5 points	.057
10 points better than 20 points	.000
5 points better than 20 points	.055

Table 2   Results of mean angle on prediction performmance

x is better than y	2-sided prob.	x is better than y	2-sided prob.
10-20/20-30	.078	60-70/10-20	.622
10-20/30-40	.874	60-70/20-30	.044 *
10-20/50-60	.198	60-70/30-40	.570
30-40/20-30	.151	60-70/50-60	.124
30-40/50-60	.261	70-80/10-20	.001 *
40-50/10-20	.005 *	70-80/20-30	.000 *
40-50/20-30	.000 *	70-80/30-40	.001 *
40-50/30-40	.008 *	70-80/40-50	.286
40-50/50-60	.001 *	70-80/50-60	.000 *
40-50/60-70	.030 *	70-80/60-70	.005 *
50-60/20-30	.895		

Testing the effect of the unit of amplitude

The 57 trends are categorized in four classes of units along the amplitude axis:
- o Class I with unit 4 or 5 pixels,
- o Class II with unit 6, 7 or 8 pixels,
- o Class III with unit 10, 11 or 12 pixels,
- o Class IV with unit 14,15 or 16.

Without differentiation to the number of visible data points the results can be summarized as: The larger the unit of amplitude, the better the prediction accuracy. The differences are not always significant. Class IV is always significantly better to induce high prediction performance than the other classes.
When differentiation is made for 5, 10 and 20 data points still class IV is superior. Differences for 5 and 10 data point are however nolonger significant between class IV and class III.

Testing the effect of the unit of time

The 57 trend presentations are classified in seven categories of units along the time axis:
- o Class I: time units of 5 and 6 pixels,
- o Class II: time units of 7 and 8 pixels,
- o Class III: time unit of 12 pixels,
- o Class IV: time unit of 15 and 16 pixels,
- o Class V: time units of 18, 19 and 20 pixels,
- o Class VI: time units of 24 and 25 pixels,
- o Class VII: time units of 28, 29 and 30 pixels.

Other time units are not experimentally tested. Without differentiation the PF-value shows no coherence in the results. Similar effects are found after differentiation.

Testing the effect of mean angle

For each trend the mean angle is calculated. This mean angle is the sum of angles between all successive data points and devided by the number of angles found. The trends are classified in seven categories with a mean angle of 10 to 80 degrees.

Without differentiation between the number of data points, a mean angle between 70 and 80 degrees results in the highest prediction performance. With all other groups except with the category 40-50 degrees significantly differences are found. The 40-50 degrees of angle category is significantly different from the remaining categories. The third place is occupied by the category 60-70 degrees, see table 2.

With differentiation to the number of data points the categorie of 70-80 degrees in the class of 5 datapoints is still superior to the other categories in that class; mostly significantly different or close to significancy.
In the class of 10 data points the mean angles between 40 and 50 degrees are inducing highest performance. Differences with 60-70 and 70-80 degrees are not substructured by the statistics.
In the class of 20 datapoints again the categorie 70-80 shows significantly better results. It has to be remarked that the categorie 40-50 degrees is missing here and the other categories do lack many observations.

Summary of the results of numerical data

(1) 10 data points of historical information in the trend, results generally in better prediction performances compared to 5 and 20 data points. 5 datapoints leads in most cases to better results than when 20 data points are visible in the trend.
(2) The larger the unit of amplitude, the better performance is measured. The categorie with the largest unit of amplitude (14, 15 and 16) is significantly better than the other categories.
(3) Variations in the unit of time has not shown significantly different results. The parameter unit of time is not a sensitive one in these experiments.
(4) The mean angle between 70-80 degrees (the largest angles, ratio 3/1) leads to the best prediction performances. 40-50 degrees of angle (ratio 1/1) is the second in the row. 60-70 degrees (ratio 2/1) although with some hesitation is the third best.

SUBJECTIVE RATINGS

Introduction

Many systems are judged on the technical part wheras the interface between man and machine is evaluated on face validity. This means that when one gets a positive impression of the information presentation, a maximization of colour coded variables on VDT for instance, one is easily apt to invest in such a system. The question to be answered here is whether one can rely on face validity or not. It has to be remarked that the investigation deals mainly with other aspects than attractiveness of VDT presentations. The subjectively estimated prediction performance with different trend presentations is requested and compared with the objectively measured prediction performance.

Perceived difficulty to predict

On a 5-points Gutmann-scale the subjects has to
indicate the level of perceived difficulty for
each predicted trend (1=very difficult, 5=very
easy).
Every trend has been investigated by means of 5
measurements. In order to be more or less subject
independent the 5 scores for each trend has been
added up. The rate of perceived difficulty now
ranges from 5=very difficult to 25=very easy.
To evaluate the perceived difficulty, the trends
are classified in a well known manner:
(1) number of data points, (2) unit of amplitude,
(3) unit of time, (4) mean angle.
The statistic applied here is the non-parametric
Kruskal-Wallis test.
The level of significance is determined on P= .05.

Testing the effect of number of data
points on perceived difficulty

The test is done with all 57 trends involved as
well as with groups of comparable units. Both
tests do not show any statistical differences in
percieved difficulty.

The effect of the unit of amplitude
on perceived difficulty

When the trends are classified according their
unit of ampltitude, the perceived difficulty shows
to be higher when the unit of amplitude becomes
larger. Differences between class I (4,5) and
class II (6,7,8) are not signifant; the same
counts for class III (10,11,12) and class IV
(14,15,16). All other comparisons between classes
are highly significant, see Table 3.

Table 3 Testing if class x is perceived
         more difficult than class y

		2-sided prob. P =
Class II more difficult than class I		.160
Class III more difficult than class I		.000 *
Class IV more difficult than class I		.000 *
Class III more difficult than class II		.000 *
Class IV more difficult than class II		.000 *
Class IV more difficult than class III		.600

Testing the effect of unit of time

The results with regard to the unit of time are
not clearcut. There is no systematic trend to be
found in the results. On the basis of statistics,
only class I is perceived more difficult than
class VI and class I more difficult than class
VII.
From these results one can deduce, but only with
hesitation, that larger units of time are in
preference to smaller ones.

Testing the effect of mean angle

The results show very clearly that the larger the
angle becomes the more difficulties one say to
have to correctly predict one data point in the
future. The angles between 70 and 80 degrees, between 60 and 70 degrees and in a somewhat smaller
extend between 50 and 60 degrees are perceived as
significantly more difficult than the remaining
categories, see Table 4.

Table 4 The effect of mean angle on
         the perceived difficulty

x easier than y	2-sided prob.	x easier than y	2-sided prob.
10-20/20-30	.576	30-40/40-50	.632
10-20/30-40	.600	30-40/50-60	.094
10-20/40-50	.241	30-40/60-70	.001 *
10-20/50-60	.048 *	30-40/70-80	.024 *
10-20/60-70	.000 *	40-50/50-60	.440
10-20/70-80	.005 *	40-50/60-70	.020 *
20-30/30-40	.709	40-50/70-80	.252
20-30/40-50	.559	50-60/60-70	.086
20-30/50-60	.195 *	50-60/70-80	.384
20-30/60-70	.000 *	60-70/70-80	.291
20-30/70-80	.010 *		

* = significantly different

Prediction stategy

In order to get insight in the subjects prediction
strategy several prefab answers are printed on the
questionnair. Space is left open to indicate other
strategies. The subject can indicate that more
than one strategy is applied during the prediction
cycle of one trend presentation. The mean score
for each trend is (712/285=) 2.50 strategies.
One can indicate that the vertical position of the
last, the last two, the last three or the last
four data points are used to make the prediction.
One can indicate that a line is interpolated
between the last two, three or four data points,
that whole picture is taken into account, that one
sees regularity in the data points positioning,
that one makes use of some sort of feeling, or
that one select the next point just independently
of all visible points
When analysing the number of stategies, a general
tendency is that prediction strategies are hardly
related to the number of data points on which a
prediction is made. Exceptions are the strategy
'drawing a connection line trough the last four
data points' and 'recognizable patterns'. The
number of times these strategies are mentioned
increases with an increasing number of data
points.
Furthermore, the results show that subjects say to
predict largely on the basis of all data points
presented ('the whole figure') with 'drawing an
imaginairy connection line between the last four
data points' and a large quantity of 'feeling'.

Summary of results of subjective ratings

(1) The number of data points has no effect on the
    perceived difficulty.
(2) The larger the unit of amplitude becomes, the
    more difficulties are met to predict.
(3) The largeness of the unit of time does not
    clearly affect the perceived prediction
    difficulty.
(4) The larger the mean angle becomes the more
    difficulties are said to be encountered in the
    prediction activity.
(5) A prediction is said to be largely based on
    the whole figure, completed with the imaginai-
    ry aid of drawing a line between the last four
    data points and the use of a substantial
    quantity of feeling.

General conclusions of experiment 1

From the results just discussed the following
three conclusions are drawn.

-1. It it not justified to think that the more
    history is given the more accurate predictions
    are made. In this experiment 20 data points
    show less performance than 10 points and even
    less than 5 data points. One can even suppose
    an overdone of information. Just for the mo-
    ment the optimum of data points is determined
    to be ten.
-2. In general, the larger the physical size of
    the trend becomes, the better prediction
    performance is found.
-3. Subjects rate the difficulty to predict higher
    when trends are increasing in magnitude. This
    result is contradictory with what one should
    expect on the basis of conclusion 2.

EXPERIMENT 2

Introduction

In the previous experiment it is shown that trends
having a large unit of amplitude are superior to
smaller dimensioned ones. One fundamental argument
against this results can be formulated as follows:
'It is quite logical that large units of amplitude
are better than small units because positioning
resolution in the first case is much higher than
in the latter case'. One pixel on the y-axis
dealing with a small unit is comparable with for
instance 2, 3 or 4 pixels on the axis with a large
unit of amplitude, while the cursor accuracy (CA)
in the latter case remains 1 pixel. In order to
treat this remark seriously the following experiment is executed.
Three different units of amplitude (UoA) are used
i.e. 5 (small), 10 (medium) and 15 (large). The
trends are predicted with a cursor accuracy of 1
pixel, 1 and 2 pixels, and 1 and 3 pixels for the
units small, medium, and large respectively.
Moreover, in this experiment the factor line
through the successive datapoints is taken into
account.
Ten different presentation forms are shaped for
experimental verification, see Table 6.

Table 6   Shaping of the presentation forms

trend number	unit of amplitude	cursor accuracy	line
1	5	1	no
2	5	1	yes
3	10	1	no
4	10	2	no
5	10	1	yes
6	10	2	yes
7	15	1	no
8	15	3	no
9	15	1	yes
10	15	3	yes

The unit of time is choosen equal to the unit of
amplitude. In this way every trend has a mean
angle of 50 degrees (actually ranging from 49.7 to
50.7 degrees).
Two statistically equal time series have been
applied. These series have been used as well as
their mirror versions (the originals with switched
poles). From the 60 predictions in one time series
the last 50 are applied to determine the RMS-error
score. Sixteen subjects participated in this
experiment. Statistical testing is executed with
SPSS.

Results of the Performance Factor data

- The Performance Factors are normally distributed
  according to the Kolmogorov Smirnov testing (Z =
  .862, P = .447). As a consequence, parametric
  testings are allowed.

- Results of the Oneway Analysis of Variance show
  that the factor 'subject' has a significant effect on the PF (F-ratio= 3.642, F-prob.= .0000).

- The factor 'time series' has also a significant
  effect on the PF (F-ratio= 4.379, F-prob.=
  .0055).

The Performance Factors are corrected for the
factors 'subject' and 'time series'. This correction is done by deviding the PF's for each subject
by the mean PF of that subject, and deviding the
PF's for each series by the mean of that series.
The result is called PFCOR with a mean of 1.0000
and a standard deviation of 0.0773. This PFCOR is
also normally distributed according Kolmogorov
Smirnov (Z = .809, P = .530).
Table 7 shows the results of the Oneway Analysis
of Variance on PFCOR in relation to the
experimental variables.

Table 7   Results of Analysis of Variance

Factor	F-ratio	F-prob.
Subject	.009	1.000
Time series	.000	1.000
Presentation Form (Table 6)	2.259	0.0212 *
Line (yes/no)	0.482	0.4885
Unit of amplitude (UoA)	4.831	0.0092 *
Accuracy cursor including UoA = 5	0.353	0.5532
Accuracy cursor excluding UoA = 5	0.479	0.4901

The 10 different presentation forms are tested on
PFCOR by a comparison with t-tests. The following
main results are found:

- UoA=5 has a higher PF-score without line than
  with a line (not significant).
- UoA=10 has a significant higher PF-score with a
  line than without a line.
- UoA=15 has approximately equal PF-score with and
  without a line.

- The UoA's 5, 10 and 15 without a line involved
  have almost equal PF-scores

- UoA=10 with line has a significant higher PF-
  score than UoA=5 with line.
- UoA=10 with line has a significant higher
  PF=score than UoA=15 with line.
- UoA=15 with line has a significant higher PF-
  score than UoA=5 with line.

- When UoA=10 and for cases with and without a
  line, effect of the cursor accuracy on the PFCOR
  is not found.
- When UoA=15 and for cases with and without a
  line, a minor but not significant effect of the
  cursor accuracy on the PFCOR is found.

Results of the subjective ratings

According to the Kruskal Wallis test, only the
Factor 'subject' has a significant effect on the
perceived difficulty (Chi^2 = 67.2, sign.= .000).

For further testing, the perceived difficulty is
corrected for the Factor 'subject' as well as for
the Factor 'time series'. Although the latter
factor is not significant, a small influence is
probably present. The perceived difficulty measures for each subject is substacted by the mean of
the subjects perceived difficulty and this result
is devided by the standard deviation of that
subject. The result is thereafter corrected for
'time series' in a similar way. The result is
called DIFCOR.

Table 8 shows the results of the Kruskal Wallis
Analysis of Variance of the different experimental
variables related to DIFCOR.

Table 8 Kruskal Wallis test on DIFCOR

Factor	Chi^2	Sign.
Subject	.294	1.000
Time series	.081	.994
Presentation form	31.321	.000 *
Line	1.804	.179
Unit of Amplitude	23.847	.000 *
Cursor Accuracy including UoA = 5	9.839	.002 *
Cursor Accuracy excluding UoA = 5	3.229	.072

The 10 different presentation forms are thereafter
compared with eachother by means of a Wilcoxon
Matched Pairs Signed Rank test with DIFCOR as
dependent variable. The results can be summarized
as follows:

- There is no difference in perceived difficulty
  between a line and no line for UoA=5, for UoA=10
  and UoA=15. For UoA=15 and AC=3 a difference
  close to significance is found between a line
  and no line; with a line is perceived more
  difficult.
- For both conditions with and without line counts
  that the larger the unit of amplitude becomes,
  the more difficulties are encountered, although
  not all differences are significant.
- For UoA=10 and no line, AC=1 and AC=2 do not
  lead to different perceived difficulties.
- For UoA=10 and with a line AC=1 is perceived
  easier than with AC=2 (P=.088).
- For UoA=15 the condition AC=1 with and without
  line is percieved equally difficult as AC=3 with
  and without line.

General conclusions of experiment 2

Objective data:
-1. From the differences in cursor accuracy, as
    investigated in experiment 2, no influences
    are found on the prediction performance. The
    critical remark which has led to this
    experiment is hereby rejected.
-2. When no line is connecting the successive data
    points, the units of amplitude 5, 10 and 15
    reach comparable prediction performances. This
    effect can be made plausible by the notion
    that a rather optimal mean angle has been
    choosen. Probably the mean angle influence
    obscures under this condition the possitive
    effect of the unit of amplitude.
-3. When however a line is drawn, UoA=10 shows the
    highest prediction preformance and UoA=5 the
    lowest performance. In relation to the
    previous concluding remark, the effect of a
    line is considerable.
-4. The presence of a line for UoA=5 leads to
    lower prediction preformance than without a
    line. The reverse result counts for UoA=10.
    Any effect is completely absent for the
    condition UoA=15.

Subjective data:
-5. With and without line the results indicate
    that the larger the unit of amplitude becomes
    the more difficulties one encounters with the
    prediction. This result is in coherence with
    the results as found in experiment 1.

GENERAL REMARKS

The best results are found with trends which have
almost- to completely- screenfilling sizes. If
smaller sizes of trends are applied, in order not
to be completely space consuming with one trend
this will automatically lead to less accurate
predictions. How worse the predictions can become
to be still tolerable is dependent on a multitude
of factors and will probably also be situational
dependent. The message drawn from this experiment
is that:
- one cannot miniaturize trend presentation in
  order to reduce the number of screens
  (which is desirable according to ergonomics
  handbooks) without a substantial loss of
  prediction accuracy,
- one cannot rely (completely) on the personal
  judgement of subjects about the accuracy of
  their predictions when evaluating different
  trend presentations. It is even possible that
  the results will become even worse when the
  subjects are not even fedback about the
  prediction accuracy.

Safety of large technical systems supervised by
men rely to a great extend on the Man-Machine
Interface. The experiments as discussed here show
that also the way of evaluating the MMI has a
substantial inpact on the answers we get. One
should therefore keep in mind that although ment
to be serious answers of people involved objective
data in combination with subjective ratings are
essential for fundamental questions about how to
display information to the human supervisor.

REFERENCES

Bösser, T., E.M. Melchior (1984). Utilization of
time-series information in an optimization
task. Proceedings of the fourth European Annual
Conference on Human Decision-making and Manual
Control, Soesterberg the Netherlands, pp.201-
212.
Koch, C.G., T.R. Edman, and R.K. Guenther (1982).
Evaluating options for CRT display of process
trend data. Proceedings of the Human Factors
Society, 26th annual meeting, pp. 49-53.
Van Bussel, F.J.J. (1980). Human prediction of
time series. IEEE Transactions on Systems, Man,
and Cybernetics, Vol. 10., no.7, pp.401-414.
Van Heusden, A.R. (1980). Human prediction of
third-order autoregressive time series. IEEE
Transactions on Systems, Man, and Cybernetics,
Vol. 10., no.1, pp.38-43.
Van Schaik, P., and J.E. Rijnsdorp (1985).
Detection of on-coming off-normals in VDU trend
presentation. Proceedings of the ninth Congress
of the International Ergonomics Association,
Bournemouth, England.
White, T.N., and A.R. Van Heusden (1984). Various
display scalings of trend information and
human's predictability. Proceedings of the
fourth European Annual Conference on Human
Decision-making and Manual Control,
Soesterberg, the Netherlands, pp.263-275.
White, T.N., and P. Van Schaik (1985).
Perceivability of trend deviations based on
static VDT-presentations. Proceedings of the
fifth European Annual Conference on Human
Decision-making and Manual Control, Berlin.

RELIABILITY ANALYSIS OF PROCEDURAL HUMAN ACTIVITIES: A CASE-STUDY

G. Heslinga

*NV KEMA, Risk and Reliability Analysis Group, PO Box 9035, NL-6800 ET Arnhem,
The Netherlands, in co-operation with the Man-Machine Systems Group of the Delft
University of Technology*

Abstract. Human Reliability Analysis (HRA) event trees are used here to study the
reliability of human activities. The study concerns itself particularly with
activities which have to be performed according to procedures. HRA event trees allow
the paths leading to certain consequences to be presented, i.e. the success
consequence if the procedure is followed correctly and varies failure consequences if
the procedure is not followed correctly.

A case-study is presented in order to analyse the benefit of applying HRA event
trees. Two different procedures, leading to similar consequences, are assessed in
this case-study. The study concerns the adjustments by an operator of a setpoint with
which he controls the temperature of a boiler.

The results show that HRA event trees can help in deciding which procedure is the
best. It is concluded that a justified decision about the design of the procedure can
only be made if all paths of the HRA event tree, leading to all the possible failure
consequences, are taken into account. The conclusion implies that applying THERP (a
generally used technique that involves HRA event trees) may not be sufficient.

Keywords. Human factors; probability; critical path analysis; reliability theory;
nuclear power stations; industrial plants; human error; event trees.

INTRODUCTION

The influence that human performance can have on
the safety of a plant (a nuclear power plant or
another industrial plant) is becoming
increasingly important. Especially in systems
where the technical reliability has been
increased by applying many redundancies, the
human factor becomes the weak component. It is
therefore becoming essential to analyse in depth
human errors that can lead to the occurrence of
undesired consequences that degrade the overall
reliability of a system.

At the Delft University of Technology a study is
performed having the aim to analyse the use of
event trees for the assessment of human
reliability. This study is performed in
co-operation with the KEMA (Joint Laboratories
and other services of the Dutch electric supply
companies), which is the sponsor of the study.
The analysis is restricted to 'rule-based
behaviour' which implies the performance of
certain activities according to procedures
(Rasmussen, 1982). There is a generally applied
technique involving the use of HRA event trees,
known as THERP (Technique for Human Error Rate
Prediction, Swain and Guttmann, 1983; Bell and
Swain, 1983). In THERP, single errors are
usually considered leading to one undesired
consequence i.e. not following the procedure.
THERP usually does not consider the different
consequences that might take place if the
procedure is not correctly followed. In the
study reported here the attention is especially
directed to the different consequences that
might take place if a human error or a sequence
of human errors occurs.

Many activities (in- or outside a control room)
are performed according to procedures. These
procedures can be regarded as sequences of
specific actions such as the adjustment of a
switch, a setpoint, the reading of a meter etc.
An important aspect of improving the reliability
of human performance is the design of 'good
procedures'. Good procedures are defined here as
procedures that are designed in such a way that
the probability is minimized that certain
undesired consequences will occur due to human
failures (taking into account the extent of
these consequences).

The object of this paper is to use event trees
to assess how 'good' different procedures are.
For that purpose a case-study is presented where
two different procedures will be considered,
each leading to the same goal. Before this
case-study is presented, this paper will start
with a description of HRA event trees.

HRA ELEMENT

A Human Reliability Analysis (HRA) element is an
HRA event tree of a specific action. An HRA
event tree is a set of events presented in the
form of a tree in which almost each event
reflects a possible human error. Figure 1 shows
an example of an HRA event tree of a specific
action, i.e. an HRA element. This example refers
to the adjustment of a switch, here switch 2 of
a panel. It should be noted that a difference
exists from the event trees used so far in the
literature, since certain events may no longer
have a binary possibility of occurring (e.g. the
treatment error in Fig. 1). The probability that
a YES-direction will be followed occurs with a
so-called Human Error Probability (HEP). Low
HEPs are applied in Fig. 1 since 'rule-based
behaviour' is involved. Most HEPs are taken from
Swain and Guttmann (1983).

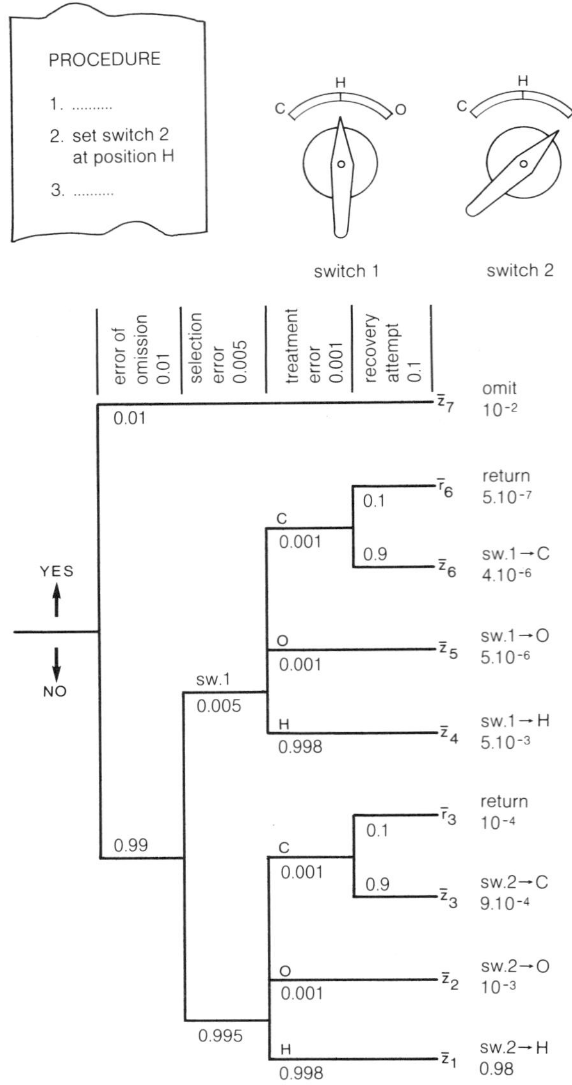

Fig. 1. HRA element and the relevant panel lay-out and procedure.

Several consequences are possible on completing the tree. Two categories of consequences will be considered: the termination consequence, $\bar{z}_q$, and the recovery consequence, $\bar{r}_j$. The second, although termed a 'consequence', is not really a consequence since it implies that the same and/or other actions of the procedure are (partly) repeated. The termination consequences can be characterized by the success consequence (the desired consequence, $\bar{z}_1$) and the failure consequences (the undesired consequences, $\bar{z}_2$ through $\bar{z}_7$). The paths leading to these are the success path and the failure paths, respectively.

The probability of a path being completed is given by the path probability, which is obtained by multiplying the relevant HEPs of the path. The path probabilities in Fig. 1 are presented at the offshoots of the tree. In Fig. 1 it is assumed that the events are not dependent, i.e. the treatment errors have equal HEPs as well as the recovery attempts. Multiplication is also possible when dependence is present. But then the HEPs have to be adapted to incorporate dependence. Dependence implies that the HEP of

an event is not only directed by the type of error but also by success or failure of a previous event.

HUMAN RELIABILITY ANALYSIS APPLIED TO THE ADJUSTING OF A SETPOINT

An example is presented now. It concerns the adjustment of a setpoint with which the temperature of a boiler can be regulated. To simplify the case, the setpoint can only be adjusted to five different positions (see Fig. 2).

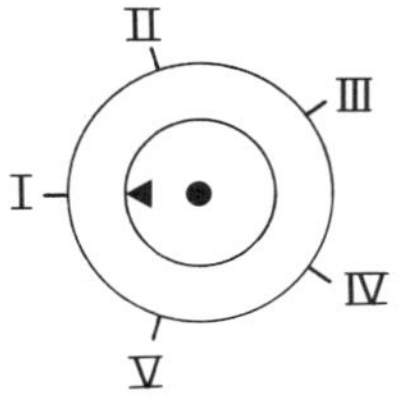

Fig. 2. The setpoint to be adjusted.

The operator is instructed to move the setpoint from its starting position (position I) to position IV. He must take into account that more than two steps are not allowed because a thermal shock would almost immediately take place if more than two steps were used within ten minutes. Due to this limitation, two procedures are possible. They are presented in Fig. 3.

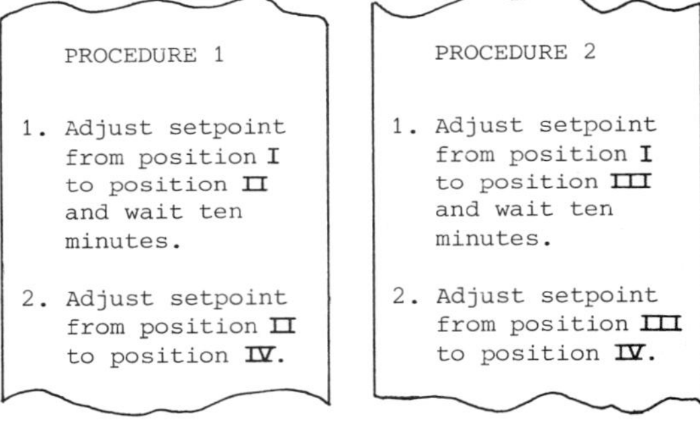

PROCEDURE 1

1. Adjust setpoint from position **I** to position **II** and wait ten minutes.

2. Adjust setpoint from position **II** to position **IV**.

PROCEDURE 2

1. Adjust setpoint from position **I** to position **III** and wait ten minutes.

2. Adjust setpoint from position **III** to position **IV**.

Fig. 3. Two possible procedures for adjusting the setpoint.

On completing one of these procedures four different termination consequences are possible:

- A thermal shock ($\bar{T}$). This will take place if the setpoint is adjusted to position **IV** within ten minutes.

- An explosion ($\bar{H}$) because of the creation of an excessively high temperature and, hence, too high a pressure. This will result if the setpoint is adjusted erroneously to position **V** and kept there. This damage will not take place immediately, so recovery is possible within a few minutes.

- Too low a temperature ($\bar{L}$) because the procedure is completed with the setpoint in the wrong positions "I", "II" or "III".

- The success consequence ($\bar{S}$) when one of the procedures is completed correctly.

The HRA element of each procedure step consists
of three human errors. The HRA element of the
first step of procedure 1 is presented in
Fig. 4. The selection error is not shown because
only one setpoint is considered for the sake of
simplicity. The treatment error implies the
adjustment of the setpoint to a wrong position,
not corresponding with the procedure step. As it
was assumed that recovery is possible when

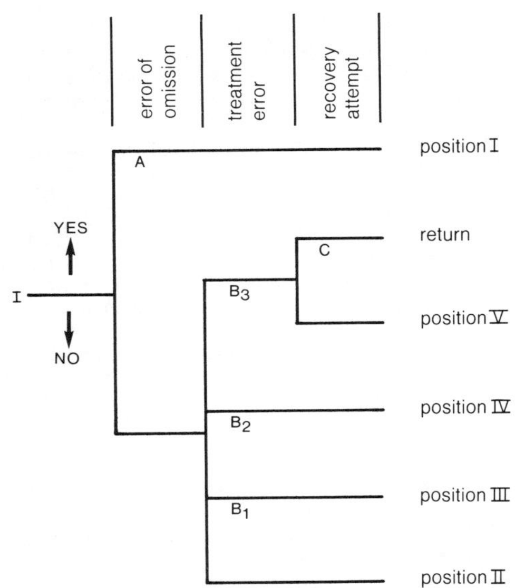

Fig. 4. The HRA element of the first step of
         procedure 1, i.e. HRA element 1.

the setpoint is adjusted to position $\overline{V}$, Fig. 4
contains this recovery attempt there.

It is agreed that the activities performed at
the setpoint in the first ten minutes belong to
the first procedure step; they are represented
by HRA element 1. The activities performed after
that belong to the second procedure step. The
HRA element of this step, i.e. HRA element 2,
has the same form as HRA element 1 in Fig. 4.

By combining the HRA elements of the two
procedure steps of one procedure, the HRA event
tree of the complete procedure is obtained. The
HRA event tree of procedure 1 is presented in
Fig. 5. Several limbs in the HRA element 2 are
combined when they lead to the same
consequences. The tree of procedure 2 is not
presented since it is nearly similar. It is
agreed that a symbol with a bar indicates a
consequence and one without a bar a probability.

Path probabilities

The following assumptions are made in order to
quantify the path probabilities S, L, T, H and R:

1 The HEP of each error of omission is P_1.

2 The HEP of the treatment error of the first
  procedure step is P_2.

3 The HEP of the treatment error of the second
  procedure step is P_2, except for the
  following cases:

   3.1 the HEP is P_3 if the setpoint is turned
  with a step size corresponding with the
  procedure but with an incorrect result
  because the setpoint was adjusted
  incorrectly in the first procedure step;

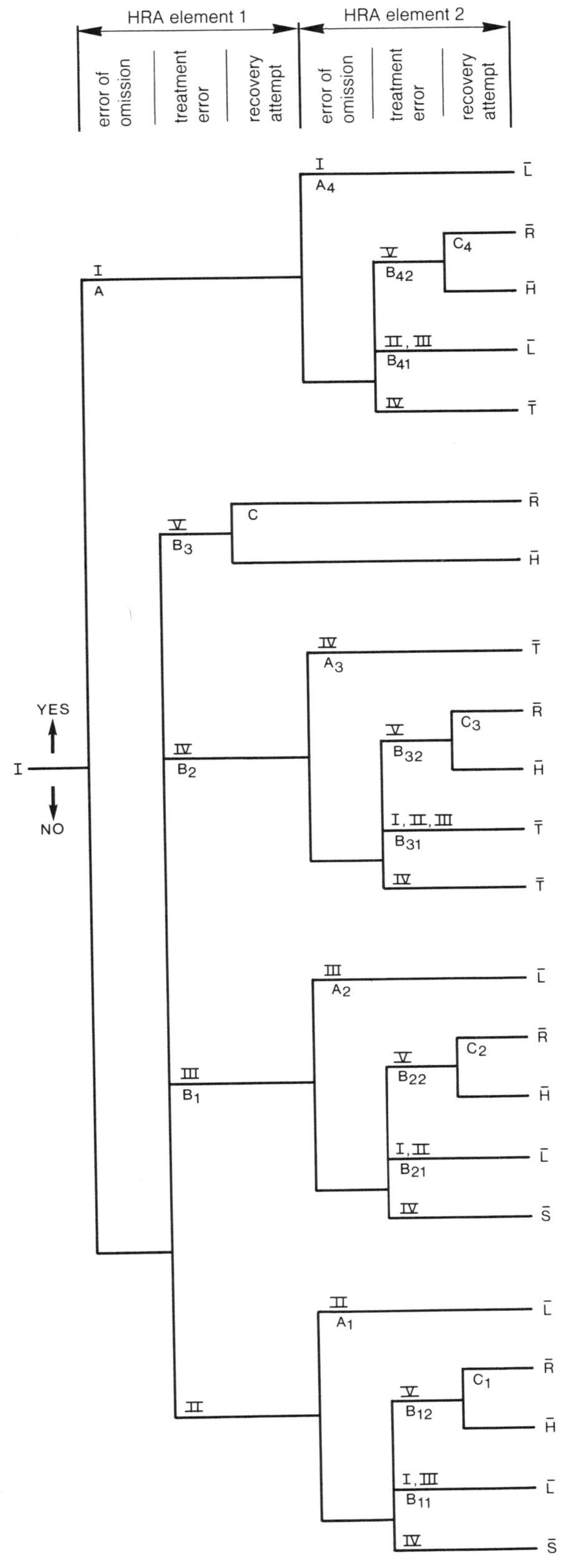

Fig. 5 HRA event tree of procedure 1. The
        positions of the setpoint are given
        above the limbs; the HEPs are given below
        the limbs.

3.2 the HEP is P_4 if the step size of the adjustment is the same as done in the first procedure step because of memory influences;

3.3 the HEP is the maximum of P_3 and P_4 if both 3.1 and 3.2 are valid.

4 The HEP of each recovery attempt is P_5.

Table 1 shows how the branch probabilities of Fig. 5 are expressed in P_1 through P_5. As

$A = P_1$	$A_1 = P_1$	$A_2 = P_1$	$A_3 = P_1$	$A_4 = P_1$
$B_1 = P_2$	$B_{11} = P_2 + P_4$	$B_{21} = 2P_2$	$B_{31} = 3P_2$	$B_{41} = 2P_2$
$B_2 = P_2$				
$B_3 = P_2$	$B_{12} = P_2$	$B_{22} = \max(P_3, P_4)$	$B_{32} = P_2$	$B_{42} = P_2$
$C = P_5$	$C_1 = P_5$	$C_2 = P_5$	$C_3 = P_5$	$C_4 = P_5$

TABLE 1   <u>The branch probabilities of Fig. 5 expressed in P_1 through P_5</u>

several limbs in Fig. 5 were combined the branch probabilities can be a combination of HEPs (e.g. B_{11}). Assuming certain values for P_1 through P_5 allows the path probabilities S, L, T, H and R to be calculated. In Table 2 this is done for both procedures 1 and 2. The HEPs P_3, P_4 and P_5 are varied since they are important for the difference in the path probabilities between procedures 1 and 2.

Consequence probabilities

A consequence probability is the probability that a termination consequence occurs. It may not equal the path probability because the recovery attempts are included in the consequence probabilities. If an operator makes errors, he might reach the return consequence $\bar{R}$, which implies here that he repeats (partly) the procedure. He might then follow the success path

when he discovers his errors. It might also be possible that he introduces other errors, follows other failure paths and ends at other or the same failure consequences. Hence, the consequence probability of a termination consequence is a combination of the path probability of the path immediately leading to this consequence, and of the probability of reaching that consequence via all kinds of recovery attempts.

We do not know at the moment which HEPs are present in the tree during the recovery attempt, i.e. when the procedure is (partly) repeated. The assumption is made here that when an operator ends at the return consequence $\bar{R}$, he will return to the top of the same HRA event tree (having the same HEPs) and complete this tree. If he again ends at a return consequence he will start the same tree for the third time etc. In Heslinga (1984) it has been shown that, when this assumption is made, the number of recovery attempts is not important and a consequence probability F_r becomes: $F_r = F/(1-R)$, where F is the path probability of the termination consequence $\bar{F}$. In table 2 the consequence probabilities S_r, L_r, T_r and H_r are calculated by applying this equation.

COMPARING THE CONSEQUENCES

Calculating only the consequence probabilities is not enough for making the decision as to which procedure is the best. Certain consequences are more undesired than others. Hence, two factors are important for the decision about the best procedure design:

1 the probabilities that failure consequences occur, i.e. the consequence probabilities, and

2 the extent of the failure consequences.

The extent of the failure consequences can be expressed by weighting factors. By combining these weighting factors with the corresponding consequence probabilities a cost factor C is obtained. If, for the sake of simplicity, a linear equation is used for this cost factor, then C can be expressed in this example by:

P_1	P_2	P_3	P_4	P_5	S	L	T	H	R	S_r	L_r	T_r	H_r	
.01	.001	.1	.01	.5	.9663	.0208	.0109	.0010	.0010	.9673	.0208	.0109	.0010	procedure 1
					.9663	.0121	.0108	.0054	.0054	.9716	.0121	.0108	.0055	procedure 2
.01	.001	.1	.01	.0	.9663	.0208	.0109	.0021	.0000	.9663	.0208	.0109	.0021	procedure 1
					.9663	.0121	.0108	.0109	.0000	.9663	.0121	.0108	.0109	procedure 2
.01	.001	.1	.001	.5	.9751	.0120	.0109	.0010	.0010	.9761	.0120	.0109	.0010	procedure 1
					.9751	.0121	.0108	.0010	.0010	.9761	.0121	.0108	.0010	procedure 2
.01	.001	.1	.001	.0	.9751	.0120	.0109	.0021	.0000	.9751	.0120	.0109	.0021	procedure 1
					.9751	.0121	.0108	.0021	.0000	.9751	.0121	.0108	.0021	procedure 2
.01	.001	.001	.1	.5	.8784	.1087	.0109	.0010	.0010	.8793	.1088	.0109	.0010	procedure 1
					.8784	.0121	.0109	.0494	.0494	.9240	.0127	.0011	.0519	procedure 2
.01	.001	.001	.1	.0	.8784	.1087	.0109	.0021	.0000	.8784	.1087	.0109	.0021	procedure 1
					.8784	.0121	.0109	.0987	.0000	.8784	.0121	.0109	.0987	procedure 2

    experimental variables   path probabilities   consequence probabilities

TABLE 2   <u>The path probabilities and the consequence probabilities of procedures 1 and 2 for several P_1 through P_5</u>

$$C = W_1 L_r + W_2 T_r + W_3 H_r.$$

In this equation W_1, W_2 and W_3 are the weighting factors expressing the extent of $\bar{L}$, $\bar{T}$ and $\bar{H}$, respectively. The procedure with the lowest cost factor should be preferred.

In practice $\bar{H}$ (explosion) will be less desirable than $\bar{T}$ (thermal shock) and $\bar{T}$ will be less desirable than $\bar{L}$ (too low temperature). Therefore, it is assumed that $W_1 \leqslant W_2 \leqslant W_3$. In table 3 the cost factor is presented for the case where $P_1 = 0.01$, $P_2 = 0.001$, $P_3 = 0.1$, $P_4 = 0.01$ and $P_5 = 0.5$ (the first case in table 2). Table 3 shows three situations with different weighting factors (W_i). It is shown that, for situations 1 and 2, procedure 1 has a lower cost factor than procedure 2 and the reverse holds for situation 3. Similar results for cost factors are obtained for the other P_i's as applied in Table 2.

	W_1	W_2	W_3	C procedure 1	C procedure 2
situation 1	0.05	0.45	2.50	.0084	.0192
situation 2	0.50	1.00	1.50	.0228	.0251
situation 3	1.00	1.00	1.00	.0327	.0284

TABLE 3   The cost factor C of procedure 1 and 2 for several weighting factors W_1, W_2 and W_3 where $P_1 = 0.01$, $P_2 = 0.001$, $P_3 = 0.1$, $P_4 = 0.01$ and $P_5 = 0.5$

DISCUSSION

The consequence probabilities of procedure 1 shown in Table 3 are the same for the three situations. The same applies to procedure 2. The Table shows, however, that different decisions can be made regarding the best procedure because of different weighting factors. If the weighting factors for the first situation in Table 3 ($W_1 = 0.05$, $W_2 = 0.45$ and $W_3 = 2.5$) are considered as realistic values the conclusion is then that procedure 1 is better than procedure 2 because of its lower cost factor.

The situation where $W_1 = W_2 = W_3 = 1$ implies that the consequences are equally weighted. This would lead to an opposite decision about the best procedure. Equal weighting happens in THERP where the failure consequences are regarded as identical since each failure consequence is "not following the procedure". This study shows that not following the procedure may lead to failure consequences, some of which might be very undesired. In that case equal weighting is not permitted since it may result in opposite decisions about the procedure design. This implies that THERP is not directly applicable to those situations where certain failure consequences are far more undesired than others. Then the complete HRA event tree should be obtained with all the possible failure paths leading to the possible failure consequences. This could be done by systematically combining the HRA elements.

The simplicity of the example presented here, the assumptions about the HEPs and the weighting factors can be discussed. This is however of little importance since the goal of this paper was to show that HRA event trees which are a combination of HRA elements can be used to choose among similar procedures. The method of using HRA elements might also be applied to more complex procedures and to situations where the influence of human factors on the safety of a plant is to be assessed.

CONCLUSION

In general, a justified decision about the procedure design can only be made if the following items are taken into account:

- all paths of the HRA event tree and the consequences that may occur,
- the probability that these consequences occur,
- the extent of their occurrence.

REFERENCES

Bell, B.J., and A.D. Swain (1983). _A Procedure for Conducting a Human Reliability Analysis for Nuclear Power Plants, NUREG/CR-2254._ U.S. Nuclear Regulatory Commission, Washington D.C.

Heslinga, G. (1984). Analysis of Human Reliability considering a Sequence of Independent Actions. _Proceedings of the Fourth European Annual Conference on Human Decision Making and Manual Control,_ Institute for Perception TNO, Soesterberg, The Netherlands, pp. 73-86.

Rasmussen, J. (1982). Human Errors. A Taxonomy for Describing Human Malfunction in Industrial Installations. _Journal of Occupational Accidents,_ _4_, 311-333.

Swain, A.D., and H.E. Guttmann (1983). _Handbook of Human Reliability Analysis with Emphasis on Nuclear Power Plant Applications, NUREG/CR-1278._ U.S. Nuclear Regulatory Commission, Washington D.C.

PRACTICAL ACTIVITY ON RELIABILITY
CONTROL OF THE COMPONENTS

T. Yamamoto

Control System D&E Department, Yokogawa Hokushin Electric Corporation

Abstract. Very high reliability is required for components used in a industrial instrument, due to enormous loss and danger in the case of its failure. To achieve such high reliability, we employed around 1975 a method to obtain a component with the best quality. Candidate components were purchased from some vendors using the same specification. The reliability of those components then were compared by utilizing general evaluation methods. And the component which showed the best reliability performance was selected for the use. Nowadays, however, the general methods became almost ineffective according to the improvement of the reliability of components and also of the evaluation techniques. Defect of a component is not easily detected by the relative evaluation, and it is revealed at the stage of mass production, or after long duration of usage. Even in this case, the rate of failure is small. Therefore, we consider a method important - we call the method "weighted evaluation" - and have emproyed it in these late years: we evaluate the reliability of component independently focusing on the weak points of the component based on our experience, while general evaluation of components is left to the manufacturer, in the following, our practical activity on reliability control of the components is described.

Keywords. Reliability of components; evaluation method; practical activity on reliability; H_2S test; pressure cooker test; CuS dendrite.

THE CHANGE OF THE EVALUATION METHOD

As methods for the reliability evaluation of component, there are such general methods as temperature cycling test, high temperature and high humidity test, vibration test, salt water spray test and so forth. Genarally these tests are conducted by the manufacturer of component, and a user can obtain the test data from the manufacturer. However, a user tends to conduct same or similar test by himself to confirm the validity of the reliability data when the user intends to apply the component to a product with a high reliability requirement. in this occasion, although there is rarely a case such that defects in a component could be found through test by user from the difference of manufacturing lot or from the time Length of the test conducted, it is basically considered that user would obtain the same data as the manufacturer's data. Fig.1 shows a distribution of working hours consumed to evaluate the reliability of components from 1974 to 1983, by a group of a department developing industrial instruments. There is no specific classification criterion applied to the data such that nature of components or number of items or the quantity of components, the result, however, shows the distribution fits to the logarithmic normal distribution. Upon development of a new product, many new components are used. It is quite difficult to conduct the reliability evaluation uniformly to all of such new components. Furthermore, it sometimes happens general reliability evaluation methods fail to detect defects of components. Followings are such three examples.

A Failure of Bonding Wire of a Photo-Coupler

Fig.2 illustrates construction of a photo-coupler,

viewed from a cross section. After the general Reliability evaluation, this component was accepted for use. In spite of the fact that the components behaved normally at the time of shipment of the products, they failed at a certain rate in the field. The failed samples were returned to the manufacturer for analysis. Although it was found that bonding other wires were broken for most of the failed components and no failure was found no definite cause was fixed. We, then, assumed a hypothesis as follows by referring to the nature of the component.

A) In a component of mold sealing, a wire tends to be weak against thermal stress, in contrast to in a component with vacant inside sealing as found in can type packaging. In the latter case a wire tends to be weak against mechanical vibration and/or shock.

B) As a photo-coupler is defferent from general IC in the point that it inherently utilizes two types of resin; tansparent and non-transparent a wire in a photo-coupler sealed by mold tends to accept force because of the different thermal coefficients of the molding materials.

C) If the breakage of wire is caused by either the expansion or the shrinkage of the materials there will be a possibility that the broken wires will recover the connection by the shrinkage or the expansion of the materials. Then, the photo-coupler with a broken wire can behave normally at some temperature range and the breakage of wire is not detected. (Fig.3)

In other word, judgement a general temperature cycle test with continuous observation is necessary.

CuS Dendrite

A printed wiring board (PWB) in recent years, has
become strong against corrosive atmosphere because
traces on the PWB have become covered by solder-
resist materials. A PWB without the solder-resist
material, however, traces are covered and protected
only by solder plating. In this case, the plating
sometimes fails to cover all the surfaces of
traces (Fig.4). By the application of hydrogen-
sulfide or sulfide-oxide atmosphere to the PWB,
sides of the traces without the plating are corroded, and copper-sulfide (CuS) grows from the
corroded locations. When the acceralation factor
of the corrosion atmosphere test is low (E.G.
hydrogen-sulfide gas, 40°C 90% RH, 96 HRS), and
gaps between the traces are wide enough, there will
be observed no short of circuit, though the CuS has
a low resistance. It is desirable, to for an instrument, to be protected from such environment
though, the protection would be quite difficult at
some sites, for example at chemical plants. For
the installation to these site, application of
silicone-coating to the surfaces of PWB can improve
the strength against corrosion atmosphere, as effective as the application of solder-resist material. From the evaluation point of view, careful
attention must be paid to the following points.
When the silicone is   the varnish type and the
coating is   a thin film whose thickness is less
than 10 um, numerous number of cracks, almost in-
visible to bare eyes, can arise on the surface of
the coating according to the elapse of time and to
the variation of ambient temperature. Upon the
existence of such cracks, humidity in the air is
going to be condensed to dew (capillary conden-
sation phenomena). The CuS grows rapidly along
these narrow cracks, and traces are shorted by them
in a short time. This is the CuS dendrite phenomena, an example of this phenomena is shown in Fig.6.
In this case, the silicone coating has brought
worse effect to the resistivity against CuS.
Without very careful attention to the evaluation
test, this phenomana will not be detected.

Failure to Arise Depending Upon the Way of Usage.

Two failure modes were found in a mold sealed type
ceramic capacitor in a product at the stage of production proto stage, which had passed general evaluation tests. The first failure mode was the degrade of resistivity against moisture. The cause of
this failure was introduced by inserting legs of
the capacitor to two vias, whose distance was
wider than the width between legs. Due to the
forced insertion to these vias, mold at the part of
the legs cracked and moisture was able to go into
inside of the capacitor.(Fig.7) The second failure
mode was the peel off the plated Sn or SnPb from
leads by sharp edge at vias. The shape of the
peeled Sn or SnPb was as if it was a whisker. The
peeled fragments shorted conductors. Manufacturers
of components do execute the evaluation tests regarding the strength of lead for pulling and/or
bending though. They seldom execute test regarding
points described above. A user of the component
would avoid such the usage, or he has to devise and
execute his original test to make sure the point
above defects would not arise to the component.

The three examples described above are the typical
ones illustrating the point of view for the tests
focusing on the weak points of components, obtained
through the practical activities for the management
of reliabilty control of components. Also they
illustrate that the weak points of components com-
prise such modes that are hard to be found through
the general evaluation tests.

The weak points of components, acquired through the
experience, are summerized in table 1. The points
in the table are also the examples we devised

originally to focus in the execution at the time of
employment of new components.

ORIGINAL EVALUATION TEST METHODS AS A USER
OF COMPONENT FOLLOWINGS ARE THREE ACTUAL
EXAMPLES

H_2S Test

Connection components such as connector, switch
contact have a possibility to be corroded through
producing copper-sulfide (CuS) by gases as
hydrogen-sulfide (H_2S) or sulfide-oxided (SO_2)
because their base material is either copper or
copper alloy. In an industrial product, which may
be used at a chemical plant, components with gold
plating for the connection are often utilized in
order to decrease the effect of the gases as above.
However the the thickness of the plating is getting
thinner for the years as gold is expensive. It is
a dificult problem for a user of connection component to find a method for the evaluation of
plating by gold.

The difficulty lies in the points as:
a. Without the acceralating factor of the test
   condition (gas density, humidity, and temperature) to the actual environment, the time for
   the test can not be determined.
b. The effect of factors are complex. As an
   example, even gas with high density corrodes
   metal weakly under dry environment. This
   implies the difficulty in measuring the real
   environment, to simulate it.
c. In general the judgement is made based on the
   measured data of the resistance through contact.
   However, this resistivity is affected by such
   factors as state of insertion, construction of
   contact, and randommess. So obtained data show
   wide variation, and make difficult to judge the
   acceptability.

Until we have clear conclusion, we employ the
method shown in table 2 to evaluate the reliability
of connection component. As shown, the method consists of three sub-tests: measurement of thickness
of plating, nitric acid (NH_3) immersion test, and
hydrogen-sulfide (H_2S) Test. At H_2S test, we
submit to the test both the connection component
samples and a reference sample at the same time.
The judgement is made by the comparison between
the connection component samples and the reference
sample. The reference samples are two types of the
contact   which have been used in the process automation field more than 10,000,000 in number, or
more than 10 years. We have a large number of
stock of the reference sample for the H_2S test.

Pressure Cooker Test

We apply the pressure cooker test for the evaluation of humidity proof of semiconductors with mold
sealing. The test condition is 121°C, 100% RH, 2
ATM, 96HRS. When no failure is observed after the
test, the semiconductor is accepted. When failure
is observed, the semiconductor will be tried for
the test under milder conditon of 85°C, 85% RH,
1000HRS. This is based upon the fact that there
are such components empirically known and confirmed through experiments that their humidity
proof rates are not adequately estimated by the
pressure cooker test (Fig.9). The time duration
under the condition provides the effect matching
to the usage of the sample at Bangkok for more
than 20 years (1). We had acquired the similar
result through test and the experience in the
field (Fig.10). In fig.11, results of pressure
cooker test (PCT) under differnt conditions are
shown. The samples used throughout the test are
the same kind of TTL IC which is a little weaker
in the humidity proof characteristic.

The conditions are:on bias, bias with low voltage (1V, 0.1mW), Bias with rated voltage (5V, 65mW), and in water (no bias). By the comparison of time at which failure starts, these conditions are arranged in the order: bias with low voltage ≒ no bias < bias with midle voltage < bias with rated voltage. It is hard to determine whether of not bias is to be applied because the application of voltage accelerates the electro-chemical corrosion, and increases the power dissipation inside which at the same time decreases the acceleration rate.

<u>Inside Visual Inspection by Decomposition</u>

Although this test looks like a test without focus and without clear opportunity, we consider it particulary important according to our actual experience. As for an example, there were two failures of variable resistor of winding wound type, after the usage in the field for several years. The failure mode was same and it was the breakage of wire. They were returned to the manufacturer for analysis. Manufacturer estimated the cause to be electro-chemical corrosion because similar traces looked like by electro-chemical corrosion were found and grease of high activity type was applied to the winding wire. The manufacturer proposed the change of the type of grease, based on the estimation. However, further investigation, by scanning electron microscopy (SEM), was conducted, and it was found that the section surfaces of the breakage did not agree. There were indentions at both surfaces. As the conclusion, the primary cause was the existence of a pin hole inside the winding wire.

The information as above are reported to the parts administration committee (Fig.12) and the   collected information are classified with respect to the kind of components, and examined to help the determination of important observation points of the inside visual inspection test.

CONCLUSION

A component to be used in an instrument with a high reliability requirement as one for industrial use must also have a high reliability. However, the reliability of the component is not necessarily estimated adequately by depending upon the reliability data of the manufacturer and/or the reliability grade assigned to the component based on reliability standard. Also the general reliability evaluation tests can not always be successful to reveal weak points of component. In order to determine whether or not a new component is acceptable for the use in an instrument with high reliability, we conduct original evaluation tests, which are proved effective, to detect weak point(s) of each of the new component based on the experiense with user claims we encountered in the past, while the general reliability evaluation is left to the manufacturer as much as possible. Also a committee named "Parts Committe" and an associated system are described. The objective of the committee is to gather information on reliability of components and make use of them to the evaluation of components which would follow. The committee could provide information to the design team for recovery even in the occasion that a component fails in the field in succession.

We consider that practical activity to increase reliability of product is such a activity which needs accumulation of experiences with continuous and steady movement. We also consider this activity is the survival issue.

REFERENCES

1. Koyama, H.K, H.M. (1979). Moisture resistance of plastic mold transistor. <u>The institute of electronics and communication engineer of Japan.</u> R79-24

TALLE 1 Major Items on Components Evaluation

Components Type		Weak Points for Reliability
Semiconductors	Diodes	Mo-Galvanic Corrosion (Glass Package)
	Diodes Zener	Voltage Drift at Lower Current (Vz)
	ICs Transistors	Purple Plague, Open-Wire (Bonding) Humidity (Plastic Package) Crack of Sealing-Glass
	Photo Couplers	Dark Line Defect on LED, Dust Open-Wire (Boundary of Resin, Bubble)
	LEDs	Dark Line Defect, Luminosity Matching Resistance to Soldering Heat
	CMOS-ICs	Latch-Up, Static Electricity Resistance to Humidity
	ZnO Varistors	Voltage Drift for Surge Current Resistance to Humidity
Resistors	Metal Film	Electoro-Chemical Corrosion (Thin Film) Whisker (Figure 8)
	Wire-Wound	Electro-Chemical Corrosion (Fine Wire)
	Metal Oxide	Resistance to Humidity (Low Power) Resistance to Surge Power
	Variable	Pin-Hole in the Wire Water-Proof, Multi-Slider Loose Contact (Lead to Resistor) Insulation Problem (Metalic Dust)
	Module	Bubble, Temperature Shock
Capacitors	ALuminuam Electrolytic	CL-Corrosion, Dry-Up Failure Failure by Large Ripple Current
	Ceramic	Insulation Problem (High Temperature) Ag-Migration
	Plastic Film	Loose Contact (Lead to Electrode)
	Tantalum	Ag-Migration, Mechanical Shock Failure by Ripple Current Possibility of Combustion by Short Circuit
	Variable	Resistance to Humidity
Electro Magnetic Components	Reed Relyays	Soft Stick of Contact Crack on Glass Tube Migration of Contact Material Dust
	Mercury Wetted Relays	H_2 Gas Leak, Mechanical Shock Crack of Glass Tube
	Relays	Si-Contamination, Dust Brawn & Black Powder Influence in Magnetic Field, Gas
others	Connectors	Fretting Corrosion Au-Cu Diffusion Pin-Hole and Crack Contact Pressure
	Switches	Crack, NH_3 -Problem, Dust Crick Action
	Fuses	NH_3-Problem, Corrosion (Fine Wire) Reliable Solderring
	Terminals	Sn-Whisker, Ag-Migration NH_3-Problem

TABLE 2 Evaluation Tests for Au Plating

Test Method	Condition	Judgemerit Criterion
Plating Thickness	Au, Ni	Specification
HNO_3 Test	HNO_3+H_2O (61%) $Cu+2HNO_3 \longrightarrow$ $Cu+(NO_3)_2+H_2 \uparrow$ $Ni+2HNO_3 \longrightarrow$ $Ni(NO_3)_2+H_2 \uparrow$	No bubble from area less than 1.5mm radius from the point of contact(30sec)
H_2S Test	$H_2S=10$ppm 40°C, 95%RH 24Hr,48Hr,96Hr (5%NaCl dipping)	Visual Check Comparison to Standard Sample

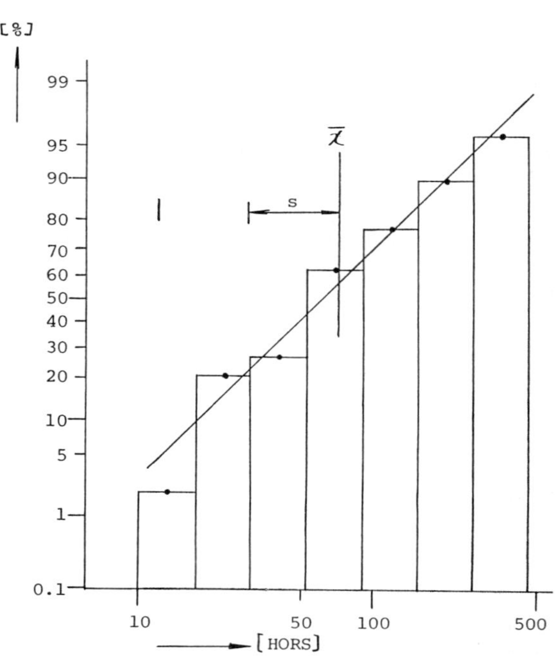

Fig.1. Distribution of Working Hours to Evaluate the Reliability of Components

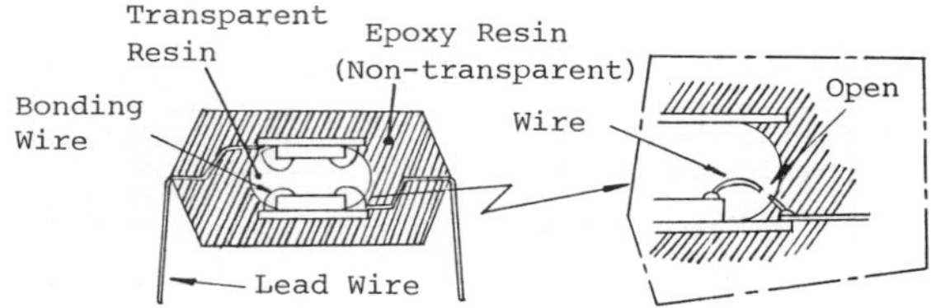

Fig.2   Construction of a Photo Coupler

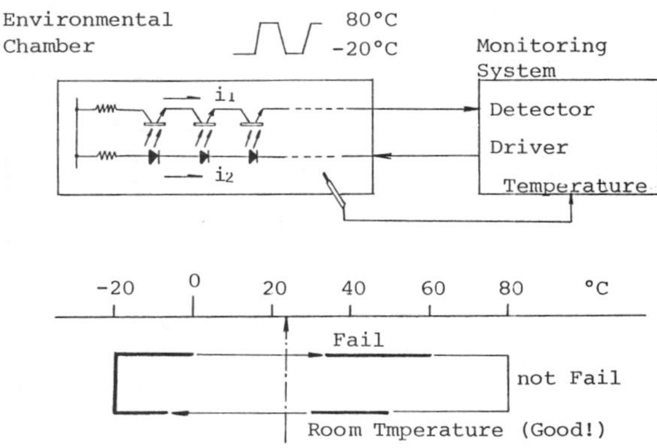

Fig.3   Test System & Test Result

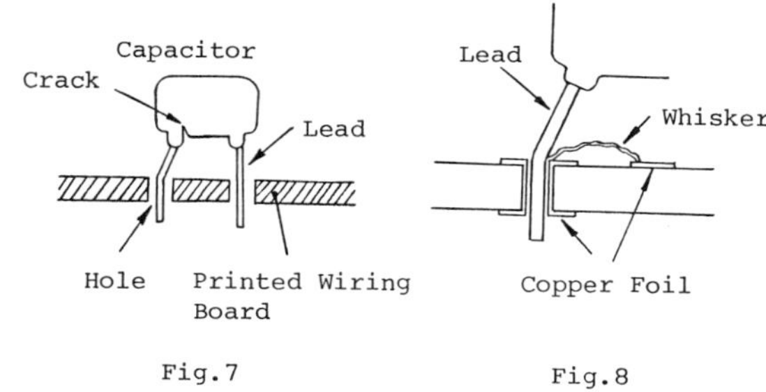

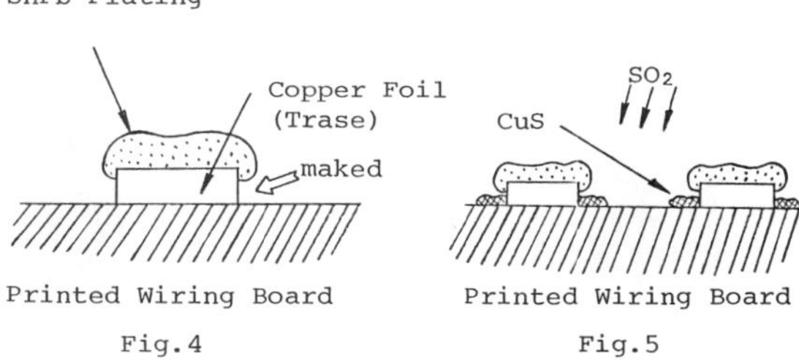

Test Condition	Resistance
Humidity Test(=1atm)	Increased
PCT (>1atm)	Decreased

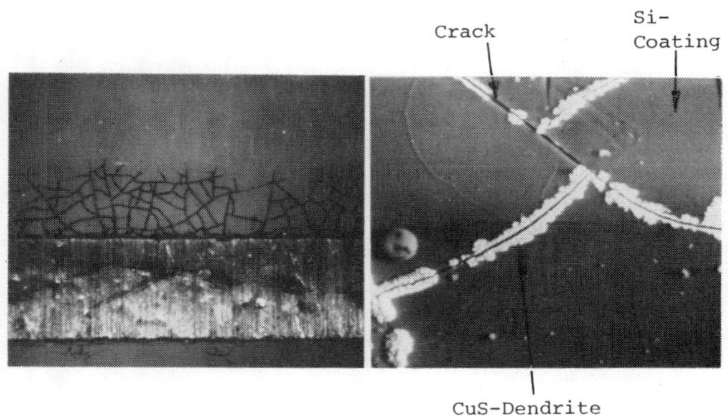

Fig.6. CuS Dendrite

Fig.9   PCT for Some Kind of Metal
        Film Resistor

RISC-L

Acceleration Factor (Reference 1)
AF=PCT 121°C/Bangkok+10^4/3≒3300

3 days×3300≒27 years.

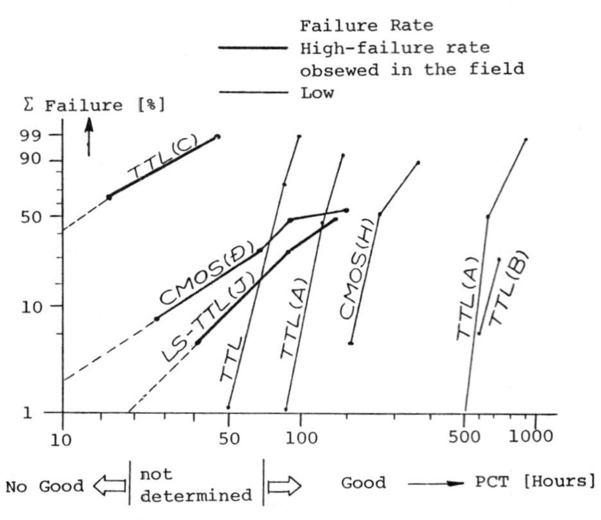

F.10   PCT Results of IC₃

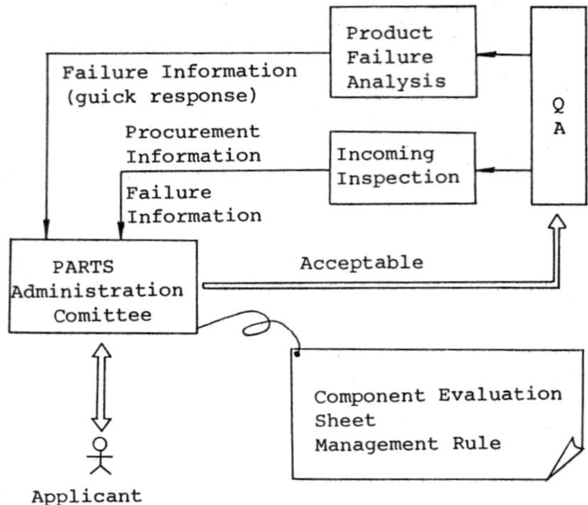

Fig.12 PARTS Administration System

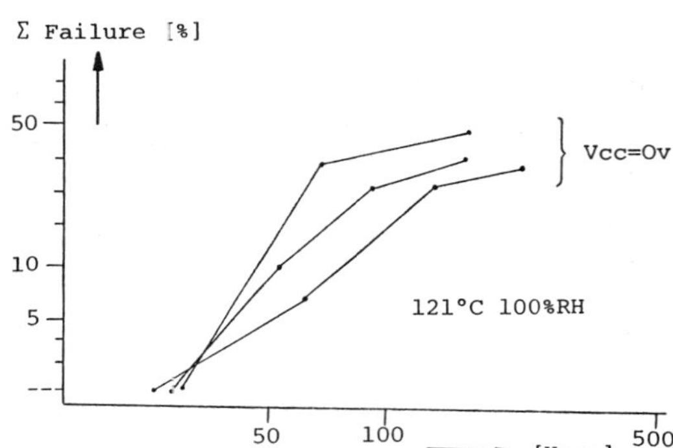

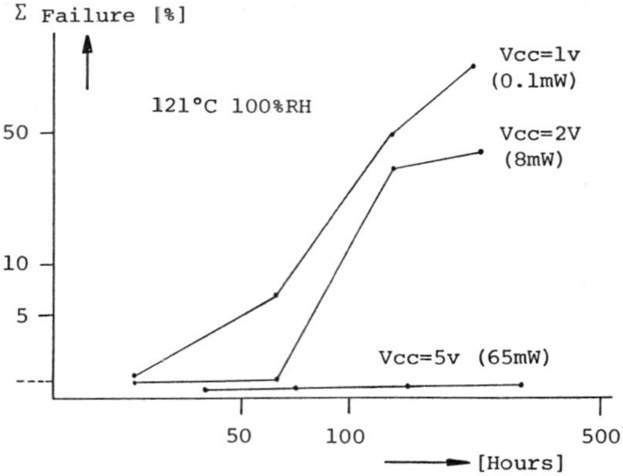

Fig.11   Result of Bias PCT

A RELIABILITY MODEL FOR THE ANALYSIS OF HAZARDS CAUSED BY INTRINSICALLY SAFE APPARATUS

J. K. Fraczek

The Silesian Technical University, Gliwice, Poland

Abstract. The paper discusses a reliability model which enables to assess the mean time between explosions (MTBE) as a measure of the safety level. The probability of the presence of explosive gas mixture, the reliability of electrical instruments (MTBF), the ignition probability assessed with the IEC testing apparatus and the number of electrical instruments located in the explosive area are taken into account in this model. Some examples are included which ilustrate various situations from the viewpoint of different safety levels and additional protective means.

Keywords. Explosion hazard; intrinsic safety; standardization; system failure; chemical industry; mining.

INTRODUCTION

Intrinsic safety is one of several techniques for preventing explosion in hazardous areas (Magison, 1978; Hutcheon, 1981). This technique is based on limiting of electrical energy of the apparatus to the level that is too low to ignite the most easily ignitable gas mixture. The safety is achieved by designing and testing of apparatus according to requirements of standards. Normal as well as probable fault conditions are taken into account. Although, under testing conditions, certifying authorities simulate the extremely dangerous conditions, additional safety factors are included. Even if safety factors are not yet the same in all countries and if there are still arguments for applying them to energy (in the USA and Canada) or to currents (members of CENELEC, Poland, USSR), we expect that the tested apparatus will be safe with a high margin of safety. It does not mean, that the margin of safety is really known and is needed. Nevertheless, such approach is necessary becouse we have not sufficient collected data from practical situations in order to change this approach.

Frączek (1977) and Magison (1980) have described propositions of the model approach to the intrinsic safety which had been met in literature. The most developed model was derived by Benjaminsen and van Wiechen (1969). The author's own model has been precisely described in 1977 and next presented in the condensed form (Frączek, 1979). The new concept of some problems and new quantities introduced are presented in this paper.

The model approach presented in this paper shows quite different approach to a safety problem. This model enables to analyse an influence of different quantities on the safety level of all system. It enables also to compare what is an additional risk of explosions caused by intrinsically safe apparatus in relation to another existing causes. In this way it is possible to propose a reasonable margin of safety for the whole system.

PROBABILITY OF IGNITION

The probability p_z that an explosion will occur is equal to the product of the probability p_w that explosion mixture is present and the probability p that sufficient ignition energy will be released. For the i-th device of the n devices existing in the same explosion area the probability of ignition p_{zi} in single sparking equals

$$p_{zi} = p_w p_i, \tag{1}$$

where the probability p_w is defined as: $T_g/(T_g+T_o)$, in which T_g is the mean dwell time of the explosive mixture and T_o is the mean disappear time of the explosive mixture.

Eq. (1) is valid in the case when device has no safeguard system with the delay time T_d. When such a system exists for faults of an obvious nature, than for a period equal to the MTBF of device T_{ui} only a period T_d is dangerous and then,

$$p_{zi} = p_w p_i \frac{T_d}{T_{ui}}. \tag{2}$$

Faults of a nonobvious nature are eliminated by periodical inspections with an inspection time T_k. In this case Eq. (2) should be modified to the following formula:

$$p_{zi} = p_w p_i \frac{T_k}{T_{ui}}, \tag{3}$$

which is valid for $T_k/T_{ui} \ll 1$ (according to the geometric distribution).

MEAN TIME BETWEEN EXPLOSIONS

Let the probability of ignition p_{zi} be constant for each sparking then, the probability of the event that $X_i = m$ explosions will occur when the number of sparks N_i is described the binomial distribution,

$$P(X_i=m) = \binom{N_i}{m} p_{zi}{}^m (1-p_{zi})^{N_i-m}. \tag{4}$$

N_i is regarded as a number of failures of a device, with the simultaneous sparking. Let the number of failures N_i be described by the Poisson distribution,

$$P(N_i=l_i) = \frac{(\lambda_i t)^{l_i}}{l_i!} e^{-\lambda_i t}, \tag{5}$$

in which $\lambda_i = 1/T_{ui}$ – intensity of failures with simultaneous sparking. From Eq. (4) and Eq. (5) the complex process is derived,

$$P(X_i=m) = \frac{(p_{zi}\lambda_i t)^m}{m!} e^{-p_{zi}\lambda_i t}, \tag{6}$$

for which the $MTBE_i$ equals

$$MTBE_i = \frac{1}{p_{zi}\lambda_i}. \tag{7}$$

For the explosive area the entire number of sparking is important, so let us introduce a new random variable

$$X_O = \sum_i X_i, \tag{8}$$

while each variable X_i is characterized by the Poisson distribution. The process of the creation of the resultant number of explosions will be Poisson process,

$$P(X_o=m) = \frac{(\lambda_o t)^m}{m!} e^{-\lambda_o t}, \tag{9}$$

with the global intensity of explosions $\lambda_o = \sum_i p_{zi}\lambda_i$, for which the $MTBE_a$ is given by the formula,

$$MTBE_a = \frac{1}{\sum_i p_{zi}\lambda_i}. \tag{10}$$

In the particular case, when $p_{z1} = p_{z2} = \ldots = p_{zi} = p_z$ and $\lambda_1 = \lambda_2 = \ldots = \lambda_i = \lambda$, then according to Eq. (2) or (3),

$$MTBE_a = \frac{1}{n} \frac{T_u}{p_w p} \frac{T_u}{T_d} \quad \text{or,} \tag{11a}$$

$$MTBE_a = \frac{1}{n} \frac{T_u}{p_w p} \frac{T_u}{T_k}. \tag{11b}$$

For expressing p as a function of circuit parameters, for example of current I in the resistive or indictive circuit, the following formula may be used, which is done from practical experiments (Frączek, 1977):

$$p = \left(\frac{p_o}{I_o{}^a}\right) I^a = \frac{1}{A} I^a, \tag{12}$$

in which p_o and I_o may be a pair respective to minimum ignition currents (MIC) according to any definition. Therefore from Eq. (11a) and Eq. (12), the final expression is obtained,

$$MTBE_a = A \frac{1}{n} \frac{T_u}{p_w p} \frac{T_u}{T_d} I^{-a}. \tag{13}$$

Depending on the values of the respective quantities in Eq. (11), the following cases may be foreseen, when no safeguard system is provided:

1) Gas is continuously present – this respecting to Zone O (Division 1) of explosive area,

$$p_w = 1 \longrightarrow MTBE_a = \frac{T_u}{np}.$$

2) Devices are not intrinsically safe,

$$p = 1 \longrightarrow MTBE_a = \frac{T_u}{np_w}.$$

Safety is determined by the reliability of devices and the probability of gas appearance.

3) Only one device present in the explosive area,

$$n = 1 \longrightarrow MTBE_a = \frac{T_u}{p_w p}.$$

It is the case of the lowest danger.

4) One device in Zone O,

$$n = 1, \ p_w = 1 \longrightarrow MTBE_a = \frac{T_u}{p}.$$

5) As in 4), but device is not intrinsically safe,

$$n = 1, \ p_w = 1, \ p = 1 \longrightarrow MTBE_a = T_u.$$

Safety is determined by the reliability of the device only.

6) Single device with a constant danger. This is the case of the open sparking and p = 1. The MTBE then depends on the mean time between sparks T_I and the probability of gas appearance p_w,

$$MTBE_a = \frac{T_I}{p_w}.$$

When $p_w \longrightarrow 1$ and $n = 1$, $MTBE_a \longrightarrow T_I$ and when n devices are present, then

$$MTBE_a = \frac{T_I}{n}.$$

This determines the case, when explosion danger is greatest and may be a reference for other cases.

7) Particular case is when tests are provided with the IEC testing apparatus. Then $n = 1$, $p_w = 1$ and $T_I = 3.125 \times 10^{-3}$ min. The $MTBE_a$ may be determined from the formula:

$$MTBE_a = \frac{T_I}{p}.$$

For the typical values of p we obtain:

$p = 10^{-8}$ $MTBE_a \approx 5208$ h ≈ 7 months
$p = 10^{-6}$ $MTBE_a \approx 52$ h
$p = 10^{-3}$ $MTBE_a \approx 3$ min
$p = 10^{-2}$ $MTBE_a \approx 19$ s.

Usually circuits are tested for $p = 10^{-3}$ or $p = 10^{-2}$.

SAFETY FACTOR

Let us presume that basing on collected statistical data in particular branch of industry, the $MTBE_f$ (fires) was determined, this regarding explosions caused by other factors than using intrinsically safe apparatus. Then $MTBE_a$ determined for the intrinsically safe apparatus from Eq. (11) or Eq. (13) should be greater than the $MTBE_f$, what can be expressed as:

$$MTBE_f \; k = MTBE_p \leq MTBE_a, \qquad (14)$$

where $k > 0$ (e.g. $k = 100$) - safety factor for the $MTBE_p$ (permissible value of the mean time between explosions). As an example, let us regard the data collected in the coal plant "Donieckugol" in the Donieck Basin (Michajłow, Kołobowskij, 1967). These are: $p_w \approx 0.029$, $T_u \approx 0.15$ yrs, $MTBE_f = 2.5$ yrs. Assuming $k = 100$, the $MTBE_p$, caused by intrinsically safe devices, is:

$$MTBE_p = 250 \text{ yrs.}$$

For the data quoted above it is possible to analyse the influence of the individual quantities in Eq. (13). Let us calculate an example for the single inductive control circuit with the parameters: $p_o = 0.001$, $I_o = 100$ mA, $a = 13$. Then, for a single device we obtain:

$$MTBE_a = (5.17 \times 10^{-9}) \text{ yrs,}$$

where I is in amperes.

Results of calculations are presented in Table 1.

TABLE 1   Results of Calculations of MTBE

p	0.001	0.01	0.99
I [A]	0.10	0.12	0.99
n = 1:	51724	4834	52
	1.6×10^{12}	–	1.6×10^{9}; $T_d = 1$ s
n = 100:	517	48	0.5
	1.6×10^{10}	–	1.6×10^{7}; $T_d = 1$ s

The most important conclusions are stated as follow:

1) For the probability of $p = 10^{-3}$, that is for current equal to the MIC and the probability of explosion gas appearance $p_w = 3 \times 10^{-3}$ as well as with relatively frequent failures ($T_u \approx 1.5 \times 10^{-1}$ yrs), the MTBE for a single device is very long.

2) Increasing of the circuit current to the value for which the probability of ignition p is near to 1 does not made situation critically worse, for $MTBE_a = 52$ yrs is considered value large enough. Bijl (1966) in his proposition introduces admissible value of the $MTBE_a$ equal to 30 yrs.

3) Introduction of the safeguard device with activation time $T_d = 1$ s causes (for $n = 100$ devices and p near 1) the $MTBE_a$ to be so high that explosion practically may be considered impossible.

4) There is a direct suggestion that determining of the limiting admissible number of intrinsically safe devices, that may be located in explosive area, is possible. In the real conditions when $k = 100$ and $MTBE_p = 2.5$ yrs, the values of n_1 (limited) are:

$p = 10^{-2}$ 10^{-3} 10^{-6} 10^{-8}
$n_1 = 20$ 200 2×10^{5} 2×10^{7}.

The above data show that the need of applying such a large number of devices in one plant regarded as an object with common danger, practically does not exist. In addition, we know that for the limiting values danger is 100 times smaller than danger caused by the devices other than intrinsically safe.

5) Basing on the last statement, next conclusion may be drown, regarding directly the admissible safety currents. Eq. (11a) and Eq. (14) show that fulfilling the condition of intrinsic safety, while T_u and p_w being constant, is possible when,

$$n \, p = \text{const.} \qquad (15)$$

This means that deciding $n < n_1$ we may accept greater value of p and then greater value of the admissible safety current. This is equivalent with substantial acceptation of the smaller safety factor refering to the current (formally being determined by standards) during attesting. Using Eq. (12) in Eq. (11a) the expression for the admissible current increase factor for $n < n_1$ may be obtained,

$$k_I = \sqrt[a]{\frac{n_1}{n}} \ . \qquad (16)$$

where k_I is the ratio of the MIC (as I_o) to current I.

In the case when the $MTBE_p$ = 2.5 yrs and k = 100, from Eq. (16) we obtain:

$$n_1/n = 1 \quad 10 \quad 10^2 \quad 10^3 \quad 10^4 \quad 10^5$$

$$k_I = 1 \quad 1.19 \quad 1.43 \quad 1.70 \quad 2.03 \quad 2.42.$$

As we can see, in extreme conditions, the admissible value of current could be at the level of the MIC.

FINAL CONCLUSION

The presented method is based on the principle of experimental control of all variables which influence on the MTBE. The method of determining the danger level as a reference was developed theoretically and experimentally. The safety level is proposed to be determined basing on the accepted safety factors. It is proposed that these factors should be defined according to generally accepted rules. Parameters considered safe should be comprised in the area of values where there is a possibility of practical verification of effects.

REFERENCES

Benjaminsen, J.M., and P.H. van Wiechen (1969). Probability factor as a guide for area classification and selection of electrical equipment. IEEE Trans. on Ind. and Gen. Applic., 5, 242-249.

Bijl, P.C.J. (1966). Numbers yield instrument safety balance. ISA Journal, Oct., 62-64.

Frączek, J.K. (1977). Assessing fundations of the safety level for intrinsically safe control and measuring apparatus when the IEC breakflash is used. Automatyka, Z. 38, Pol. Śl., Gliwice. 75 pp. (In Polish).

Frączek, J.K. (1979). A probabilistic model for the analysis of hazards coused by intrinsically safe apparatus. Archiwum Górnictwa, 24, 189-195. (In Polish).

Hutcheon, I.C. (1981). Intrinsic Safety Rules OK for Process Instrumentation. TP1051, Measurement Technology Ltd., Luton. 12 pp.

Magison, E.C. (1978). Electrical Instruments in Hazardous Locations. Third Ed., Instrument Society of America, Research Triangle Park, North Car.

Michajłow, W.A., and A.K. Kołobowskij. (1967). Norm of intrinsic safety for mining automatic control. Izw. WUZ Gorn. Ż., 10, No 9. (In Russian).

DISCUSSION

<u>Session:</u> Industrial Problems.

Mr J. Järvi of Kemika Engineering introduced the safeguarding of exothermic Batch Reactors.

Certain types of batch reactors can develop inhomogeneities in their liquid reaction mixture. This may be due to unequal cooling or heating, malfunctioning mixing, too fast adding of some ingredients or other causes. Inhomogeneities may lead to locally increasing reaction rates and over-heating in the exothermic case, which in turn may lead to a runaway reaction and an explosion.

Batch reactors are usually controlled by measuring temperatures and manipulating valves of the heating, cooling or reactant medium according to temperature values or time profiles. Alarms and interlocks are provided for actual values and rate of change of critical process conditions as well as for mixer faults. These functions are nowadays normally accomplished by a digital instrumentation system.

Each of the primary sensors, transmitters, the digital system and control valves may fail. Consequently there exists a need to safeguard the reactor from runaway conditions with an independent,redundant system. Temperature transducers can be duplicated and a logic should be provided at least to open cooling valves and shut heating and feeding valves when a temperature or a temperature differ-ence get too high. Maybe valves to be opened should have redundancy in parallel and valves to be shut redundancy in series. In practice the critical temperature can be near 100°C and the critical temperature difference only a few degrees. To shut the system down on the temperature difference other conditions of the batch sequence must be monitored as well.

The problem is, what is the most reliable safe-guarding system for this type of a reactor at a reasonable cost.

From the discussion no single solution can be formulated. Mr J. Pearson of U.K. Health and Safety Executive offered the following guidance:

1. Identify Hazard and Event leading to the hazard using a procedure for failure mode and effects analysis.
   Refer e.g. to the publication 812 of 1985 of the International Electrotechnical Commission.

2. Identify the safety systems.

3. Decide on the risk level acceptance: Some guidance may be:
   Machine 10^{-2} - 10^{-3} per year.
   Chemical 10^{-4} - 10^{-6} per year.
   Nuclear (Individual risk) 10^{-5} per year.
   Nuclear general 10^{-7} - 10^{-9} per year.
4. Calculate accident frequency.

5. Decide whether the safety or integrity is adequate, improve weak links introducing better designs, diversity or redundancy.

6. For the exothermic Reaction protection system the following points could be considered:
   (a) a single control system for the "Temperature-Time profile seems adequate.
   (b) For Critical inputs however redundancy (2 out of 3) or a continuous comparison of several temperatures should be applied with alarm indication on a single discrepancy.
   (c) Monitoring of the agitator speed using e.g. pulses of a proximity device and of the agitator shaft using nitrogen purge pressure.
   (d) Monitoring the Temperature/Pressure characteristic.
   (e) Apply an exotherm test by detecting temperature rise before continuing to add further material to check that reaction has started (only if chemistry allows this).
   (f) Apply if necessary a relay type or other trip system on separate temperature inputs to prevent common cause failures.
   (g) If critical, duplicate equipment e.g. cooling pumps or apply a gravity based system.
   (h) Apply the "fail safe" state for air and/or power failure (agitator excepted!) e.g. cooling on, heating off, inhibitor agent on!

The importance of the environment on the reliability of instruments was discussed. To obtain data Mr. Yamamoto offered suggestions for accelerated tests.

For the reliability analysis of distributed digital control systems the Technical Research Centre of Finland started in 1980 a project whose aim was to develop reliability analysis tools for distributed digital control systems. Now the project is finished and one of its results is RELVEC computer code for complex systems reliability analysis.

The main idea of modelling with RELVEC is based on *tasks*. The entire control system hardware is modelled first and then any number of tasks may be assigned to the control system. The computer code and the modelling technology are very user-friendly and efficient.

The most complex model of a control system analysed so far included more than 3000 components and over 230 tasks (control loops, alarms etc.).

RELVEC is available for VAX-11 series, for VAX-8600 and for IBM PC/AT/XT.

One of the least explored areas is the reliability aspects of the man/machine-interface and task allocation. During the discussion following points were highlighted:

(a) Automation that is carried too far leads to the paradoxical situation that if breakdowns occur these are less manageable due to the scarcity of their occurrence.

(b) Nowadays, in order to avoid misfits in the operator-VDU system, two approaches are necessary:
    - the ergonomic approach; "fit the machine to the man."
    - the selection & training approach. "Train the man for the Task."

(c) The ergonomic approach starts *in* the control room with a so-called "*situation analysis*". This analysis consists of three parts: process analysis; task analysis; error analysis. "Human errors" made in the control room are mostly due to designers' errors rather than the operators' errors.

(d) In the situation analysis, the designer should make use of the *operator's expertise*. This kind of user participation is not only necessary from a human point of view, but above all from a technical point of view. It may widen the scope and re-define the tasks.

(e) The plant designer's view of the process does not necessarily coincide with that of the operator.
    Ignoring the operator's view will invariably lead to suboptimal process control and possibly even to increased safety risks.

(f) Depending on process conditions, the operator usually needs only a small subset of all information potentially available.
    Presenting anything beyond this subset is not only inefficient, but it can also be harmful when the superfluous part masks vital information.

(g) Different user groups (such as: staff, experienced operators and trainees) will in some cases have such widely differing information needs pertaining to the same process part, that each user group may actually need its own specific way of information presentation.

(h) The *manufacturers of the present VDU-based* process-control systems could have prevented most of the user problems with regard to information presentation, if they could have added two sources of input into their design process:
    - a proper task analysis of the *panel*-operator's job, and
    - existing knowledge from the field of *cognitive psychology* regarding the differences between *parallel* and *sequential* information presentation.

(i) In the near future, process-control operators should be trained and retrained in a similar way as commercial aviation pilots with realistic, full scale simulation facilities and periodical tests of emergency procedures.

Based on "Integrated Instrumentation Systems: the role of reliability engineering when evaluating a system" by J. P. Jansen (ref. Journal A Vol. 21 No. 1 and 2) the Working Group on Instrument Behaviour (WIB) has made reliability assessments of 15 systems over the last 8 years.

These assessments have shown clearly the critical system parts, redundancy concepts, revealed and unrevealed faults, diagnostic strength etc. They also have led to practical tests to check the tolerance of a system for occurring faults and errors.

From discussions with manufacturers they in particular had learned that once a system is conceived it is

difficult to increase its availability in a simple way.

Mr. J. Nyland from Jakarta University stressed the last point by bringing to the discussion a reliability check list based on his experience:

1. Maintenance.

    - Using standard components with high reliability.
    - Availability of components.
    - Interchangability of items.
    - Safety-conditions.

2. Accessibility.

    - Openings of adequate size.
    - Openings easy to open and to close.
    - Frequent maintenance-areas have best access.
    - Replacing parts without disturbing others.

3. Adjustments.

    - Factory-warranty-adjustments sealed.
    - Clockwise increase.
    - Protection against accidental movement.

4. Cables.

    - Good identification.
    - Adequate clamping.
    - Long enough to remove connected components for testing.
    - Good electrical scheme at manual and/or behind panel doors.

5. Connectors.

    - Standard.
    - No possible mis-connection.
    - Easy in/out.
    - Plugs cold/receptacles hot.
    - Moisture-prevention if needed.
    - Labeled.

6. Panels.

    - Sealed against foreign objects.
    - Easily opened and replaced.
    - Good protection against moisture, sand and chemicals.

7. Diagnostics.

    - Every fault detected and isolated.
    - Condition monitoring on all major inputs and outputs.
    - Never more than two signals can be observed simultaneously.

8. Environment.

    Instrument protected against:
    - Hot and cold temp.
    - High and low humidity.
    - Corrosives.
    - Liquids.
    - Electrical static.

9. Fasteners.

    - Few in number.
    - Standard.

10. Lubrication.

    - Use standard oil/grease.
    - Mention oil-scheme at back of panel door and/or at manual.
    - Marking lubrication-points at instrument conform manual.

11. <u>Operational tasks</u>.

    - To build in checklist.
    - Related controls together.
    - Fail-safe.

12. <u>Packaging</u>.

    - Efficient package-material.
    - Stacking components avoided.
    - Plug-in replaceable components.

13. <u>Parts and components</u>.

    - Mean-time between maintenance known.
    - Labeling with part number.
    - Delicate parts protected.
    - Standard.

14. <u>Personnel involvements</u>.

    - Weight (portable) max. 16 kg.
    - Male or female.
    - Special clothing needed.

15. <u>Safety</u>.

    - Protection against:
      - hot areas.
      - radiation.
      - high voltage.
    - Corners and edges round.
    - Adequate fuses.
    - Guards on moving parts.

16. <u>Test points</u>.

    - Functionally grouped.
    - Clearly labeled.
    - Protected against physical damage.

17. <u>Tools and test-equipment</u>.

    - Standardized.

18. <u>Transport and storage</u>.

    - Integrated moving.
    - Practical methods for moving.
    - Weight not above 30 kg.

The point of intrinsically safe instrumentation for
use in plants where explosion hazard is a possibility
such as petro-chemical plants was brought up by
Mr J. Fraczek, who's approach was received with
great interest.

AUTHOR INDEX

AKASHI: Control Science and Technology for the Progress of Society, 7 Volumes

ALONSO-CONCHEIRO: Real Time Digital Control Applications

ATHERTON: Multivariable Technological Systems

BABARY & LE LETTY: Control of Distributed Parameter Systems (1982)

BALCHEN: Automation in Aquaculture

BANKS & PRITCHARD: Control of Distributed Parameter Systems (1977)

BARKER & YOUNG: Identification and Systems Parameter Estimation (1985)

BASANEZ, FERRATE & SARIDIS: Robot Control "SYROCO '85"

BASAR & PAU: Dynamic Modelling and Control of National Economies (1983)

BAYLIS: Safety of Computer Control Systems (1983)

BEKEY & SARIDIS: Identification and System Parameter Estimation (1982)

BINDER & PERRET: Components and Instruments for Distributed Computer Control Systems

BRODNER: Skill Based Automated Manufacture

BULL: Real Time Programming (1983)

BULL & WILLIAMS: Real Time Programming (1985)

CAMPBELL: Control Aspects of Prosthetics and Orthotics

Van CAUWENBERGHE: Instrumentation and Automation in the Paper, Rubber, Plastics and Polymerisation Industries (1980) (1983)

CHESTNUT: Contributions of Technology to International Conflict Resolution (SWIIS)

CHESTNUT, GENSER, KOPACEK & WIERZBICKI: Supplemental Ways for Improving International Stability

CHRETIEN: Automatic Control in Space (1985)

CICEL: Automatic Measurement and Control In Woodworking Industry—Lignoautomatica '86

CICHOCKI & STRASZAK: Systems Analysis Applications to Complex Programs

CRONHJORT: Real Time Programming (1978)

CUENOD: Computer Aided Design of Control Systems*

DA CUNHA: Planning and Operation of Electric Energy Systems

De GIORGIO & ROVEDA: Criteria for Selecting Appropriate Technologies under Different Cultural, Technical and Social Conditions

DI PILLO: Control Applications of Nonlinear Programming and Optimization

DUBUISSON: Information and Systems

ELLIS: Control Problems and Devices in Manufacturing Technology (1980)

ELZER: Experience with Management of Software Projects

FERRATE & PUENTE: Software for Computer Control (1982)

FLEISSNER: Systems Approach to Appropriate Technology Transfer

GEERING & MANSOUR: Large Scale Systems; Theory and Applications (1986)

GELLIE & TAVAST: Distributed Computer Control Systems (1982)

GENSER: Control in Transportation Systems (1986)

GERTLER & KEVICZKY: A Bridge Between Control Science and Technology, 6 Volumes

GHONAIMY: Systems Approach for Development (1977)

HAASE: Real Time Programming (1980)

HAASE: Software for Computer Control (1986)

HAIMES & KINDLER: Water and Related Land Resource Systems

HALME: Modelling and Control of Biotechnical Processes

HARDT: Information Control Problems in Manufacturing Technology (1982)

HARRISON: Distributed Computer Control Systems (1979)

HASEGAWA: Real Time Programming (1981)*

HASEGAWA & INOUE: Urban, Regional and National Planning—Environmental Aspects

HERBST: Automatic Control in Power Generation Distribution and Protection

ISERMANN: Identification and System Parameter Estimation (1979)

ISERMANN: 10th IFAC World Congress

ISERMANN & KALTENECKER: Digital Computer Applications to Process Control

JANSEN: Reliability of Instrumentation Systems for Safeguarding and Control

JANSSEN, PAU & STRASZAK: Dynamic Modelling and Control of National Economies (1980)

JOHANNSEN & RIJNSDORP: Analysis, Design, and Evaluation of Man-Machine Systems

JOHNSON: Modelling and Control of Biotechnological Processes

KAYA & WILLIAMS: Instrumentation and Automation in the Paper, Rubber, Plastic and Polymerization Industries (1986)

KLAMT & LAUBER: Control in Transportation Systems (1984)

KOPACEK, TROCH & DESOYER: Theory of Robots

KOTOB: Automatic Control in Petroleum, Petrochemical and Desalination Industries

KUZUGU & TUNALI: Microcomputer Application in Process Control

LANDAU: Adaptive Systems in Control and Signal Processing

LARSEN & HANSEN: Computer Aided Design in Control and Engineering Systems

LAUBER: Safety of Computer Control Systems (1979)

LEININGER: Computer Aided Design of Multivariable Technological Systems

LEONHARD: Control in Power Electronics and Electrical Drives (1977)

LESKIEWICZ & ZAREMBA: Pneumatic and Hydraulic Components and Instruments in Automatic Control*

MAFFEZZONI: Modelling and Control of Electric Power Plants (1984) (1986)

MAHALANABIS: Theory and Application of Digital Control

MANCINI, JOHANNSEN & MARTENSSON: Analysis, Design and Evaluation of Man–Machine Systems (1985)

MARTIN: Design of Work in Automated Manufacturing Systems

MARTOS, PAU, ZIERMANN: Modelling and Control of National Economies (1986)

McGREAVEY: Control of Distillation Columns and Chemical Reactors

MILLER: Distributed Computer Control Systems (1981)

MUNDAY: Automatic Control in Space (1979)

NAJIM & ABDEL-FATTAH: Systems Approach for Development (1980)

NIEMI: A Link Between Science and Applications of Automatic Control, 4 Volumes

NORRIE & TURNER: Automation for Mineral Resource Development

NOVAK: Software for Computer Control (1979)

OLLUS: Digital Image Processing in Industrial Applications—Vision Control

O'SHEA & POLIS: Automation in Mining, Mineral and Metal Processing (1980)

OSHIMA: Information Control Problems in Manufacturing Technology (1977)

PAUL: Digital Computer Applications to Process Control (1985)

PONOMARYOV: Artificial Intelligence

PUENTE: Components, Instruments and Techniques for Low Cost Automation and Applications

QUIRK: Safety of Computer Control Systems (1985) (1986)

RAMAMOORTY: Automation and Instrumentation for Power Plants

RAUCH: Applications of Nonlinear Programming to Optimization and Control*

RAUCH: Control of Distributed Parameter Systems (1986)

RAUCH: Control Applications of Nonlinear Programming

REMBOLD: Information Control Problems in Manufacturing Technology (1979)

RIJNSDORP: Case Studies in Automation related to Humanization of Work

RIJNSDORP, PLOMP & MÖLLER: Training for Tomorrow—Educational Aspects of Computerized Automation

RODD: Distributed Computer Control Systems (1983)

RODD & MULLER: Distributed Computer Control Systems (1986)

ROOS: Economics and Artificial Intelligence

SANCHEZ: Fuzzy Information, Knowledge Representation and Decision Analysis

SAWARAGI & AKASHI: Environmental Systems Planning, Design and Control

SINGH & TITLI: Control and Management of Integrated Industrial Complexes

SKELTON & OWENS: Model Error Concepts and Compensation

SMEDEMA: Real Time Programming (1977)*

STRASZAK: Large Scale Systems: Theory and Applications (1983)

SUBRAMANYAM: Computer Applications in Large Scale Power Systems

SUSKI: Distributed Computer Control Systems (1985)

SZLANKO: Real Time Programming (1986)

TAL': Information Control Problems in Manufacturing Technology (1986)

TELKSNYS & SINHA: Stochastic Control

TITLI & SINGH: Large Scale Systems: Theory and Applications (1980)

TROCH, KOPACEK & BREITENECKER: Simulation of Control Systems

UNBEHAUEN: Adaptive Control of Chemical Processes

VALADARES TAVARES & DA SILVA: Systems Analysis Applied to Water and Related Land Resources

WANG PINGYANG: Power Systems and Power Plant Control

WESTERLUND: Automation in Mining, Mineral and Metal Processing (1983)

WITTENMARK: Adaptive Systems in Control and Signal Processing

van WOERKOM: Automatic Control in Space (1982)

YANG JIACHI: Control Science and Technology for Development

YOSHITANI: Automation in Mining, Mineral and Metal Processing (1986)

ZWICKY: Control in Power Electronics and Electrical Drives (1983)

*Out of stock—microfiche copies available. Details of prices sent on request from the IFAC Publisher.

IFAC Related Titles

BROADBENT & MASUBUCHI: Multilingual Glossary of Automatic Control Technology
EYKHOFF: Trends and Progress in System Identification
ISERMANN: System Identification Tutorials (*Automatica Special Issue*)